Techniques in Physical Geography

Nyland Hill seen from the southern slopes of Mendip (taken from N.G.R. ST(31) 480520)

TECHNIQUES IN PHYSICAL GEOGRAPHY

JAMES HANWELL

Head of Social Studies Department,
Blue School, Wells

and

MALCOLM NEWSON Ph.D.

Research Hydrologist,
formerly of Bristol University
Geography Department

MACMILLAN

First published 1973

Published by
MACMILLAN EDUCATION LIMITED
Basingstoke and London

The Macmillan Company of Australia Pty Ltd Melbourne
The Macmillan Company of Canada Ltd Toronto
St Martin's Press Inc New York

Companies and representatives throughout the world

Printed in Great Britain by
ROBERT MACLEHOSE AND CO LTD
The University Press, Glasgow

Acknowledgements

The authors wish to thank Barbara Glover, Valerie Newbury and Cathy Whiting for help with the illustrations and Stephanie Jones for advice on the text. We are indebted to the Hydraulics Research Station for Plate 7.2; to the Director General of the Meteorological Office, Bracknell, for the information used in figs. 2.10 and 2.11; to Michael Chaddock and Dennis Kemp for data in figs. 3.17 and 3.18; to Trisha Jordan for fig. 4.18; to the Institute of British Geographers for permission to use figs. 6.11, 6.18 and 7.3; to Alan Carr for help with figs 7.3 and 7.6 and for other advice; to Professor Kidson for permission to use fig. 7.6; to Professor Steers for fig. 7.2 and to Dr Ken Gregory for fig. 6.13. Dr Alastair Pitty kindly let us use fig. 6.13, while D. Ingle Smith is the originator of the 'pebble-ometer' shown in fig. 7.5; fig. 7.4 (left) is reproduced from the Ordnance Survey map with the sanction of the Controller of the H.M.S.O. (crown copyright reserved), reprinted by David & Charles; fig. 7.4 (right) is reproduced from the Ordnance Survey map with the sanction of the Controller of the H.M.S.O. (crown copyright reserved).

We also wish to thank many friends at the Universities of Bristol and Exeter, whose Departments of Geography have had major influences on the material contained, to F. J. Davies for advice concerning many theoretical aspects in chapter two and to Anna and Judy for encouragement and sustenance.

Wookey Hole, Somerset, 31.7.72

Contents

Introduction

> 'Learning without thought is labour lost.'
> Confucius

Man is becoming much more concerned about his relationship with the physical or natural environment. Quite rightly, subjects other than geography are sharing this concern and responding to the challenge in their own ways. Many are considered more suited to tackle the problems involved, and even new disciplines are emerging for the specific purpose of studying the environment. Does physical *geography* as such still have a positive and special contribution to make beyond the scope of these emergent subjects? We are certain that it has and must.

In this book we do not seek to present a comprehensive text or work of reference concerning the increasingly sophisticated field of physical geography. Our chief concern is to highlight easily neglected concepts and techniques in the main branches of the subject. At the same time, we endeavour to link each topic within a *geographical* framework. Whilst addressing ourselves mainly to sixth formers and undergraduates, we trust also that their teachers will find scope for reappraising syllabuses in danger of flagging or even drifting away from the mainstream of geography during its present flood of development in human affairs.

New methods of approach are freeing geographers from their former preoccupation with position, form and their description. A great amount of published material exists to champion and justify these changes. We do not propose to amplify this here, although the true story that follows helps us to explain our support for such developments and so the philosophy behind this book. It concerns the curiosity of a young child about the landforms near his home in Somerset (see frontispiece).

In answer to insistent queries concerning the hill, the child was told that a Mendip giant had concluded a feud with his lowland adversary by kicking an enormous clod of earth into him: the valley being the 'rut' and the hill the 'divot'. The story, suitably elaborated to capture a young imagination, was accepted with awe. Had the child understood, it would certainly have agreed with Aristotle's view that 'wonder is the first cause of philosophy'. Nowadays we would term this the **intuitive** or **prelogical** phase following the widely accepted findings on learning by the Swiss psychologist Piaget.

On later rambles the child progressed to collecting all sorts of local information. Appealing titbits were hoarded and haphazard ventures at written records were attempted. When a little more knowledgeable, and certainly less credulous, doubts about the giant began to dawn. Piaget would deem this as the **concrete operations** phase during which simple conclusions may be drawn from selected objects or situations.

In this case, similar dry valleys were observed as not having corresponding hills. Furthermore, the respective sizes of valley and hill were seen to be at odds; the latter being much larger. Identically constituted bedrock found in both required the rejection of the idea that the hill comprised of loosely compacted debris from the valley. Painful attempts at kicking clods of earth produced unlikely shaped features and a realisation that other explanations must exist. *Position* and *form* alone could not provide the answer.

Only by evaluating the likely processes operative, after following comparative studies of valley and hill formation elsewhere, did an explanation approximating to the truth eventually replace that of the irascible Mendip giant. This was not possible, however, until the facility to *reason* on the basis of a hypothesis had developed at a later age. Briefly, this **formal operations** phase heralds the ability to formulate a relation *between* relations: a vital part of any scientific enquiry. Not everyone aspires to this phase, although it is clear that most sixth formers, especially geographers, should do so.

Despite obvious scale differences in time and space, we suggest that there are marked parallels between an individual's intellectual development and the historical growth of the *body* of knowledge called physical geography. It is always difficult to tear oneself away from the past, yet we too must not be afraid to have doubts and seek new explanations. Quantitative approaches are now basic to most acceptable theories, often replacing those limiting and misleading descriptive studies reliant on position and form. Pursuing our analogy with Piaget's phases, it is claimed that the subject has at last 'come of age'. Just as relationships are essential to any formal operation, so we see the future organisation of physical geography being centred around the points where its different topics can be linked. We return to this theme in the concluding chapter of the book. In short, we follow Einstein's great dictum that the formulation of a problem is essential to its solution:

With the help of physical theories we try to find our way through the maze of observed facts, to order and understand the world of our sense perceptions. We want the observed facts to follow logically from our concept of reality.

Thus, by adopting the view that the world comprises of various integrated **systems,** we provide ourselves with a means of organising this book and its users with a methodological framework to which they can relate the techniques discussed. As with the child in our story, we are concerned with how to solve particular relationships in physical geography. A knowledge of the subject by itself is not enough.

In stressing the increasing importance of *techniques* over knowledge in the development of science, Emile Boirac listed four steps to be followed; observation, hypothesis, experimentation and induction. Fig. i.i translates a modification of this into recognised learning situations and distinguishes the more philosophically sound and exciting inductive path from its less rewarding deductive counterpart. This point is amplified in the first chapter. Readers might explore both routes concerning the techniques of other academic subjects they are studying. We suggest this, not only to clarify our approach to physical geography, but also to amplify three important consequences affecting the choice of material for this book. (You may care to return to this figure after completing the book to see how the various techniques discussed can be applied.)

Firstly, fig. i.i bears out our intention to stress how to *think* and what to do rather than facts to be remembered. Secondly, it infers that physical geography must be of more *practical* value concerning man and his environment and, thirdly, it demonstrates that inductive approaches demand we initiate enquiries at a *smaller* scale level. Just as the beauty of a great painting is only revealed in the texture of the brushwork and the mixing of the paints, so we take a more 'close-up' view of patterns and processes in physical geography. By standing too far away from the canvas our view is scaled down, intricate and fundamental information is lost and we only register the vaguest impressions of its composition. Why be content with such a picture gallery view when most of the detail is to be found in the air we breathe, our gardens, parks, streets, the nearest river and woods?

Geography is concerned with three-dimensional space and with time. An approach through *scale* provides the obvious focus for both spatial and temporal aspects. Unaccountably, however, scalar factors are often taken for granted or merely ignored by geographers. The adoption of systems theory here necessitates a rational evaluation of scale on the lines suggested in fig. i.ii.

We take advantage of this illustration to give the reader a foretaste of the type of exercise commended throughout each chapter. It is not so much intended as a vintage aperitif as a mild stimulant! Whilst this one is designed to be completed indoors, however, most others will involve some observations in the field.

1 The reader will require squared paper, mathematical tables or a slide rule; then, by studying the explanation below as to how fig. i.ii has been prepared, an attempt should be made to reproduce the graph and to complete the questions that follow.

2 Both the x axis (or abscissa) and the y axis (or ordinate) show two alternative scales of *time* (X and x) and of *space* (Y and y) respectively. The X-time scale is called a *denary* one because it is divided by appropriate powers of ten, viz. 10^{-3}, 10^{-2}, 10^{-1}, 10^{0}, 10^{1}, 10^{2} and so on to 10^{7}. As *any* number to the power zero is equal to one, 10^{0} has been chosen to represent *a year* on the scale; thus, 10^{-3} at the lowest end represents about 8 hours and 10^{7} about 10 million years. On the other hand, the x-time scale is called a *binary* one because its intervals are in appropriate powers of two, viz. 2^{-10}, 2^{-5}, 2^{0}, 2^{5}, 2^{10} and so on to 2^{40}. Here 2^{0} is taken as the equivalent of *a day* which means that 2^{-10} represents about 1 minute and 2^{40} the approximate age of the earth. Thus it is possible to embrace vast ranges of values by such techniques.

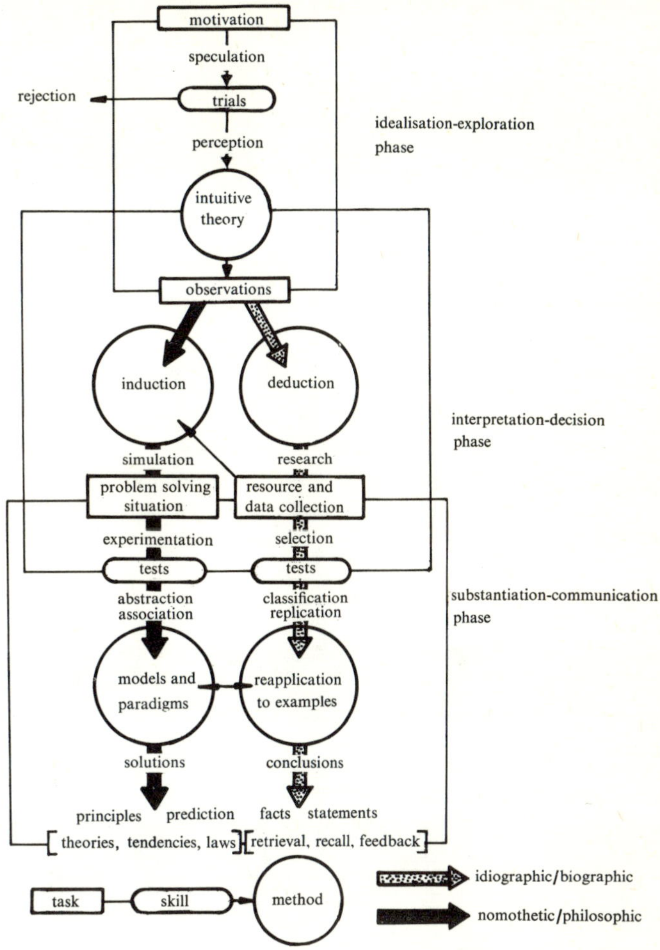

Fig. i.i Alternative paths to learning and research

On the graph, this is done by taking *logarithms* of the base number used; hence, $-3\text{Log}10$, $-2\text{Log}10$ to $7\text{Log}10$ for the denary scale, or $-10\text{Log}2$, $-5\text{Log}2$ to $40\text{Log}2$ in the case of the binary scale. Since we have deliberately chosen ten class intervals for both scales, each can be then be expressed as 0, 1, 2 and so on to 10, e.g. on the x or binary scale, let $-10\text{Log}2$ be 10, $-5\text{Log}2$ be 9 so that $40\text{Log}2$ becomes 0.

A similar treatment has been given to the Y and y space scales; the former being a logarithmic denary scale from 10^{-10}

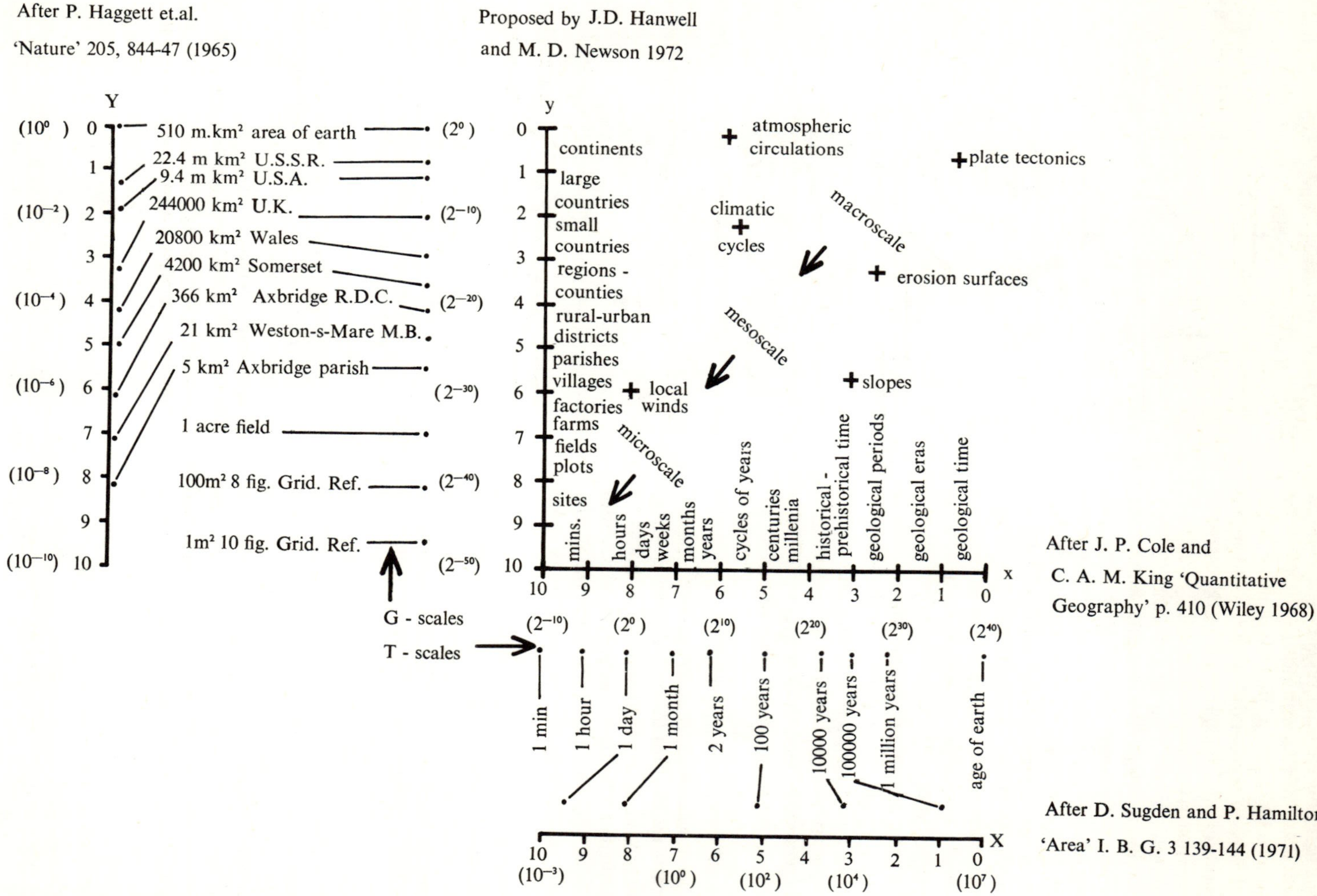

Fig. i.ii A rationalisation of scales

to 10^0, where 10^0 represents the area of the earth, and the latter a logarithmic binary scale from 2^{-50} to 2^0 also based on the area of the earth. The reader should prepare these scales and calculate the smallest areas represented by each, viz. 10^{-10}, and 2^{-50} of the earth's area.

3 Having prepared two graphs, one denary using the X and Y axes and the other binary using the x and y axes, the reader should attempt to 'locate' selected topics in physical geography with respect to time and space. Various examples have been shown in fig. i.ii as a guide. The topics throughout this book can be located on the graphs. Suggest reasons why the binary scales are to be preferred. Can other branches of geography be recorded in the same way to show their scalar relationships?

It will be found that most of the topics dealt with in this book tend to fall within the *micro-scale* portions of the graph. This is not only a neglected area in physical geography, but one where less marked divisions exist between say topics in climatology and geomorphology. Also, there is clearly more scope to apply a varied range of field and laboratory techniques by seeking greater *integration* with allied subjects. Physics, biology and chemistry are obvious examples, and we feel certain that many will have already consulted colleagues reading mathematics concerning the exercise above! We encourage such co-operation. After all, mergers are becoming a more obvious feature of our lives whether academic, economic, political or whatever. They are manifestations of the need to break down outmoded barriers where links prove more fruitful. Geography has an honourable tradition here, though it is hardly one which can be expected to persist unchanged. We believe that our approach provides sixth formers with a more stimulating view of physical geography than the repetitious observe-describe-classify-recall one.

Geographers cannot content themselves with words however elegant and, although no authors of a book would dare to suggest that they lacked feeling for the written word, we are sure that mathematics provides the only logic to *aid* any study of natural phenomena. (See chapter 1 on systems and their measurement.) However, to avoid elaborating upon many of the *statistical methods* and *mapping devices* to which we refer, the appropriate accounts by P. Toyne and P. Newby in *Techniques in Human Geography* (Macmillan, 1971) will be found equally valid in physical geography. We must give more attention to the greater problems involved in the *measurement* and *inter-relationships* of physical data. For example, counting children rushing home from school is quite a different proposition to calculating the energy they expend in doing so! Many of the techniques discussed in physical geography call for experiments using *instruments,* therefore. Some are simple enough to make but others are costly. Yet, we venture to encourage their use in the knowledge that many education authorities and industries are establishing resource centres for expensive equipment for loan or hire. Technology and techniques are inseparable.

We believe that, by blending the geographer's interest in the natural environment with the techniques of scientific enquiry, this book may focus attention on the stimulating future of our chosen subject. It is a subject of great importance in the classroom concerning our increasingly crowded world. Just as the child outgrew its wonder in a fictitious Mendip giant, we hope that readers will be infected by the same spirit which prompted the poet R. W. Emerson to observe that:

> Science surpasses the old miracles of mythology.

1 Natural systems and their measurement

'You will not learn from me philosophy, but how to philosophise, not thoughts to repeat, but how to think. Think for yourselves, enquire for yourselves, stand on your own feet.'

Kant

Our aim in this chapter is to provide a methodological background to the work which the reader may do in physical geography, whilst encouraging him to examine critically the work of others.

The philosophy of science is a complex subject, not usually tackled by researchers until they are fairly senior and in a reflective mood. It may well be, therefore, that the reader will choose to read this chapter last, but we trust he will also read it first. The strict methodology stressed in this chapter is not intended as a hindrance to experiment or discovery and a warning about being reluctant to 'have a go' is given in section 1.8.

1A The nature of systems

It would be difficult to imagine that the objects of study for the physical geographer were totally unrelated, that clouds, rain, soil, slopes, vegetation and rivers were all unique elements of a disjointed world. As scientists searching for pattern and as geographers concentrating on spatial patterns, we may comfort ourselves with the knowledge that they are there! The world we perceive is orderly and consists of a series of objects linked by flows of energy and mass; it constitutes a **system**. Within the global system are legions of others making up the study theme of the various natural sciences. The indefinite numbers of systems present opportunities for investigation at an almost indefinite variety of scales.

In the Introduction we implied that the concept of **scale** is an important part of the methodology to be followed in the study of natural systems and we will return to this shortly.

Systems have been defined by Harvey as constituting:

a) a set of elements defined with some *variable* attribute of objects (i.e. they are measured);

b) a set of relationships between the attributes of objects (i.e. they are linked);

c) a set of relationships between those attributes of objects and the environment (i.e. they are not independent of their surroundings).

1A 1 Organisation

The picture may become clearer if we consider certain familiar man-made systems, such as that which provides the home or school with hot water. In this way we may introduce an important technique in the investigation of systems: the **flow diagram**. Fig. 1.1 shows the domestic water system in normal diagram form. Obviously the elements are the tanks, taps and boiler whilst they are related by the flow of water in the pipes. While conforming to our definition of a system the concept of scale in systems is perhaps better demonstrated by reference to electricity. Domestic electricity supply begins with a network of generating stations linked by the National Grid of wires and pylons. At a slightly smaller scale are the supply wires down our own street, linked to a 'main' in each house. Within each house are a series of appliances, each with a very densely-packed network of wires. The electricity supply system is shown in fig. 1.2 as a flow diagram. The diagram consists of boxes (representing (a) in Harvey's definition) and arrows variously linking the boxes ((b) in the definition); (c) is

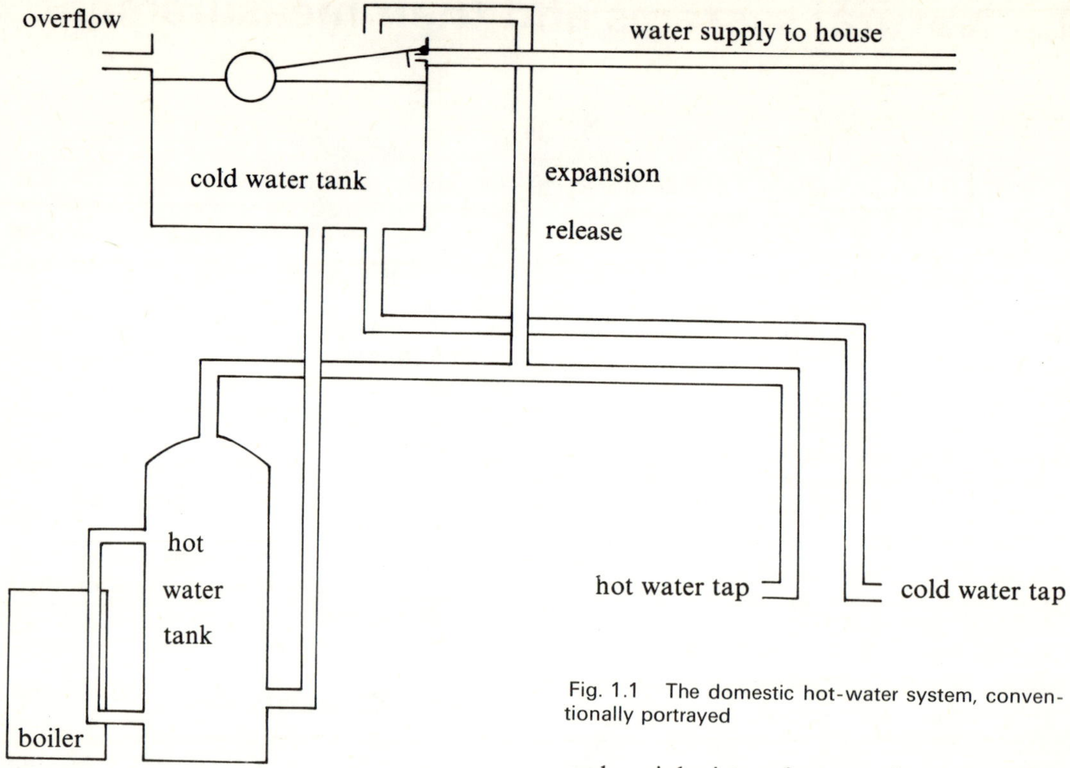

Fig. 1.1 The domestic hot-water system, conventionally portrayed

partially represented by the input of coal, water, etc.

Taking an individual appliance such as a television set we may demonstrate **closed systems**. The television engineer is unlikely to want to study the National Grid or the main circuit in the house when he calls. Rather he isolates the system in the set for study and action: he adopts a kind of closed system approach. Most systems in nature may be 'closed' for study—that is their relationships with their *environment* (the larger systems of which they are part) are subordinated to those internal relationships which are of interest. The **open systems** approach to natural science is ever mindful of the environmental relationships of the system under study, particularly from the point of view of energy flows, just as the television engineer takes for granted the relationship of the set to the mains and the Grid. The open system approach is of demonstrable value in solving some of man's current problems of population, resources and pollution. Natural scientists and sociologists refer to such an approach using the term **ecosystems** (see Chapter 4). Figs. 1.3 and 1.4 show open and closed systems.

It is important for us to know, when we do an experiment or collect any data in physical geography, the degree to which we are closing the system under investigation. Despite the ability of modern statistical techniques to make possible the consideration of many variables affecting a phenomenon, closure to some degree is inevitable. For this reason the construction of flow diagrams before embarking on a project will help us define its scale and thereby its environment.

Systems analysis and the construction of flow diagrams provides a very *efficient* basis for working. In the field of electronic engineering, communications and business systems analysis (or *cybernetics*) has become an important aid to financial success through efficiency. We all tend to regard being *systematic* as a virtue! Many readers will have followed the steps encountered in *programmed learning* or used teaching ma-

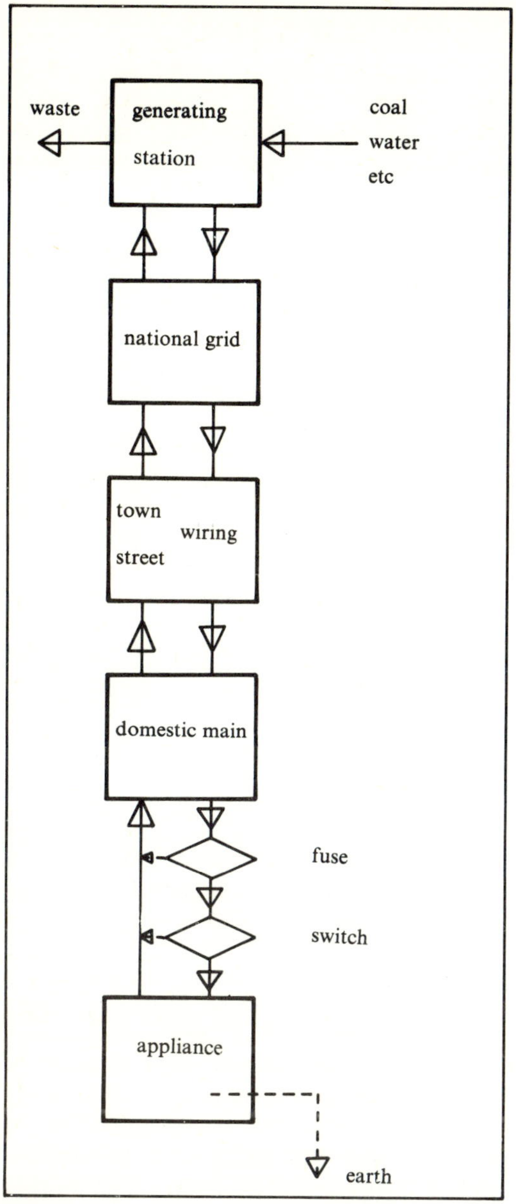

Fig. 1.2 The electricity supply system, portrayed as a flow diagram

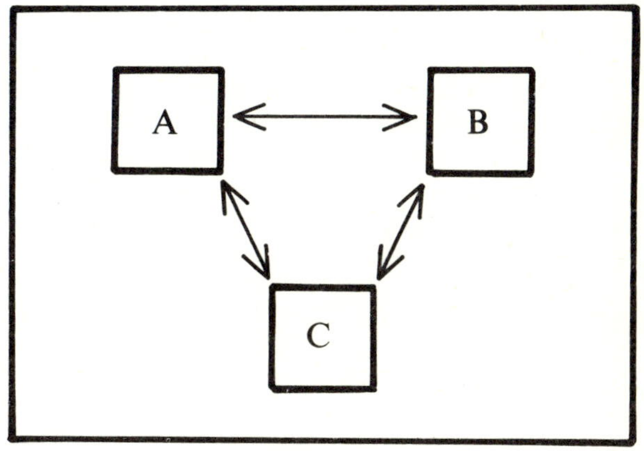

Fig. 1.3 A closed system

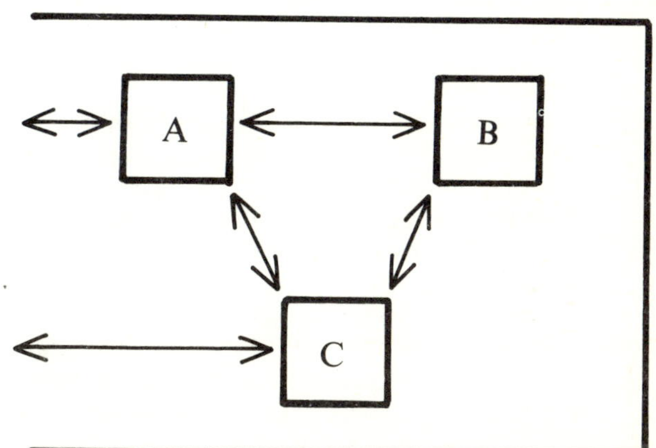

Fig. 1.4 An open system

chines. Both employ logical steps similar to those in flow diagrams and the fact that computers work in a similar way has made flow diagrams an essential prerequisite of writing a computer program. The reader ought to exercise his logic in writing a simple program to unambiguously describe the steps by which, for example, a rainfall measurement is made.

It is hard to consider natural systems as demonstrating efficiency in business terms. However, even such complex systems as slopes or stream channels generally show such harmonious adjustment of their elements and flows that a **steady state** exists. In a steady state, opposing forces in the system, although active, are balanced. The open systems in nature can, however, perform work under such steady conditions which would make the organiser of human affairs green with envy! The equilibrium of the steady state condition is a **dynamic** one.

1A 2 The scale of systems

Geographers are introduced to scale through maps. That we have chosen flow diagrams as a basic technique in physical geography need not distinguish us from map makers or users. The main difference between maps and systems is that the latter have an even greater range of scale than published maps (fig. i.ii). In case the reader feels that maps are too static to represent systems (whose dynamism has been defined) let him try inserting arrows upon a chosen map to indicate any flows of energy or materials which he feels may be going on between the objects represented there by their static spatial attributes; for example, the movement of water and sediment in the drainage basin, or the movement of pupils after the break-up of school assembly!

Maps are also conceptually useful for systematic analysis since they represent **models** of the real world, no part of the earth's surface being as flat as a map, nor arbitrarily disconnected from its surrounding 'sheets'. Models are discussed later as part of the process of deriving theories but we have already seen how the artificial closing of an open system for the purposes of study is creating a model of that system, an idealized version of reality.

Part of the progress of modern geography has been the change in the basic scales of study, a point touched on in our Introduction. During the period when geography was concerned with discovery and description on the *world* scale, travel was a basic geographical technique and the globe hung in every classroom. The later preoccupation of geographers with *regions* qualified the period when they began to examine the inter-relationships between objects at another scale. Basic objections to regional geography stem from the fact that the degree to which the region was considered as a closed system was often indefensible and that boundaries to human interactions were frequently given in physical terms. Too often regions such as 'the Central Lowlands' and 'the Coast Ranges' were considered separately even if their coalfields, railways and cities overlapped or interacted. Recently we have been tending towards measurement and, as a result, smaller units of study. However, by correctly processing the results of such small-scale or local-scale measurements we may often successfully draw conclusions for other scales of application. Moreover, the present phase is not wholly preoccupied with minute analyses—for there is increasing interest in natural resource management, hydrology and meteorology at national scales and in remote sensing of the global situation by artificial satellite. The hierarchical organisation of scales should be familiar to those in schools where classes, forms, houses, etc. are arranged on scales of operation.

1A 3 Investigation

It is important to examine and choose a scale of investigation in physical geography for the following reasons:

a) It will enable the researcher and his critics to assume the extent to which internal and external influences have been consciously (or even accidentally) ignored. Thus a report on the climate of the British Isles is unlikely to include, or be invalidated by, the peculiarities of climate in a small valley in Yorkshire.
b) The scale of investigation will largely decide the precision necessary in instruments and their recordings. Properties with a wide variability may be measured with fairly 'coarse' equipment. One would not, however, expect to learn much about the variation of velocity through a stream profile at a site using a floating orange, although the same method might be useful to compare flood and drought flows of several different rivers. Similarly, if the difference between two recordings of

temperature was expected to be less than one degree, we should need thermometers reading to at least one decimal place.

Once the choice of system and scale has been made it is wise to determine how many points in the system will be used for measurements and how many properties will be measured. Do instruments exist to make these measurements? We may qualify two basic situations, most familiar to meteorologists and hydrologists (see chapter 5) the **black box**—for which measurement is possible of the gross **inputs** and **outputs** to the system, namely rainfall and streamflow, and the **white box**, describing the situation in which recordings are also made of intermediate processes (e.g. soil moisture and groundwater movement, or evaporation).

It should be added that, as well as measuring objects and flows in a system, we also are increasingly quantifying the **network** along which the flows occur and the objects are linked, for example the stream network. Thus the subject of **topology** has entered physical geography, particularly in the study of drainage basins. Topology deals with the connectivity of a geometrical figure rather than its shape or size; it employs a rather more abstract, mathematical treatment of landscape systems than does the traditional **topography**, which delineates the ground features of a locality.

1B Measurement

Though the application of scientific principles to research and learning has united the study of human and physical geography, they are at their most dissimilar in the realms of measurement. The aim of this chapter is partly to stress this very point. The part played by instruments and the problems of designing true experiments in the behavioural sciences is small; much of the data used has been recorded officially in censuses, trade figures and so on, or may be gathered by questionnaires (see chapter 1 of Toyne and Newby, 1971). The data is in the form of large numbers of people, commodities or transport movements. It is, therefore, not surprising that human geography took the lead in the statistical analysis of large amounts of data while physical geography still seems to be preoccupied with data-collection. Thus, while national agencies have made meteorological recordings (at the macro-scale) for many years, only recently have river flows been consistently measured, water chemistry analysed and soils mapped, by any but interested individuals or researching academics. Without this data-collection in the dynamic sections of the subject, physical geography has largely been concerned with historical interpretations (see section 1 D). Thus, measurement is one of the central problems of physical geography and deserves an examination of some of its basic concepts.

1B 1 Its rules

Beneath the everyday word 'measurement' is concealed a large body of firmly-established scientific rules. For those who think they do not wish to know these rules we have a testing question! What is temperature? If you can reliably answer this you may know something about measurement and should go on to define length, mass and so on!

The problem of temperature is helpful in illustrating the principles of measurement. Firstly it is an **associative measurement**, as opposed to the **fundamental measurements** of mass, length and time interval, which may be measured with the appropriate standard. One could hardly preserve the original degree Celsius in a French museum for future *calibration* (scaling) of instruments, though this is the fate of the standard metre rule! In associative measurements we choose some property associated with that to be measured and rely on a constant relationship or **correlation** between the two to provide our measurements. The correlation must be of high degree and unvarying for the reliable use of such measurements. In measuring temperature we rely on the state of thermal equilibrium between the test material and the glass of the thermometer. We also rely on the unvarying expansion of the contained liquid to the appropriate place on the chosen scale (see pages 26–30).

All measurement is based upon *units* and *scales* which are rigidly standardised by accepted convention. Standardisation was present even when it lacked rigidity, for example the former use of human dimensions such as spans and cubits for length measurement. The effects of changing standards are well demonstrated by the recent change to decimal currency in Great Britain. In the near future Système International (Metric or SI) units will be officially introduced for length and other scales. Except for reference to map and survey dimensions which may occur in SI and Imperial units for some time, the remainder of this book tries to adopt the SI convention. If readers are not yet agile in conversion they may like a gentle introduction to graphs and equations by preparing a compendium of such devices to aid them in conversion (fig. 1.5).

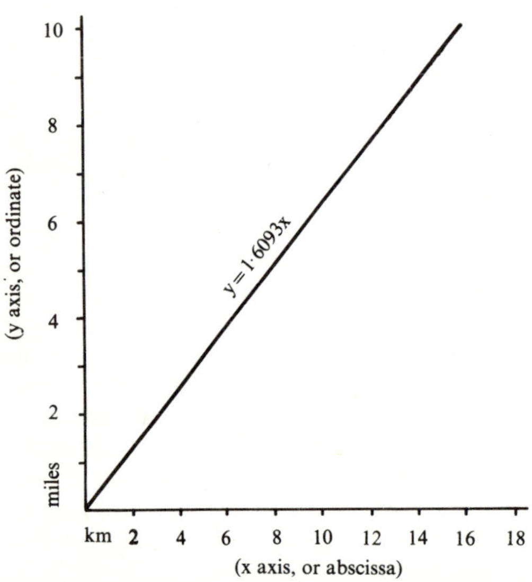

Fig. 1.5 Graph and equation for relating miles to kilometres

Units are classified by their name and the measured property to which they refer. There is a rather more important division of scales, according to the way in which they divide up the acknowledged range of properties. This is now explained.

1B 2 Words and numbers

Nominal scales are based on word classifications. In most cases they come close to breaking the rules of measurement by being virtually impossible (except in the case of the most commonplace or by international convention) to standardise. Thus the intention of the scientist who uses such nominal scales as 'drumlin', 'meander' or 'podzol' may be badly misinterpreted by the readers of his work. The authors are well aware that, in spite of careful editing, there are verbal pitfalls in this text! The philosopher Bentham's view was that, 'Error is never so difficult to be destroyed as when it has its roots in language.' What scientists try to do is to develop a language which has its own internal logical rules. This is inevitably mathematics. Not only is it universally standard in meaning but it also allows us to manipulate our observations and measurements in a way not possible with words. This is not to say that at the descriptive level (maintained by geographers for so long) there is not both power and delicacy in a rich vocabulary, compared with mathematical symbols. Technical writing can, and ought to (as Sir Ernest Gowers points out) become very refined and concise. On a more artistic level, few English-speaking geomorphologists would fail to acknowledge the beauty and perception of the topographical descriptions in classic novels by Hardy, Lawrence and the Brontës. Most of those opposed to mathematical geography hold this view because they are convinced that geography is descriptive, not analytical in the scientific sense. For those convinced that the topographical style of writing is geography (even physical geography!) at its purest the following extracts would provide excellent ammunition:

The time was not long past daybreak, and the yellow luminosity upon the horizon behind her back lighted the ridge towards which her face was set—the barrier of the vale wherein she had of late been a stranger—which she would have to climb over to reach her birthplace. The ascent was gradual this side and the soil and scenery differed much from those within Blackmore Vale. Even the character and accent of the two peoples had shades of difference, despite the amalgamating effects of a roundabout railway;

from *Tess of the d'Urbervilles*
by Thomas Hardy

They wound through the pass of the South Downs. As Siegmund, looking backward, saw the northern slope of the downs swooping smoothly, in some great, broad bosom of sward, down to the body of the land, he warmed with sudden love for the earth;

from *The Trespasser* by D. H. Lawrence

Before dealing with mathematical scales we may, therefore, summarise their benefits:

a) Mathematical scales are better fitted than nominal scales to the range of natural phenomena. How many ambiguous words have we for the surface landforms of the earth?
b) The language of mathematics is universal, as are its internal logical rules and applications.
c) In a move from pure description to prediction, the logical nature of mathematics is a stronger framework for guiding our conclusions which, if made by breaking mathematical rules, must remain substantially invalid.

Mathematical scales are divided into **ordinal scales**, along which objects are ranked (even in words) according to the degree to which they possess certain characteristics, **interval scales**, in which the exact separation of these objects is quantifiable and **ratio scales**, similar to interval but with the additional property of being directly comparable by their ratio. Most ratio scales are based on a zero value. Examples of scales are given in table 1.1. The type of scale available for measurement largely affects the statistical methods used to process the results. The scales defined above represent themselves an order of increasing potential for the more powerful **statistical tests** (we shall deal with these below) from nominal through ratio. The parametric tests make assumptions about the statistical distribution of data, while the weaker non-parametric ones do not. There is again a difference here between physical geography which uses the more powerful parametric tests and human geography which often cannot.

Table 1.1 Scales of measurement and their use with statistics

Scale	*Example*	*Statistical Tests*
nominal	drumlin esker	a group of *'non-parametric'* tests (See Toyne and Newby, pp. 60 and 61)
ordinal	1st order, 2nd order, 3rd order streams cobble, pebble, sand, silt, clay	
interval	0°K–100°K	*'parametric'* statistics which make more rigorous assumptions about the range of data.
ratio	2 grammes, 4 grammes etc.	

Throughout this text the words **variable** and **parameter** are used to describe the elements we measure. Thus variables in a study of soil may be the percentage of silt, pH, wet strength and so on. Though we may sometimes call these soil 'parameters' the latter term is best reserved for the measured properties of statistical distributions (and even then only when the distribution is that of the whole population, and not a sample). Thus we speak of the parameters of the flow duration curve in hydrology; they are the **mean** flow, **modal** flow, **variability** and so on. Variables are essentially instrumentally measured; parameters are statistically derived. A further word, **index**, is often used, meaning a variable measured to indicate the behaviour of another which cannot be directly measured.

1C Designing an experiment

'Doing experiments' has not been thought of as being part of a school or university geography course until quite recently. Yet experimentation is a basic method of science as any dictionary will tell you. It really means the measurement of certain chosen variables **within a conceptual, or problem-based framework** and under conditions largely controlled by the experimenter. The word also implies that the experimenter has begun with a **hypothesis** of the possible results and will collect and process his measurements or **data** with a view to its proof or disproof. As such, even taking measurements from a map constitutes an experiment.

1C 1 Careful forethought

The following are properties of a good experiment:

a) that there should be no *systematic* unmeasured errors (for example, the influence of iron on a compass);

b) that the remaining *random* errors (often caused by lack of skill or precision) should be reduced by increasing the number of experimental units (or recordings), though precision should still be aimed at with as few as possible;

c) the conclusions should have a wide range of *validity*; they should not apply just to that experiment;

d) the experiment should be *simple* in design and analysis;

e) a proper *statistical* analysis of the results should be possible, including a calculation of the uncertainties of the results.

They apply particularly to the type of experiment performed by geneticists and botanists, applying different treatments to different species under controlled conditions (you may have seen the trial plots used by experimental horticulturalists). Such controlled experiments are not common in physical geography although we shall use them in laboratory tests in which outdoor conditions are simulated artificially. Human geographers find such experimentation virtually impossible.

It is unusual for the physical geographer as a field scientist to be able to control any variables in his experimental work by operating a truly closed-system approach. Similarly, he cannot hope to measure every property at every location in the system. He is bound to follow, therefore, a routine of **sampling** or making measurements at points *in space and time* so as to obtain data on the most important variables and their changes of property. The sample is a subset of the total **population** of cases open to investigation. We are all restricted to one temporal sample—that provided by our 'three score and ten' lifespan—and this fact is important when some elements of physical geography, such as erosion and climatic change, may take thousands of years. There are other unavoidable factors restricting us in a choice of sampling routine; for example, daily rainfall is always measured at 0900 hrs GMT and BST because this is the most convenient time for the observers. Luckily the advent of.automated recording apparatus is gradually freeing data collection from the basic human routine of sleeping and eating! Spatially, the sample we choose will obviously be related to the

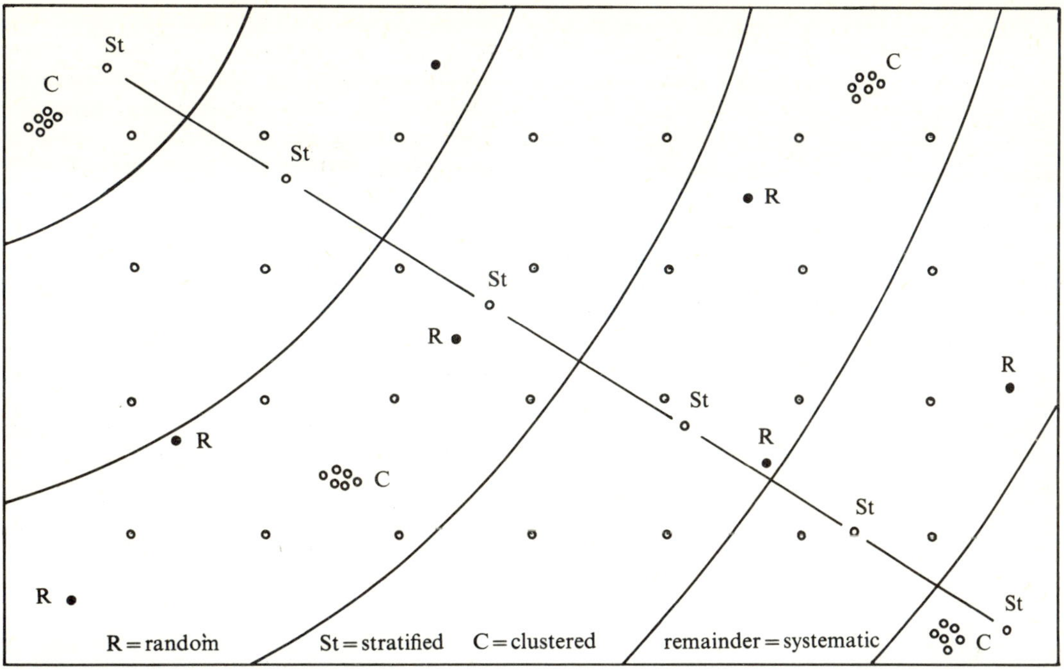

Fig. 1.6 Basic spatial sampling designs (the curved lines are contours)

spatial scale of the system being investigated. Again, we must have some prior intuitive knowledge of what this will be.

There are four basic types of sampling, all dependent on a **probability** (or chance) relationship with the total population. We have to assess the *chance* that our sample is a **representative** one without knowing directly what the results would be if derived from the population as a whole. If you imagine that you know nothing about the properties of a pack of cards, sampling is like deciding how many cards to draw (and from where) in order to reliably learn that there are four suits, Jacks, Queens, Kings and so on. You may like to try this as a game! Fig. 1.6 shows **random, systematic, stratified** and **cluster** sampling applied to soil moisture measurements on a hillside. Random sampling is of most value in drawing conclusions or *inferences* from the resulting data. At present there are restrictions on the number of inferential tests which may be performed on non-random probability samples. Generally speaking large samples are most valuable but this is an entirely subjective judgement and, as usual, the scientist must have at least some intuitive knowledge prior to measurement to know the best sampling routine to choose, from the inherent variability of the property being investigated. Sampling in human geography is dealt with in Toyne and Newby (1971), chapter 1 C, and their section is essential reading matter for the physical geographer too.

1C 2 Data collection and processing

It is not easy to judge at the outset of an experimental investigation just what sort of data will be useful at the end. Clearly, measurements of the major variables, based on the working hypothesis, will be the most important. However, it will almost inevitably happen that the conclusions of a piece of work will need other results to aid their proof: general 'diary' type data which reveals the proper working of the instruments, the dates of important changes in conditions, recalibration and synchronisation of instruments, measurements of some of the variables linking the system to its environment and measurements which require little extra effort but which may aid the research project of another group. All these appear less tedious considerations if the experimenter thinks of his work as the setting up of a **data bank** to which he and others afterwards refer. Even in schools, such comprehensive efforts are possible and no data ought to be thoughtlessly discarded. In complicated experiments there will be little chance for modifying hypotheses at intermediate stages—so the methods had better be mindful of contingencies from the start. As Louis Pasteur once said, 'Chance only favours the prepared mind.' Many major discoveries have been made by scientists who were prepared for the unexpected during merely routine investigations.

In the absence of automatically-recording instrumentation, the recording of readings by hand in a strong laboratory- or field-notebook is an essential part of data collection. Such data may then be transferred to ledgers for office storage and thereafter converted into the form in which it will be analysed, i.e. it will remain tabulated for mental arithmetic or desk calculator work, or be punched on to cards or tape for computer use. One major advantage of the automatic recording equipment now available is that the recordings are usually made in a form immediately usable by a computer—generally paper or magnetic tape. In cases where the recording is by more traditional pen and ink on a continuously revolving chart there is now a means of producing a computer-compatible tape from the chart by using pencil-following or digitizing equipment, so called because it converts the trace of the pen to digits in the form of pairs of coordinates for points at small intervals along the penned line.

For those equipped to use electronic methods of calculation there are obvious advantages in the vast amounts of data which can be used simultaneously and the speed with which complex arithmetic may be done. However, some degree of **cost-benefit analysis** is usually required by professional data processors before embarking on schemes involving a lot of computing. Obviously their use depends on the likely importance of the conclusions. It is important that error checks be made regularly in all forms of data processing—one only has to receive a faulty bill from a company using a computer accounting system to realise that the programmer's data (and very occasionally computers) are not infallible.

Much of the work to be done on the data collected during experiments must be statistical if the experimenter admits any of the following:

a) that there are possible inaccuracies of data;
b) that the data are only a sample of the population under investigation;
c) that the data are not in a form which renders their full meaning readily interpretable.

The reader may be appalled that statistics must therefore enter much of the work he is ever likely to do! It is not our purpose here to provide an introduction to statistical methods. Toyne and Newby provide a clear survey in their chapter 2. Many statistics books are today written in simple styles for those, without a formal training in mathematics, who wish to process data from experiments. Examples are quoted at the end of chapter 9. Before complimenting oneself on acquiring a knowledge of statistical methods one should heed the warning of W. W. Sawyer that: 'In statistics more than in any other branch of mathematics, it is easy to do something that looks entirely reasonable and turns out to be a first-class howler.' He goes on to urge those without a long experience of using statistical techniques to consult a statistician *before* embarking on experimental work so as to ensure that the data collected are compatible with the most powerful available technique for drawing conclusions. In schools without statistics expertise, one of the simpler books recommended is obviously an alternative of which Sawyer would approve.

Statistical methods fall basically into **descriptive** and **inductive** categories. Descriptive statistics solve some of the problem (c) above in that they present ways of expressing the data as an entity rather than as separate items. They summarise the behaviour of the separate variables being measured, usually by describing parameters of their **distribution** or dispersion about some average value. Maxima, minima, averages, and standard deviations are all descriptive statistics. Inductive statistics are used to sort out the meaningful variations from '**noise**'—or background information such as is provided by random inaccuracies, assess the representativeness of the samples we have taken and quantify the relationships between the variables. A further branch of statistics, again based on probability or chance, allows us to quantify the *uncertainty* which must attend use of the obtained relationships, for example **significance tests** and **confidence limits.** The use of probability again serves to test our inductive logic against the more valid mathematical rules. Inductive statistics are, therefore, a good deal more 'vital' than those 'vital statistics' (which are, incidentally, not statistics but 'raw data') of which we hear and see so much!

For the purposes of the work described in this text we may further subdivide inductive statistics on the basis of the number of variables whose relationships are studied. We shall consider the **bivariate** case, in which two variables (often the **independent** variable or control variable and the **dependent** variable) are considered and the **multivariate** case where more than two variables are analysed simultaneously. The basic test of relationships between variables will be **regression analysis** and the calculation of **correlation coefficients.** Bivariate regression and correlation are easier to contemplate conceptually because both have a convenient pictorial representation by means of a conventional Cartesian graph. We may consider our earlier exercise of graphing Imperial against SI units as one resembling regression, although of course the correlation is unusually perfect (fig. 1.5) and we know the relationship is constant.

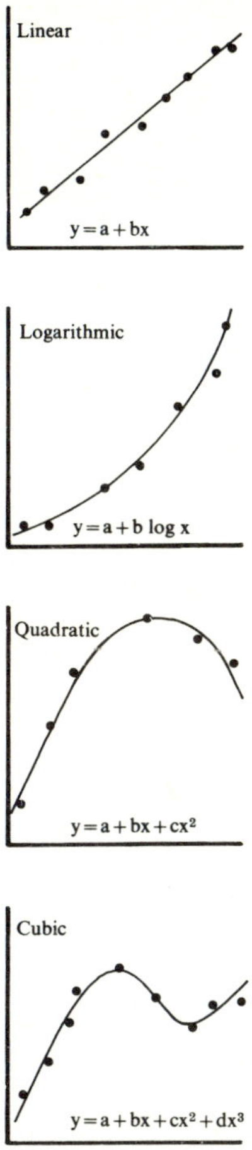

Fig. 1.7 Some mathematical relationships between two variables (i.e. bivariate)

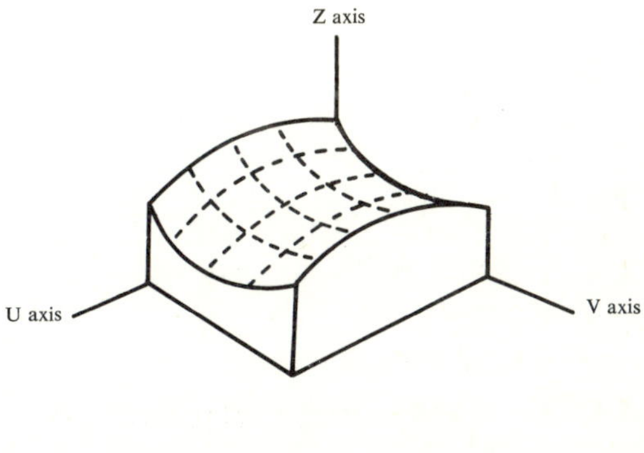

$Z = a + bu + du^2 + euv + fv^2$

Fig. 1.8 The trend surface corresponding to the bivariate quadratic case. The surface is indicated by the deformed rectangular grid (dashed lines)

Table 1.2 Examples of the use of statistics in physical geography

Mean Values	Temperature
Standard Deviations	Sizes of stones in glacial deposits
Coefficient of Variation	Streamflow duration
Regression Analysis	Stream channel slope and flood flows (bivariate). Trend surface analysis of relief and slopes (multivariate).
Correlation Analysis	Relationships between drainage basin characteristics

In most of the work we shall suggest there will inevitably be an imperfect relationship involving a considerable **scatter** of points on such a graph (see fig. 4.16). Multivariate relationships must be considered in terms of **trend-surfaces** linking values on as many axes as there are variables. They require three-dimensional thinking! In each example, both bivariate and multivariate, the mathematical form of the relationship between variables may be simple or complex (see the bivariate examples in fig. 1.7). Most of the relationships encountered in schools will be bivariate. However, fig. 1.8 shows the concept of a trend-surface since it crops up later in the book (pages 151 and 186).

Correlation analysis expresses the degree to which the data points are scattered around the line we have 'drawn' (i.e. calculated the formula of) to plot their best-fit relationship. Correlation is an important property of natural systems in dynamic equilibrium. The close relationship between variables is the means by which these systems acquire their properties of self-regulation. Table 1.2 gives a brief introduction to the ways in which both descriptive and inductive statistics may be used in physical geography.

1C 3 Drawing conclusions

There is a cynical quotation at the beginning of some university notes on statistics: 'Statistics are a means of being precise about matters of which we will remain in ignorance.' It, of course, implies that we need to draw conclusions from our analysis beyond the expression of statistical results. Our work, from sampling to analysis has been based on chance and consequently our results are *only the most probable solution* to the problem investigated. However, we have established **empirical laws**, derived from measuring the properties under investigation, rather than deducing them from theory; empiricism is the philosophical doctrine that experience is the only source of knowledge. Every science aims to develop a body of theory whose constituent laws are completely logical and whose empirical validity has been proven beyond doubt. Though the theoretician in science may be considered 'the thinker', by the very nature of theoretical advance it is necessary to carry out empirical investigations too; the theoretical and empirical branches of science, therefore, hopefully cooperate and advance together. It is probably worth mentioning now that there is virtually no theory of physical geography—our theories are basically those of the physicist. (Fig. 9.3 in our final chapter returns to this theme.) The field for empirical study, therefore, is very wide.

We need to spend a considerable time on the **causes** behind the **effects** we have demonstrated in our systematic analysis. Rainfall may be related statistically to river flow, but how? There is probably no other branch of the philosophy of science which has caused more controversy than cause-and-effect reasoning. The general causal principle is:

If C Happens, then (and only then) E is always produced by it. *Or* C ⟶ E

It may be that we can immediately assume the mechanism of causation in our results from theory or other laws. For example, the correlation between slope failure and relief has gravity as its major explanation. However, it is extremely easy to make faulty causal inferences from our results and the early use of correlation analysis by geographers has shown that quite strong statistical relationships may occur by chance between variables which have non-existent **causal** connection. For example, it is possible to demonstrate 'fake' correlations between rainfall and coal mining, or beer drinking. If we do make conclusions about cause-and-effect we must not assume that the correlation we have discovered is a simple relationship. Are there any intervening stages? Are there **covering laws** stating the conditions under which causation occurs? Does a reversible situation exist where **feedback** occurs; in other words, does C ⟶ E ⟶ C? Feedback is returned to in chapters 3, 4 and 8.

There has been much talk amongst geographers about **models.** Basically most of our work will be modelling; since we are mainly working empirically we can only assume our results to be idealisations of reality (and this is a brief definition of models). Our laws are statistical models because our empirical experience can never be total. Models are built up as experimentation proceeds, the gaps in the conclusions from one investigation forming the hypotheses for investigation in the next. Thus, currently, the model of runoff which stresses water movement down the surface of a hillslope (see chapter 5) is being adapted and modified and a new model of runoff has appeared stressing throughflow. Though models tend to grow bigger and more complex it must be remembered that, as Hempel says, 'science aims at a parsimonious description of the world', dealing basically in simple principles of general application. What is the everyday meaning of parsimonious?

1D Hindsight and foresight

1D 1 The past

Our treatment of experimental technique and scientific principles above has revolved around present-day functioning systems: the atmosphere, the river basin, the soil, and so on—at a variety of scales. The restricted temporal nature of our work has been touched on in connection with sampling. There are, however, many physical geographers studying the historical behaviour of these systems by applying the same scientific techniques to the traces left from a former time. The most obvious example is in the strong links between geomorphology and the geology of the recent past. We may divide these *historical* investigations from *process* studies. Though the general move is towards process studies and a systems treatment incorporating dynamic as well as static elements, historical studies add the qualifications which we cannot detect in our short lifespans. We may muse that there will, in the future, be less need for historical aspects of physical geography thanks to the records which are being taken during process studies today. For the present, however, historical studies remain, are of great value and are dealt with in this text where appropriate (especially chapter 8).

1D 2 The future

Much of the 'profit' in conducting process studies and discovering relationships by regression and correlation analysis is the ability to *predict* the future behaviour of one of the variables following changes in the other. In chapter 9 we hope to show the role that the physical geographer may come to fill in society, one of revealing such relationships and predicting effects, particularly those resulting from changes caused by man himself. The more relationships which can be quantified, the better systems are understood and the greater can be our preparedness for the future.

1D 3 A warning!

Most of the material in the following chapters presents the results of work already performed by others. You will have a chance in the exercises to prove (or dispute!) the findings for yourself. Though the strict methodology stressed in this chapter will not frequently be referred to, the basic elements of care and skill in data collection, processing and making conclusions will be required if you undertake work yourselves. We trust that you will not be deterred from experimental work because some of the terms in this chapter have been new to you. There is a considerable amount of methodological licence exercisable in science and the amount of time spent following the rules depends largely on the importance of and your ambitions for the results! The basic step of measuring two variables, graphing the data and thinking about the relationship should not be beyond any of you. However, to be respected for a scientific contribution towards predicting the future behaviour of important environmental systems you had better adopt a stricter attitude!

2 Ground Level Meteorology

Should it be imagined that technology has the capability to master or control atmospheric processes, then we urge that greater thought be given to the innumerable ways even subtle weather variations influence our outlook and actions. The reader may care to log personal reactions to weather from say the mental effort required to get up and about on a chilly winter morning to the prospect of a cosy evening by the fire, and the resultant fuel bills! Any day's routine will do. Even in our temperate climate we must take elaborate precautions to deal with 'unexpected' hazards. Consider the capital expenditure on council rates for maintaining rarely-needed snow clearing equipment, or the extra cost of designing water supply systems which can cope with prolonged droughts. On a global scale, it is difficult to appreciate that the world faces annual weather damage totalling some £15 000 million for the want of reliable forecasts alone! Yet, such estimates do not include countless side-effects or possible benefits arising from a more efficient utilisation of weather 'resources'. We pay dearly for our ignorance compared with the cost to acquire the information that would enable society to *co-operate* with the weather.

Whilst former claims by geographers for the dominance of weather controls cannot be supported, there is a danger that in the retreat from so-called climatic determinism we devalue the significance of weather in our lives. A simple 'shopping list' of everyday household items used largely to *combat* weather will give some idea of its impact on our purses.

Unfortunately, familiar proverbs like 'out of sight, out of mind' have more than an element of truth concerning the appeal or otherwise of meteorology. Most of the processes at work remain tantalisingly invisible, and it is a bold person who dares to isolate any part of the atmosphere as an entity or system for detailed study. We dare to do just this, however, by limiting ourselves to the processes near to the ground. Not only is this part of the system, the one where we can feel and experience the effects of atmospheric processes, but it is also the obvious one concerning the sort of techniques which readers can use effectively. Whilst being well aware that this takes us directly into what many consider as the 'deepest and most uncharted waters' of physical geography, we urge the reader to take the plunge with us. By studying meteorology first we have the logical beginning to guiding the reader along the 'paths' of *energy* and *mass* movements which link the various physical systems examined in later chapters.

Although atmospheric sciences are older than most aspects of physical geography, our inability to escape from the ground for reliable measurements in *depth* seriously hampered their development until recently. Indeed, it can be argued that until we were able to escape from the atmosphere entirely, a full and fruitful appreciation of its complexity was impossible. Nevertheless, in human terms, it is through such earthbound experiences as 'rain stopped play' that the immediate effects of weather are made apparent. Occasionally the experience is not so mundane of course.

Nearly every nation has an official organisation like our own Meteorological Office to compile records, undertake research and provide various weather services. The World Meteorological Organisation (WMO), World Weather Watch (WWW) and its associated bodies like the Global Atmospheric Research Programme (GARP) are the international agencies vital to meteorology. It is pertinent to consider why meteorology should have pioneered such close world-wide links before other sciences. In fact, between the humblest enthusiast who is a registered rainfall recorder and the most sophisticated weather satellite lies a hierarchy of management systems well worth studying in its own right.

Ever since Aristotle's *Meteorologia* about 350 B.C., studies of 'things above' have been deeply rooted in superstition and myth. Terms like 'natural' and 'supernatural', 'forecasting' and 'fortune-telling' are easily confused in dealing with such a dynamic and largely invisible medium as the atmosphere. Whilst the collection and testing of local weather sayings add much to the lore of weather, we wish to share with the reader an enthusiasm for the fundamental laws governing the physics of the lower air layers and the enjoyment which can be gained from their measurement in the field. We hasten to add that the latter is not the reason for neglecting precipitation processes aloft in favour of studying how the atmosphere is heated!

All the energy that drives the atmospheric 'heat engine' and makes life possible comes from the sun. Inputs of heat generated within the earth and that from fossil-fuels can be neglected since they only amount to some 5×10^{-2} watts per square metre—barely enough to melt a 5 mm thickness of surface ice a year. You will be familiar with the watt as a unit of power in the rating of electric light bulbs. Because we assume for the purposes of calculating averages that the present atmospheric system maintains its state on a global scale over standard periods such as the 35 years chosen for temperature averages, the energy inputs and outputs must balance.

It is equally true that balances occur at *local* and *micro-scale* levels over given intervals of time. Thus, whilst global aspects cannot be ignored, our acknowledged intention of concentrating on what can be *observed* near the ground leads us to emphasise those techniques valid to *a place*. This is not just a matter of convenience, however, for the key exchanges of both energy and mass in the environment are controlled by the state and structure of the ground. As geographers we prefer to be accused of having our feet firmly on the ground rather than of having our heads too far into the clouds! Consequently, we concentrate upon processes at the ground-air interface and deal in much less detail with those processes involving places far afield. With the latter, the 'field' is the globe itself; an approach which hinges nowadays upon the *remote* techniques of space technology. Although satellite information and the like is now available to provide us with stimulating and instructive 'overviews' of atmospheric motions, we feel that putting a satellite into orbit would strain the financial resources of schools!

2A Energy inputs and outputs

Radiation is the key to understanding the heat balance equation. Domestic heaters and the more imaginative science fiction writers have popularised the terms radiator and ray to the extent that many misconceptions exist concerning the real role of radiation in energy transfers. This is hardly surprising when we recall that it was not until Max Planck published his revolutionary *Quantum theory of radiation* in 1901 that so-called classic theories in physics were revised. Our awareness of such difficulties leads us to consider radiation more fully than is usual in geographical texts. Life itself is the most complex and beautiful manifestation of this energy.

2A 1 Solar radiation

Thermal radiation does not require a material 'carrier' as do the other better-known forms of heat transfer, viz. conduction, convection and the evaporation-condensation process. Its presence can only be detected by its effects upon matter, which includes the vital process by which we perceive light with our eyes. Solar energy (insolation) travels through space at a speed of 3×10^5 km s^{-1} (the speed of light) from the sun's surface. Like the individual in a ring of people surrounding a bonfire, the fraction of energy the earth receives from the sun is a function of the distance from the emitting source. We must imagine the earth

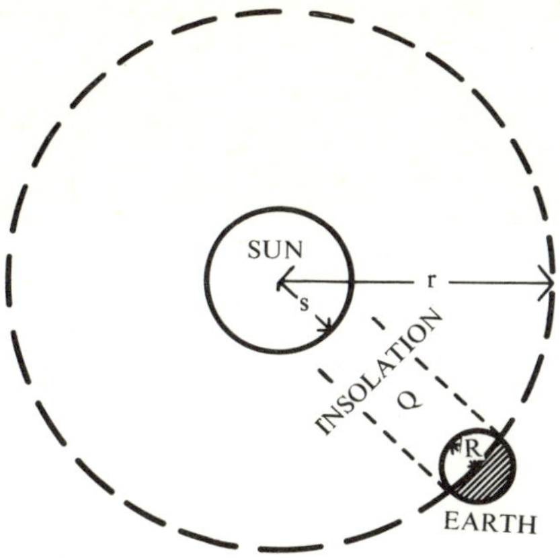

$r = 150 \times 10^9$ metres Not to scale and atmosphere ignored

$s = 69556 \times 10^4$ metres

$R = 6371 \times 10^3$ metres

Fig. 2.1 Solar radiation to the earth

as a small point on the surface of a sphere with the sun at the centre. If the total energy leaving the sun is Q then, at the earth, the radiation intensity is

$$\frac{Q}{4\pi r^2}$$

where r is the distance between sun and earth (150×10^9 metres). The reader will recognise that $4\pi r^2$ is the surface area of a sphere radius r. In short, solar radiation intensity decreases with the square of the distance travelled across space in the same way that the heat intensity from a bonfire diminishes as we move away from it. Physicists refer to this principle as the **inverse square law,** and we will shortly require it to calculate the intensity of solar radiation Q_s reaching the roof of the atmosphere (see fig. 2.1).

In 1888 Clerk Maxwell propounded the electromagnetic wave theory of radiation. Pictorially, this can be visualised as a train of waves, the distance between successive peaks or crests being the wave length (λ). Since solar radiation contains a whole range or *spectrum* of different waves, some short and others long, the various forms can be distinguished by their wave lengths. These are measured in micrometres (μm being 10^{-6} metres) and nanometres (nm being 10^{-9} metres).

We encounter the shorter waves on being X-rayed and the longer ones when tuning in to radio stations. The unpleasant and pleasant events we associate with each have real physical significance too, for prolonged exposure to short waves is lethal! This is why radiologists now keep records of the number of times we have been X-rayed. However, under controlled conditions, these rays can be used to kill unwanted tissue in tumours. *Thermal* radiation lies approximately within the range 0·2μm to 100μm. The human eye only resolves the considerably smaller range 0·36 to 0·76μm as the visible spectrum of colours from violet to red respectively. From 0·2 to 0·36μm *ultraviolet* (u.v.) waves occur, and from 0·76 to 100μm *infra-red* (i.r.) waves, all of which are invisible. Below 0·2μm are the shorter wave X-rays and emissions from radioactive substances such as gamma rays. Above 100μm occur the now equally familiar radio waves like VHF, the so-called radar bands and long electric waves. Those not conversant with the electromagnetic spectrum will find it useful to construct a linear scale indicating the ranges mentioned. The main 'bands' are illustrated on fig. 2.2.

We all accept that red is a 'warm' colour and violet a 'cold' one. Such notions stem from the relationship between the temperature of a body and the wavelengths of radiation it emits. Actually, the colour-wavelength relationship is very complex since a body emits a whole spectrum of wavelengths rather than only one. This situation can be simplified if we consider radiation from so-called *black bodies*. Such bodies are the most efficient emitters of radiation known and, by the same token, the best absorbers in that they 'accept' all incident radiation.

Empirical observations led Stefan to state that the *total* energy (Q) emitted per unit surface area per unit time over *all* wavelengths is proportional to the fourth power of the *absolute* temperature (T) of a black-body, i.e. $Q \propto T^4$. Boltzmann arrived at similar conclusions on theoretical grounds, and so the **fundamental radiation law** valid for black-bodies is:

$$Q = \sigma T^4$$

Where: Q is now measured in watts per metre^2 (W m^{-2}), T is measured in degrees kelvin (0 °C = −273 K) and σ is the Stefan-Boltzmann constant now stated as

$$5{\cdot}67 \times 10^{-8}\ (\mathrm{W\ m^{-2}\ K^{-4}}).$$

You will find radiation intensities measured as gramme-calories per square centimetre per minute in many texts; a unit often contracted to langleys per minute. Those concerned about their weight will be familiar with the calorie as a unit of energy. Indeed, our calorie intake at meals is one of the main ways that *we* convert solar energy. Many of us take much more 'fuel' than we need in energy balance terms! Too much, like too little, can adversely affect our capacity to do work. Some fuels like sugars have a high calorific value compared with their *mass*. In fact, sheer bulk is a bad indicator of energy, even if it fills our stomachs! Thus, the gramme-calorie and langley are associative measurements based upon the heat energy required to raise the temperature of a gramme-mass of water through 1 °C. However, this is specifically from 14·5 °C to 15·5 °C, and is experimentally a difficult standard because other dependent properties of water become involved. It is better to think in terms of the energy needed to perform a known amount of mechanical *work*, such as the 'exertion' in providing the *force* to accelerate a mass of one kilogramme a metre per second squared (i.e. kg m s^{-2}) if this 'exertion' or force moves through a distance of one metre, the mechanical work performed is one *joule*. This mechanical 'work' energy unit is related to the 'heat' energy unit by 1 calorie being equivalent to 4·18 joules. Dieticians base the daily food requirements of people in different jobs upon such energy values.

Whilst it is tempting to delve deeper into such basic aspects of physics, we must content ourselves here with the fact that newtons, joules, and the watt mentioned already, refer to the physical quantities of force, energy and power respectively. The joule (J) is a product of the newton (N) and the metre (m), whilst the watt (W) is a measure of joules per second; hence

1 calorie/sec = 4·2 joules/sec = 4·2 watts.

The class exercise below explains how calculations for the intensity of solar radiation reaching the earth or other planets in the solar system can be made as a function of their distances from the sun and their surface areas. Listing the results in terms of standard fluorescent lights per unit area is of interest! The reason for choosing fluorescent lights as opposed to any others for this analogy is made apparent in the discussion which follows the exercise.

CLASSWORK

Evaluation of solar radiation reaching the atmosphere

1 With reference to fig. 2.1 and, taking the sun as a black body radiator with a mean surface temperature of 5793 K (usually rounded to 6000 K), apply Stefan-Boltzmann and the inverse square laws to calculate the intensity of solar radiation Q_s(W m^{-2}) incident upon the outer surface of the atmosphere.

2 Why do the dimensions of the earth and atmosphere not enter into this calculation?

3 The earth's orbit of the sun is not exactly circular: in our winter we are 147×10^9 m away (the perihelion) and 152×10^9 m in our summer (the aphelion). Thus, what is the resultant percentage variation of radiation intensity reaching the atmosphere?

4 What is the area of the earth's surface presenting itself perpendicular to the main beam of solar radiation at any time? So, what is the energy received by the earth?

5 Find out the distances from the sun and dimensions of other planets in the solar system to calculate the solar energy they receive.

The second radiation law basic to meteo-

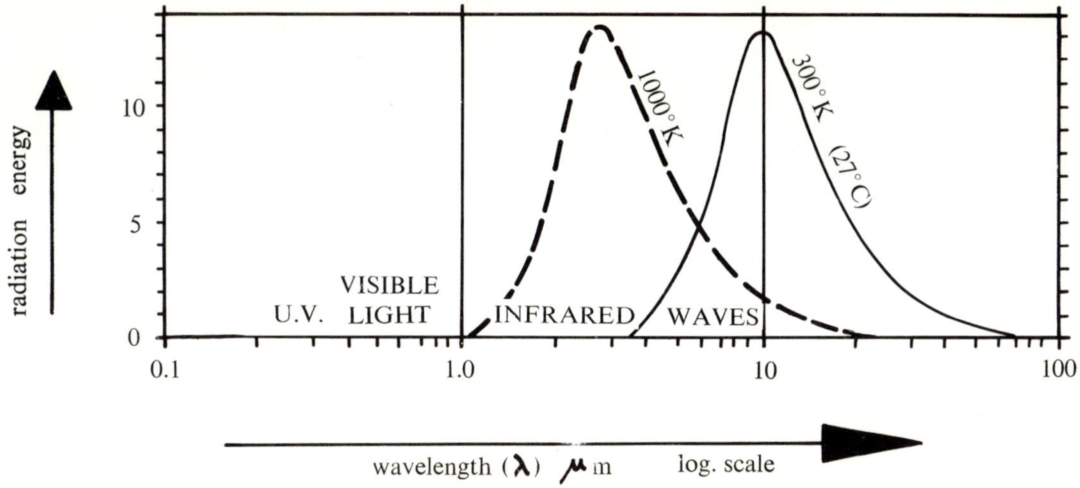

Fig. 2.2 Radiation emitted from black-bodies

rology was proved by Wien. It states that the product of the absolute temperature (T K) of the radiating body and the wavelength of the most intense radiation (λ_{max}) is a constant, i.e. $\lambda_{max}\, T$(m K). Fig. 2.2 illustrates this law graphically: two curves are shown for the radiation emitted from appropriate black bodies with surface temperatures of 300 K (27 °C) and 1000 K (727 °C). They show that the higher the temperature of a body the more its maximum radiation shifts towards the shorter wavelengths. Hence the relation is known as **Wien's displacement law**, being:

$$\lambda_{max}\, T = 2880 \quad \text{or} \quad \lambda_{max} = \frac{2880}{T}$$

Where: T is measured in degrees kelvin (K), and λ^{max} in micrometres (μm).

CLASSWORK

Nature of solar radiation

1 Using Wien's displacement law

$$\lambda_{max} = \frac{2880}{T},$$

calculate the maximum wavelength $\lambda(\mu m)$ of the most intense radiation emitted from the sun with a surface temperature of 6000 K. Compare the result with those from surfaces at 300 K and 1000 K, as in fig. 2.2.

2 Draw a graph like fig. 2.2. Using the result calculated above, locate the maximum wavelength of solar radiation at 6000 K, and construct a curve about this point identical to the two already drawn. What does this new curve show?

3 In what range does the most intense solar radiation fall?

4 Why is solar radiation called *short wave* relative to that emitted from other bodies?

5 For what reason was it suggested earlier that the power rating of fluorescent lights made a better analogy with solar radiation as opposed to others?

Having considered the intensity and nature of insolation at the roof of the atmosphere, we are now in a position to examine briefly its passage to the earth's surface. Since we will be dealing with invisible processes which can only be illustrated and quantified mathematically, readers are commended the animated film loop on the subject directed by Clark and Allen (see page 220). As we shall refer to various layers of the atmosphere and their properties, attention is drawn to fig. 2.3 showing the concentric 'shells' of the atmospheric system.

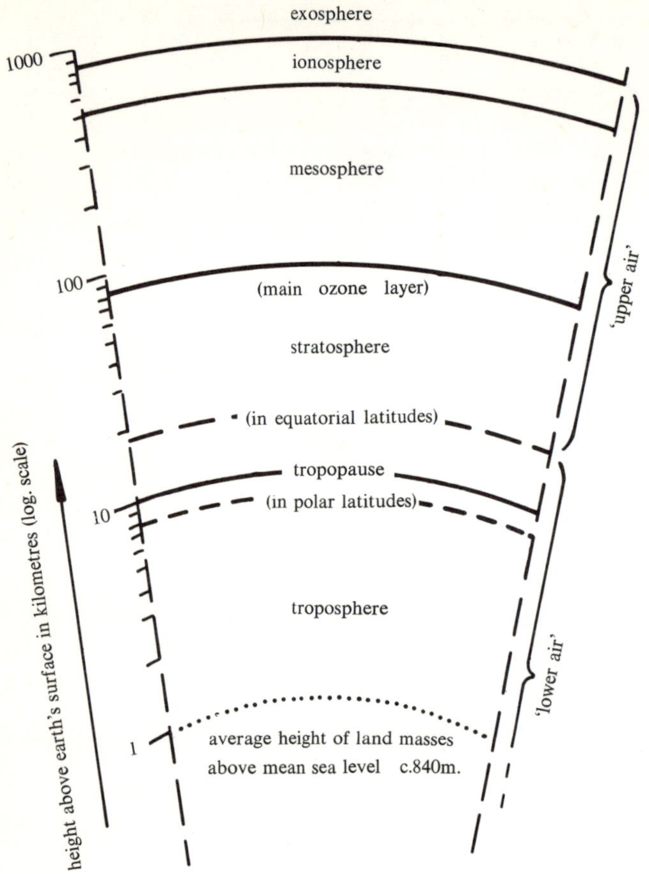

Fig. 2.3 The structure of the atmosphere

Less than half of the amount of insolation received (Q_s) is transmitted to the ground (τ); the remainder is either reflected and scattered back into space at various levels (ρ) or absorbed *en route* (α). Here it is important to remember that, although energy can be converted from one *form* to another, it cannot be created or destroyed. Once again, we are accustomed to this principle in terms of the food energy we utilise to maintain body heat and to do work. So, if a system maintains **thermal equilibrium**, it is clear that its energy inputs and outputs must balance. Such principles are embodied within the law of the **conservation of energy**; a vital concept in science, and in dealing with captured and released energy in any layer of the atmosphere at suitable time scales. Thus, processes whereby energy exchanges occur across established interfaces between each layer of the atmosphere can be represented by simple equations, e.g. Q_s is clearly equivalent to the sum of transmitted (τ), reflected (ρ) and absorbed (α) radiant energy throughout the atmosphere. We will return to the question of energy balance equations shortly.

Much as the argument above shows that absorption and emission are reversible processes, there are common features between the transmission and reflection of radiation in quantum terms. It is a matter of certain wavelengths of radiant energy being 'accepted' or 'rejected' by different constituents in the air, much as different people and organisms select different forms of food energy. Thus, each layer of the atmosphere acts like a screen or filter. For example, considerable reflection and absorption occur at the ionosphere so that all dangerous short-wave radiation is 'screened out' and only that which gives us a suntan penetrates. The screening is significant in the photo-chemical breakdown of ozone (O_3) and the ionisation of oxygen (O_2) at the roof of the atmosphere. These complex processes are known to be vital concerning intercontinental telecommunications, and biologists consider their effects as having far reaching controls over genetic mutations and so of evolution itself. To the meteorologist and climatologist, periodic variations of the processes appear to exercise significant influences upon weather patterns and climatic changes. Those fortunate enough to have seen the displays of the aurora in the northern skies have witnessed the great beauty of electrical 'disturbances' aloft. As in the volcanic eruption or earthquake, the atmosphere occasionally gives us glimpses of its complexity and power.

Further reflection and absorption take place in the transitional mesosphere and the stratosphere. Respectively, the surfaces of volcanic and meteoric dust bands called noctilucent clouds and concentrations of ozone are responsible. However, as these exchanges are illustrated later (see fig. 2.5) and our chief concern is with processes at the ground surface, we move on to consider the fate of the transmitted fraction of solar radiation.

2A 2 Terrestrial radiation

The ground will absorb incident radiation transmitted (τ) less that immediately returned to the atmosphere because of surface reflection. That reflected can be given as a percentage ratio known as the reflection index or **albedo** (A). For example, the total reflection from a mirror-like surface would be 100 per cent (A = 1·00) and the complete absorption of a black body 0 per cent (A = 0). We explain how albedos can be measured in the field later (see page 43). At any time on a global scale the cross-sectional area of ground surface normal to the direct beam of transmitted insolation is πR^2. So, the *flux* of direct radiation received is $\tau(1-A)\pi R^2$. However, the earth *emits* radiation over its whole surface area of $4\pi R^2$ according to Stefan-Boltzmann's law, viz. $4\pi R^2 \sigma T^4$ where T is the mean surface temperature of the earth's surface (K) and σ the Stefan-Boltzmann constant already given.

Over a suitable period, thermal equilibrium is maintained on the ground and so imputs and outputs balance. The following expression, therefore, is a statement of this balance:

$$\tau(1-A)\pi R^2 = 4\pi R^2 \sigma T^4$$

or

$$\tau(1-A) = 4\sigma T^4$$

CLASSWORK

The contribution of direct solar radiation

1 By substituting for the Stefan-Boltzmann constant in the expression above, show that the mean surface temperature T of the earth through *direct* solar radiation alone might be expressed as:

$$\log T = 1{\cdot}66 + 0{\cdot}25 \log \tau (1-A)$$

2 Using this equation, compute values for T when τ equals say 10, 100 and 1000 $W\,m^{-2}$ and A equals say 0, 0·1, 0·2 and so on to 0·9.

3 Prepare a graph with the abscissa divided according to the logarithm of T (from 0 to 300 K) and the ordinate according to the logarithm of τ (from 0 to 1400 $W\,m^{-2}$). On the graph plot lines showing the relationships between T and τ with selected albedos of say 0, 0·2, 0·5 and 0·9.

4 Study the graph carefully regarding the extent to which *direct* solar radiation does not account for the heating of the earth. What other radiation process must account for the deficiencies?

This exercise shows beyond doubt that *direct* solar radiation alone is not responsible for heating the atmosphere. We have emphasised the point in this way to destroy any misconceptions to the contrary. Furthermore, the results plotted indirectly indicate how much additional energy is needed to maintain the mean temperature of the earth's surface at 14 °C (or 287 K). Clearly it would be a very frigid world if we had to rely upon direct solar radiation *alone*! The deficiencies are largely made up by the absorption of *terrestrial* radiation returned to the troposphere, and so of radiation from the air itself back to the ground. This process is aptly termed as *counter-radiation* (c). We shall see that variations in the degree of absorption of terrestrial radiation in the lower air layers are crucial in local climatology (see page 58) and in helping to explain major climatic changes (see page 178). In effect, the atmosphere is heated from the ground layers upwards making it evident why air temperature normally decreases with altitude. Techniques for studying the nature, extent and variations of terrestrial and counter-radiation near the ground are crucial to both meteorology and climatology.

CLASSWORK

The nature of terrestrial radiation

1 Using Wien's displacement law and fig. 2.2, plot a graph to illustrate radiation emitted from a surface with a mean temperature of 14 °C (287 K).

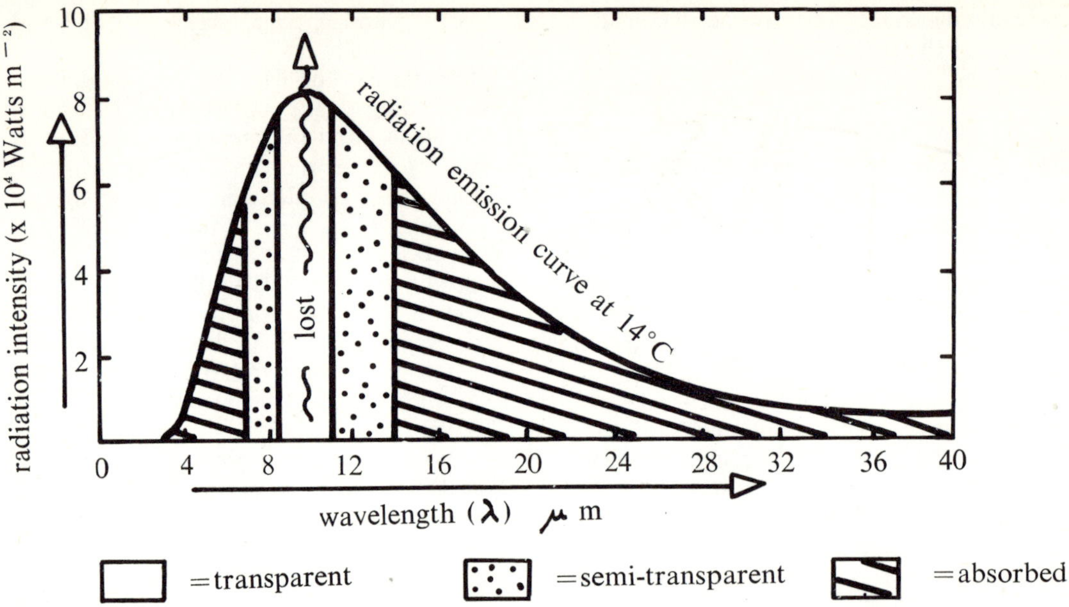

Fig. 2.4 The absorption of terrestrial radiation by water vapour and carbon dioxide

2 What is the wavelength of the maximum intensity of radiation emitted, and within what range of the electromagnetic spectrum does the emission curve fall?

3 Why is terrestrial radiation termed as *long-wave* radiation compared with direct solar radiation?

4 Redraw the emission curve of terrestrial radiation at 14°C using a linear scale for wavelengths along the abscissa instead of the logarithmic scale of fig. 2.2. Compare the result with fig. 2.4.

The extent of terrestrial radiation absorption in the lower troposphere is mainly due to the presence of *water vapour* and *carbon dioxide* in varying concentrations. Fig. 2.4 indicates the mean proportions absorbed and lost. Using standard strip, square, planimeter or even weighing techniques for measuring irregular areas, the reader will be able to calculate the percentage of outgoing radiation lost through the transparent and semi-transparent 'windows' shown in the graph, compared with the total area beneath the emission curve. We hope that the mathematically-minded will seize upon the opportunity to confirm the results obtained by using the equation of the curve and integral calculus.

In summary, we have a situation where long-wave terrestrial radiation is partially absorbed in the troposphere giving an additional rise in temperature to that through the absorption of insolation. The resultant emission of counter-radiation from the troposphere, therefore, is very critical in boosting ground level temperatures. An analogy with the **storage capacity** of a greenhouse is appropriate and we may also draw a useful parallel in hydrology with the storage of water in the soil (see page 110). It is worthwhile pondering upon the role of our atmosphere, particularly the troposphere, in conserving heat and so the likely consequences of possible changes in its water vapour and carbon dioxide constituents. Geographical variations are very important also. Other planets and satellites with 'thicker' or 'thinner' atmospheres clearly have different heat economies. Radiation-heat situations encountered by astronauts during space walks and lunar missions provide useful contrasts in this respect.

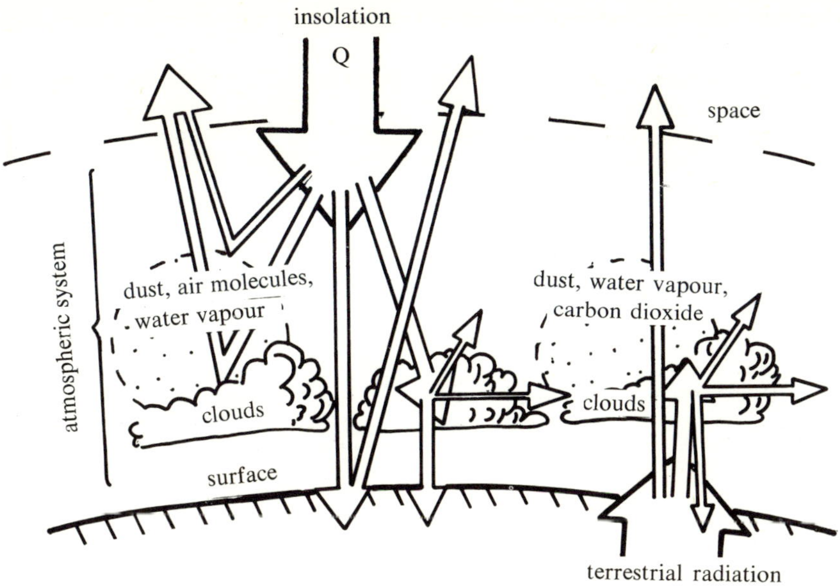

Fig. 2.5 Radiation exchanges in the atmosphere

2A 3 Radiation energy exchanges

The analogy of radiant energy transfers with wave motions permits us to picture its various forms in terms of conventional flow diagrams. Whilst these flows are complicated by interference and diffraction patterns, like those which can be generated in ripple tanks, the model is usually simplified by examining the passage of a single 'stream' or 'beam' of energy. This visual approach allows us to differentiate the hierarchy of systems (or subsystems) within systems which characterises such complex processes. Since energy cannot be destroyed, we can assign a value of 100 units to any given input of direct solar radiation, and trace the resultant transfers within the atmosphere in terms of the different *percentages* 'streamed off'. In some ways, the technique is similar to plotting the homeward routes of pupils given that 100 per cent leave school at a specified time. However, we feel sure that over half will not return to school immediately, nor will any remain 'stored' in the system between school and home! So, although it is tempting to compare the urgency of pupil departure yet slowness of their return with the respective frequencies of solar and terrestrial radiation, we will not pursue the analogy further.

Fig. 2.5 shows the gross radiation energy exchanges in the atmosphere. The transfers indicated will be used later as a means of evaluating local heat budgets (see page 38).

CLASSWORK

Main radiation energy exchanges

1 Using fig. 2.5 and assuming that Q_s represents 100 units of radiant energy, plot the following main radiant energy exchanges by drawing flow lines to scale, e.g. 1 mm to 10 units. Since each exchange is subject to considerable variation geographically, all values are rounded to the nearest five per cent for convenience. **Solar radiation** (namely, insolation Q_s = 100).

a Reflected at various levels in atmosphere, $\rho = 25$

b Absorbed at various levels in atmosphere, $\alpha = 25$ (of which 10 is returned to space and 15 to ground as counter-radiation—see below)

c Transmitted but reflected through surface albedo, $\tau A = 5$

d Transmitted and absorbed by ground, $\tau(1-A) = 45$

Terrestrial radiation, comprising $\tau(1-A)$ above plus *all* sources for counter-radiation; hence 45+15 (via b above)+55 (via f below)+25 (from mass transfers not detailed here) = 140.

e Lost directly to space through atmosphere = 10

f Absorbed at various levels in atmosphere = 105 (of which 50 passes on into space and 55 returns to ground as counter-radiation—see above)

Note: the additional 25 units from mass transfers is considered later under conduction, convection and evapotranspiration.

2 Consider possible modifications to the exchanges plotted during an ice age with marked increases in the surface albedo and decreases in the water vapour content of the air.

3 Convert the percentage values shown to $\mathrm{W\,m^{-2}}$ on the basis of the solar constant Q_s determined in the first exercise. Relate the mean output of terrestrial radiation to that from appropriately rated electric storage heaters.

As might be expected in a global study based upon mean data, some controversy surrounds the percentages assigned to the various transfers. One of the main problems here lies in the difficulty of measuring total reflectivity and its complex geographical and seasonal variations from fresh snow surfaces (*c.* 90 per cent or 0·90) to arable and water surfaces (*c.* 5 per cent or 0·05). The hazard of severe sunburn when skiing in alpine areas or the influence of the colour of our clothes on personal comfort during heatwaves are well known examples. However, further complications arise for, although surfaces like fresh snow are highly reflective to incident short-wave radiation, they absorb most diffuse long-wave counter-radiation. Those who have over-exposed photographs of snow scenes by not appreciating the improvement of visible light under such conditions will be aware of the former point. With regard to the latter, it is of interest to examine weather situations which increase the levels of counter-radiation and so the rates of ablation of snow and ice. It should also be noted that the wetting of a surface changes the albedo. For example, using the measuring techniques outlined later (see page 43), examine the albedo (A) of wet and dry playground surfaces.

If the value of A varies, then most other values must be altered accordingly since the **net radiation balance** (R) at any *place* on the earth's surface can be represented as:

$\mathbf{R_g}$	=	$\tau(\mathbf{1-A})$	+	$\mathbf{c}$	−	$\sigma \boldsymbol{T}^4$
(Net radiation balance at the ground)		(Direct solar radiation absorbed)		(Diffuse counter-radiation absorbed)		(Terrestrial radiation lost)

Similar equations can be devised for net radiation balances at other critical interfaces as the roof of the atmosphere (R_s), e.g.

$R_s = Q_s$	−	$(\tau A + \rho)$	−	q
(Direct solar radiation absorbed)		(Radiation lost through reflection)		(Radiation lost direct from ground and via atmosphere)

At any chosen time, R_g or R_s can prove either positive or negative; for example, consider the terms in the first equation at a place on a calm clear night: no direct solar radiation is being received, there may well be little water vapour or carbon dioxide in the air layers near the ground to absorb terrestrial radiation losses and so counter-radiation is low. Thus, R_g will be negative. It is hardly surprising, therefore, why the incidence of ground frost is higher under cloudless conditions, particularly during long winter nights. Low level clouds, and even fogs, act as 'blankets' by absorbing relatively more outgoing radiation to boost counter-radiation back to the ground. The same principle applies to the blankets on our beds; some readers will know of the claims made for light down-filled quilts as opposed to conventional blankets, whilst those who have ventured to bivouac on camping holidays may be aware of the virtues of the so-called 'reflective space blanket'. We return to the topic of balances later.

2B Heat storage and transport

Whilst radiation exchanges are crucial, other forms of energy and mass transfers comprise the **gross heat economy** of the atmosphere. These are losses from the ground into the troposphere by the *conduction* and *convection* of *sensible* heat and the transfer of *latent* heat by *evapotranspiration*.

Here we are dealing with energy transfers which, unlike radiation, require material 'carriers'. These carriers are vibrating molecules of whatever substance is involved. Much as the people in large crowds are referred to as being 'packed solid', so we envisage the distinction between the *solid, liquid* and *gaseous* states of substances in terms of the freedom of movement of the molecules; when only vibrations are possible the substance is solid but, when free movement is unrestricted, we have a gaseous state. What would be the definition of a liquid state? Thus, just as a few very enthusiastic spectators at an overcrowded game run the risk of transmitting a dangerous surge of people by jostling their neighbours on the terraces, inputs of energy to the ground are transformed and transferred by the resultant 'excitement' of molecules there.

This preamble would *not* be complete without mention of the common failing of confusing *heat energy* with *temperature*. The two concepts are quite distinct in scientific usage. Popular notions of temperature arise from the fact that we detect sensations of relative 'coldness' or 'heat'. However, these sensations are subjective and misleading. Most of us have experienced the feeling that the water in an open-air swimming pool is much colder on a warm sunny day than in dull cool weather. Yet, if we check the water temperatures, we find that our responses have been unreliable. In fact, temperature is a measure of a body's particular *molecular activity*. It must be quite clear that, when we take our own body temperature, the amounts of heat energy between the thermometer and ourselves have not been equalised. Rather, a **dynamic equilibrium** has been reached between the molecular activity in the skin cells and the 'indicator' substance in the thermometer, e.g. mercury. The reader will meet the distinction between heat energy and temperature many times throughout this book.

Molecules of *air* and *water* within and immediately above the soil are the respective 'carriers' of sensible and latent heat energy. The latter arises because of the three states in which water readily exists under natural conditions, viz. ice, liquid and vapour. Each change of state involves heat energy transfers. Clearly the energy to melt a frozen puddle and its subsequent drying-out through evaporation has not been destroyed but 'captured' within the water vapour molecules removed. This 'hidden' or latent heat energy is released when the water vapour condenses back to a liquid and ultimately refreezes. The fact that this usually occurs once the vapour has been lifted aloft, therefore, is important regarding the transport of heat energy upwards. Some measure of the amounts of energy involved can be determined from the fact that to evaporate a kilogramme of liquid water to an equivalent mass of water vapour at the same temperature requires 226×10^4 joules, i.e. nearly 20 minutes over a 2 kilowatt heater. The reader would prove most unpopular if this were tested with an electric kettle! Under controlled conditions, however, it is instructive to note the respective times to melt and then evaporate the ice equivalent of a kilogramme of water. How much more heat energy is required for evaporating the water than melting the ice? We outline the main energy flows involved regarding atmospheric conduction, convection and evapotranspiration processes below. To simplify the issue, and remain true to our brief, we ignore horizontal transfers over the earth's surface for the reasons stated at the beginning of this chapter.

2B 1 Conduction

Since the nature of sensible heat energy is the property of molecules moving about at random, the more 'excited' the motions at one place the greater the *flow* of heat away from that place. A warm body in contact with a colder one gradually loses energy to the latter because the *condition* of its more mobile molecules transfers across the interface until a balance or *thermal equilibrium* is reached between both. Place your hand against a cold surface and feel the loss of heat energy, or the reverse when stirring a hot drink with a spoon: energy is being transferred but not the molecules. This *conduction* of heat energy is not unlike the infectious laughter which spreads throughout a comedian's audience; much as the spontaneity of the response and the degree of uproar can be measured against each person's willingness and ability to appreciate the joke, so the flow of heat depends upon the *thermal conductivity* of the substances involved. Common examples of this property are the clothes we wear and the materials which insulate our homes compared with those which facilitate heat flow. Broadly, large masses of lightweight materials restrict conduction and *vice versa*; for example, few would prefer a metal spoon to a large wooden one when stirring a boiling soup! So it is with the conduction of heat energy in soils and the less dense air; the process is very important in soils but much less so in air. Only when the lower air layers are comparatively dense and undisturbed, as in very calm conditions, will conduction be of any significance here. Thus, we can concentrate first upon soil warming and cooling processes in response to inputs of insolation and subsequent energy losses upwards.

As heat flow is effected by temperature differences (see page 29), in a homogeneous *soil* the heat flow (Φ) is proportional to the rate of change of temperature (y) with depth (x). This can be expressed as:

$\Phi = Ak\frac{\delta y}{\delta x}$, where $\frac{\delta y}{\delta x}$ is the temperature gradient, k the thermal conductivity and A the cross-sectional area of the soil involved. For those not conversant with calculus, fig. 2.6 shows the important mathematical principles upon which this expression is based. Fig. 2.6 (a) may be envisaged as a section in which the decrease of temperature with depth and height appear 'in profile'. The depth or height at which observations have been made were initially known, or **dependent**, whilst the tempera-

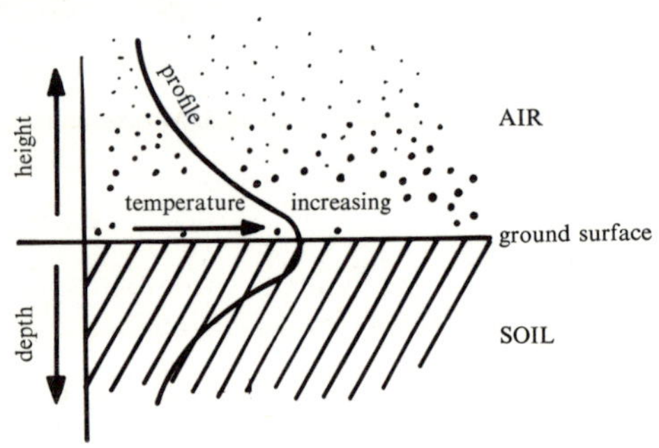

a) Typical temperature profiles recorded above and below the ground surface (not to scale)

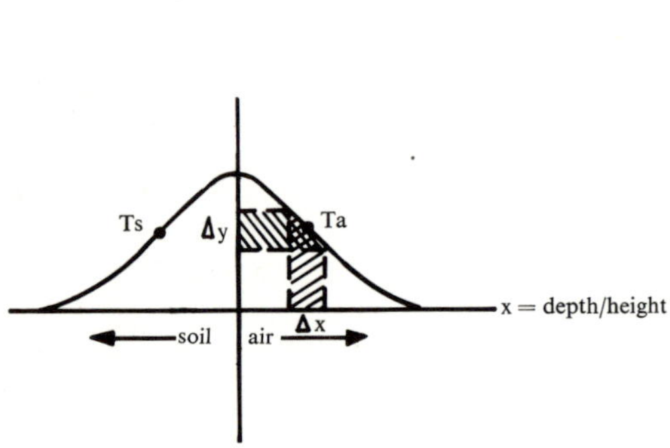

b) Temperature curve from a) above transposed to a graph for mathematical purposes (see text for explanation)

Fig. 2.6 Temperature profiles and heat flow

tures recorded at each level were unknown, or **independent**. However, when we come to plot the temperatures as a curve for mathematical purposes, as in fig. 2.6 (b), it is very important to follow the convention of scaling the dependent variable along the **abscissa** and the independent variable along the **ordinate**. As for the gradients of a hill, the temperature gradient of the curve can now be shown as a tangent to the curve. Because the curve does not have a uniform slope, however, the gradient at a point like T_a (i.e. the temperature gradient in the air) can only be realistically expressed by creating a very small triangle about it whose vertical and horizontal dimensions are Δy and Δx. Clearly, if these dimensions are considered to be *infinitesimally small differences,* the more accurate is the expression for the gradient at T_a. The notations Δy and Δx, or δy and δx, are understood to mean this in mathematics, being called **differentials.** Thus, $\frac{\delta y}{\delta x}$ is an expression for the temperature gradient at a point like T_a. The same applies, of course, to the gradient in the soil at T_s; but, why should the sign be negative, viz. $-\frac{\delta y}{\delta x}$?

Thermal conductivity (k) is the amount of heat energy (in joules) which can flow between opposite faces of a cubic metre of a particular soil in one second given a temperature difference of 1 °C at either side ($J\ m^{-1}\ s^{-1}\ K^{-1}$). As it has been mentioned already that watts are a measure of joules per second, it is possible to refine this statement of thermal conductivity to the more familiar heat energy units of watts $m^{-1}\ K^{-1}$. So, the actual heat flow (Φ, in watts) with a temperature gradient of $-\frac{\delta y}{\delta x}$ must be Ak times $-\frac{\delta y}{\delta x}$.

substance	density ($kg\ m^{-3}$)	specific heat ($J\ kg^{-1} K^{-1}$)	thermal conductivity ($W\ m^{-1} K^{-1}$)	thermal capacity ($J\ K^{-1}$) $\times 10^6$	thermal diffusity ($m^{-1}\ s^{-1}$) $\times 10^{-6}$
soil air	1.3	1008	0.02	0.0013	15.40
ice	920	2100	2.50	1.9	1.30
fresh water	1000	4200	0.60	4.2	0.14
sea water	1025	4195	0.65	4.3	0.15
sand	1500	650	0.20	1.0	0.20
loam	2000	675	0.30	1.3	0.23
clay	2200	700	0.40	1.5	0.27
limestone	2600	715	1.70	1.8	0.94
granite	2700	840	3.40	2.3	1.48
metal ores	8000	440	90	3.5	2.57

properties of typical surface materials
(per cubic metre)

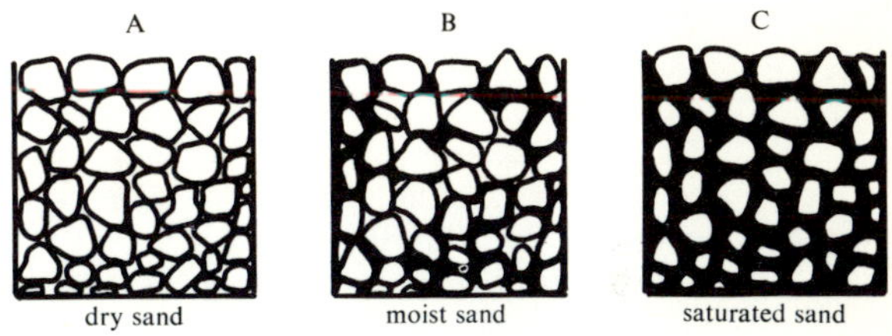

Fig. 2.7 Factors affecting soil warming and cooling

LABORATORY WORK

Measuring thermal conductivity

1 Determining the thermal conductivity of natural materials like soil poses many problems. The experiments that most readers will have met whilst studying physics involve measuring the conductivity of various metal rods, which are good conductors of course. The rod is wrapped in insulating material, one end projects over steam (100 °C) and the other into an ice and water mix (0 °C). By plotting temperature at points along the rod it is possible to calculate the conductivity of the particular metal. Poor conductors necessitate more complicated apparatus and calculations, however.

2 Rough but useful comparisons can be made by filling insulated lunch boxes with selected soils, embedding thermometers into the side of each and exposing them beneath photoflood bulbs simultaneously. The increases of temperature with time for each will give some indication of relative conductivities regarding the absorption of radiant energy. Care must be taken to provide similar masses of soil and present black surfaces horizontally beneath the bulbs.

3 The third column in fig. 2.7 gives typical conductivities for ten common materials. Using this information, calculate the heat flow per square metre for selected soils given various temperature gradients per metre.

Unfortunately, it must not be assumed that k is a day-to-day constant. For instance, it is clear that variations of soil moisture will greatly influence the thermal conductivity of the ground just as wet clothes facilitate heat losses from our bodies; the same clay is at least three times as efficient a conductor when saturated as when dry, whilst the corresponding increase for a wet sand might be eight times. This demonstrates that the voids in a soil are at least as significant as the solids. The relationship between voids and solids is clearly a function of the **relative density** of the soil as well as being an indicator of permeability in hydrological terms.

The heat energy required to raise the temperature of a kilogramme of a substance by 1 °C is called its specific heat capacity. However, since the *mass* of a soil varies with its moisture content, it is better to think in terms of raising the temperature of a unit *volume* such as a cubic metre—effectively the heat capacity per unit volume, usually contracted to **thermal capacity**. Commonplace facts like birds fluffing their feathers in cold weather, keeping the walls of buildings dry with damp courses and cavities, wearing well 'aired' clothes and aerating soils by cultivation are good examples of attempts to lower thermal capacities by decreasing relative densities. In fact, the volume thermal capacity is a product of the specific heat capacity (c_v) and density (ρ) of a substance such as soil.

Determining the specific heat of a material involves experimental procedures similar to those suggested in the previous exercise. In this case, however, it is also necessary to be able to measure the heat energy inputs. Since this makes the experiment much more elaborate, and we cannot afford the luxury of explaining too many techniques in experimental physics, the reader should refer to the appropriate data in fig. 2.7. Regarding density, however, crude but effective measurements can be made with cold tap water and a few kitchen utensils. Hand specimens of a soil can be sealed in thin polythene bags and weighed on kitchen scales. Each one in turn is then gently immersed into a mixing bowl brimming with water. If this bowl is placed inside a washing-up bowl, the water displaced can be caught and then weighed. Since the weight of a cubic metre of water at 4 °C is 1000 kilogrammes (see fig. 2.7), it is possible to evaluate the density of each sample relative to the weight of water it displaces. The density of gases like air, however, is a very different matter. We are sure that we have already introduced enough physics to make the reader aware of where he can find the answers to such problems.

The thermal capacity ($c_v\rho$) is important concerning the *retention* of warmth with time (t) and depth (x). Using differentials once again, this may be expressed as $\frac{\delta \Phi}{\delta x} = A c_v \rho \frac{\delta y}{\delta t}$, where $\frac{\delta \Phi}{\delta x}$ is the rate of change of heat flow with depth and $\frac{\delta y}{\delta t}$ the rate of change of temperature with time. Now, by substituting the previous expression for Φ, a relationship between the changes of temperature with time and with depth can be found, viz:

$$\frac{\delta y}{\delta t} = \frac{k}{c_v \rho} \cdot \frac{\delta^2 y}{\delta x^2}$$

The coefficient $\frac{k}{c_v \rho}$ is called the temperature conductivity or **thermal diffusity** of a substance, being the *thermal conductivity* divided by the *thermal capacity*. Fig. 2.7 shows appropriate data which readers might derive for themselves given the density, specific heat and thermal conductivity of each substance. It also illustrates 'dry', 'moist', and 'saturate' sands as a basis for discussing the factors effecting soil warming. Here we can put the different concepts of heat energy and temperature to a useful practical test. The dry sand has a high percentage of air voids per unit volume. Thus, since still soil air has a low thermal conductivity but a relatively high thermal diffusity, it follows that such soils are poor conductors of heat energy but efficient temperature transmitters. Its capacity to store heat is also very low. The reader should now attempt to determine how these characteristics change with the increase in soil water shown in the moist and saturated sands. The principles discovered can be applied widely to most soils and checked by simple laboratory or controlled field experiments. Physical geographers must be constantly aware of the significance of soil water in all branches of their work. It is a topic that recurs frequently throughout this book.

So far as the meteorologist is concerned, the conduction and storage of heat energy beneath different surfaces provides a key to understanding many climatic features. One of the best known manifestations, of course, is the extreme thermal characteristics of continental interiors in comparison with the more uniform regimes experienced in oceanic or maritime regions. Similar principles can be applied at quite a small scale level. The reader should attempt to assess the influence of suitable lakes, ponds and large rivers upon their immediate surrounds. The temperature traverse techniques discussed in the next chapter (see page 70) are particularly suited to such investigations.

The same principles are applicable to transfers from the ground into the lower *air*. It must not be forgotten that the soil is effectively 'breathing' in much the same way that we lose heat energy by respiration. As the air can transmit temperatures better than soils, far greater fluctuations are experienced. However, *true* heat conduction is less important in the air than a *pseudo*-conduction process called *eddy diffusion*. The difference between the two processes is that the former involves a relatively uniform progress of the temperature wave parallel to an interface like the ground surface whereas the latter is turbulent. Even if we cannot see the motions going on we can invariably feel some degree of air movement whether outdoors or indoors. Only with cold dense air do we find situations where 'stratification' can be said to exist; in effect, the molecules are being constrained. Some idea of the factors involved can be gained by dropping coloured ink into clear liquids of different densities like water, white spirit and bromoform in separate containers. What patterns does the ink trace in each liquid and how quickly does mixing occur? What is the difference in the patterns and mixing speeds when the liquids are agitated prior to dropping in the ink?

Whilst mathematical treatments of random motions in eddy diffusion are fairly complex, many visual illustrations like the ink experiment in liquids are well known to physicists in experimental techniques dealing with kinetic theory, e.g. smoke cells. When we see swirling smoke or airborne dust on a calm day, it is as well to remember that many other elements which we cannot see are similarly in motion. Smoke, French chalk or talc tracers can be introduced to study eddy diffusion outdoors. If such tracers are released near the ground on a warm summer's day, it will be found that they billow in a very turbulent manner. On a cold calm winter evening, however, the tracer will spread out and 'lie' over the ground. Herein is the explanation of the summer heat haze and the low-lying winter smog. In such experiments, however, great care is needed to distinguish between eddy diffusion and those motions generated by gusts of wind, draughts and even the experiment itself!

In the case of eddy diffusion and heat flow upwards, we are concerned with the *vertical* components of the complex swirling motions. The magnitude of upward motion is largely determined by the **potential temperature** (T_0) gradient with height (x) much as heat flow in the ground was seen to be controlled by temperature profiles in the soil. Potential temperature is defined as the temperature air would have at any level if reduced to a standard pressure of 1000 millibars—effectively mean pressure at sea level. Thus, it is independent of the environment. If the actual pressure is below 1000 mb, compression is necessary and so T_0 must exceed the actual air temperature. The reverse holds true for air pressures above 1000 mb. Further consideration is given to potential temperature concerning larger scale convective motions. *Near* relatively warm ground there is usually a decrease in T_0 with height to promote upward eddy diffusion. Only in air layers immediately above cooled surfaces will T_0 increase with height to minimise eddy diffusion and increase stratification. Herein lies the reason for the diminution of eddy diffusion at night. In practice of course, the difference between T_0 and the *actual* air temperature near the ground is so small that the latter is good enough for experimental purposes in the field. By taking simultaneous readings of air temperatures at several levels up to a metre above the surface, it is possible to calculate the mean temperature gradient in the lowest air layers. Controlled smoke or talk tests throughout the day and night in calm conditions will reveal the characteristics of eddy diffusion above selected surfaces in relation to temperature gradients.

Above a warm surface, the net energy flow (Φ) across the potential or actual temperature gradient $\frac{\delta y}{\delta x}$ is given by $\Phi = AMc_\rho \frac{\delta y}{\delta x}$ where M is the so-called **Austausch coefficient** and c_ρ the specific heat capacity of air at constant pressure. The Austausch coefficient was derived by W. Schmidt in 1917 as a measure of 'eddy conductivity' or *eddy diffusity*. Like thermal diffusity, it is a variable dependent, among other factors, upon fluctuations of air density. Although this is a similar law to that already given for conduction in the soil, it is much more difficult to pin down precise values for the Austausch coefficient in particular weather situations. Near the ground, in comparatively cool stable conditions, M will be about 10 (kg m^{-1} s^{-1}) whereas in warmer unstable weather it can exceed 10^4. Simple calculations using these values, and a general figure for the specific heat of air as 1000 (J kg^{-1} K^{-1}), will impress the reader with the magnitude of heat flow from the ground surface. As with conductivity in the ground, the relationship between changes of temperature with time $\frac{\delta y}{\delta t}$ and with height $\frac{\delta y}{\delta x}$ is:

$$\frac{\delta y}{\delta t} = \frac{M}{\rho} \cdot \frac{\delta^2 y}{\delta x^2}$$

For those who remain undaunted by such mathematical approaches, the physics and experimental procedures involved in the determination of Austausch coefficients are discussed at the end of the classic book on meteorology by H. R. Byers (1944). Suffice it to say here that eddy diffusion in the air is many thousand times more efficient than molecular or true heat conduction in the ground; hence the difficulties in the techniques of its investigation. Complex multivariate analyses are essential. We avoid these as we feel sure that, by now, readers are sufficiently aware of the problems facing the meteorologist wishing to devise techniques for studying the conduction of heat energy.

2B 2 Convection

Eddy diffusion is clearly a small scale 'intermediate' process between true conduction in fluids and the equally familiar convection process in fluids. Many similar factors are involved. However, the larger scale of convectional motions require different techniques of enquiry. In effect, this arises from the differences between excited molecular 'bubbles' of air *near the ground* and larger cells or 'parcels' of discrete air in motion *aloft*. As heat energy is transferred in these parcels, there can be no escape from considering the fundamental Gas Laws and the laws of thermodynamics. Most readers will have met the Gas Laws already in a number of science experiments concerning heat energy, whilst we have already used the laws of thermodynamics.

We are all familiar with the principle that hot air parcels are 'buoyant' and rise, while cold air parcels subside. It is useful to list and make careful observations of everyday events providing visual indications of these processes, e.g. smoke from bonfires and chimneys, 'fair weather' cumulus clouds, fogs and so on. The flight patterns of soaring birds, insects, gliders, or the characteristics of visibility all provide indirect evidence of convective updraughts often called *thermals*. When we see swallows and swifts at great heights on summer days, we can be sure that the flying insects they are seeking 'on the wing' are there too because of rising air currents above the warmed surface. Conversely, when these birds rake the ground in cooler weather, we know that little convection is taking place. In calm weather, it is often worthwhile blowing soap bubbles across different surfaces and recording their degrees of lift before bursting. A great deal can be learnt about the mobility of the air through such observations. Other information can be gained from studying the operation of hot air balloons and the like.

The General Gas Law expresses the relationship between pressure (P), density (ρ) and temperature (T) of a gas; namely $P = r\rho T$, where P is now measured in newtons per square metre ($1000\,\text{mb} = 10^5\,\text{N}\,\text{m}^{-2}$), ρ is in kilogrammes per cubic metre and T in degrees absolute (K) and r is the gas constant, being $2{\cdot}87 \times 10^{-3}\,\text{J}\,\text{kg}^{-1}\,\text{K}^{-1}$ for dry air. The density of *dry* air at 0°C and 1000 mb is $1{\cdot}276\,\text{kg}\,\text{m}^{-3}$; however, in the atmosphere, the air is mixed with water vapour. Thus, actual atmospheric densities vary according to the amount of water vapour present as well as variations of pressure and temperature. When a parcel (which is virtually a heated 'soup' of air and water vapour molecules) lifts off the ground it loses contact with its main heat energy and water vapour supplies. Thereafter, its characteristics depend upon internal circumstances rather than upon any outside sources of energy and vapour. As the pressure of the air surrounding the parcel decreases with altitude it grows in size. This expansion enables the air and water vapour molecules to spread out; a process which must use up some internal kinetic energy. Thus, the density of the air decreases and a fall in temperature occurs. Since nothing has been added or taken away from the parcel, it is clear that the amount of water vapour present has remained constant. However, there is a limit to the weight of water vapour that any parcel can hold proportional to temperature. Consequently, if lifting, expansion and cooling continue, there will come a time when the temperature is too low for the parcel to hold the water vapour present. The vapour in the parcel is then described as being 'saturated' and further cooling causes the vapour to condense into water droplets. At this stage, latent heat is liberated as mentioned at the beginning of this section. Before discussing techniques by which this form of heat energy transfer can be studied, it should be added that saturation and condensation can also be induced by *adding* water vapour to air parcels which move across suitable surfaces, e.g. seas, lakes and woodland. Indeed, we are doing just this when we breathe into cold air and create localised 'fogs'. One way of expressing the amount of water vapour present is to compute its contribution to the total pressure of the 'mixture'. This independent or partial value is called the water vapour pressure, usually contracted to vapour pressure (e) and measured in either newtons per metre squared or millibars. It is much the same as quoting the *weight* or mass of water vapour present. This leads to the concept of **humidity** as a measure of the water vapour present. Most readers will have met the term **absolute** and **relative humidities:** the former measures the mass of water vapour in a parcel of given volume and temperature, whilst the latter is the percentage ratio between the mass of water vapour present in a parcel to the amount to achieve saturation at the same temperature; namely, what water vapour *is present* against what *could be present*. Both are related in the following formulae derived from the Gas Laws:

$$\textbf{Ab. H.} = \frac{0{\cdot}622\text{e}}{\text{r}T}. \qquad \textbf{Rel. H.} = \frac{100\text{e}}{\text{E}}.$$

(Where r = 2·87 when e is measured in millibars)

(Where E is the vapour pressure at saturation in millibars)

Fig. 2.8 illustrates the relationships, and the reader is invited to redraw it and inter-

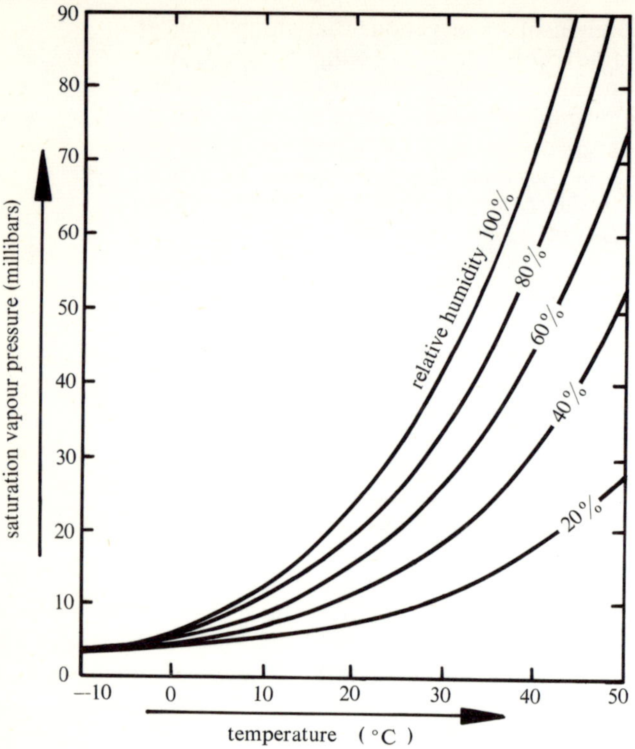

Fig. 2.8 Relative humidity, temperature and vapour pressures at saturation

polate relative humidity curves between the 20 per cent and 100 per cent ones shown at suitable intervals. Consider, for example, an air parcel with a temperature of 20 °C whose relative humidity is 60 per cent. What would be its saturation vapour pressure, and at what temperature would condensation occur? It is from such relationships as shown in fig. 2.8 that hygrometric tables (published by the Meteorological Office) are prepared. These are essential to many of the climatological techniques discussed in the next chapter.

To evaluate amounts of heat transported by convection above warmed surfaces, the *First and Second laws of thermodynamics* must be invoked. We strongly recommend both as essential concepts in *all* aspects of physical geography, and many fields of human and economic geography for that matter; both laws summarise many of the behavioural patterns apparent in systems.

The following expressions may be derived to determine the rate of change of pressure with height $\frac{\delta P}{\delta x}$ and that of temperature $\frac{\delta y}{\delta x}$:

$$\frac{\delta P}{\delta x} = -g\rho \quad \text{and} \quad \frac{\delta y}{\delta x} = \frac{-g}{c_\rho}$$

where g is the acceleration owing to gravity (*c.* 9·8 m s^{-2}), ρ is the density of the air at a place, and c_ρ the specific heat capacity of the air at constant pressure (*c.* 1000 J kg^{-1} K^{-1}).

If the numerical values for g and c_ρ are inserted in the latter expression, it can be determined that temperatures decrease at 9·8 °C every 1000 metres in altitude (virtually 1 °C per 100 m) until saturated. This is called the **dry adiabatic lapse rate** (DALR) and applies to all parcels of unsaturated air that are lifted. A useful exercise is to tabulate from fig. 2.8 the condensation or **dew point** temperatures of particular air parcels of given temperatures and relative humidities at ground level, and then to plot graphically the height at which saturation occurs for each. Fortunately, exchanges between the air parcel and the surrounding environmental air are small enough to be ignored. Föhn winds are ideal for studying these relationships in practice. Most meteorological texts explain the mechanisms involved and give examples of such winds.

Saturation in the parcel causes the water vapour to condense. If the ascent continues, the motions will be apparent in cloud formations. So, still photograph sequences or time-lapse cine film are ideal for studying the form of the processes at work, especially in conjunction with pictorial guides explaining different cloud formations. Plate 2.1 shows examples of simple time lapse cloud pictures and how they can be interpreted. This is a useful exercise which can be done on most days anywhere. Furthermore, the techniques are basically similar to those used in the study of cloud patterns on satellite photographs,

Plate 2.1 One-minute time lapse photographs of 'fair weather' cumulus clouds late on a hot summer's day. Can you explain why the cloud in the centre appears to 'dissolve'?

viz. nephanalysis. Owing to the liberation of latent heat of condensation, the so-called **saturated adiabatic lapse rate** (SALR) is not linear, the rate of cooling being slowed initially. This is because the amount of water vapour decreases as condensation increases. In effect, the air parcel is no longer discrete, as evidenced by the boiling motions at the margins of cumulus clouds and entrainment of surrounding dry air. It is important to realise that water vapour and droplets coexist in clouds and fogs and that the term 'saturation' must not be taken literally. Consequently, convection is a significant process in transporting heat upwards and releasing it aloft. We have only to observe the extent of cumulus cloud development during a warm day to appreciate its diurnal variations and cycles.

The Second Law of thermodynamics explains the *rate* at which convection occurs. Briefly, it states that the entropy within a closed system, like the unsaturated air parcel, remains constant or increases but can never decrease. Since potential temperature (T_0) is directly related to entropy, we can expect it to remain constant during an ascent until saturation occurs. After all, the very definition of potential temperature (see page 29) involves bringing an air parcel to its mean sea level pressure; what applies on the way down must also apply in reverse on the way up. Therefore, the DALR is effectively a line or isotherm of equal potential temperature. Any technique which supplies information concerning a decrease or increase of potential temperature within the environmental air will enable us to apply the same principles already used in examining the magnitude of eddy diffusion.

Radiosonde and rocket ascents from meteorological stations provide routine data up to about twenty km and seventy km respectively, and there are eight such stations scattered throughout the British Isles with a ninth on the Faeroe Islands. Seven further stations exist in northern France and the Low Countries. Codified information from these stations appears in the Daily Aerological Records published by the Meteorological Office. They enable temperature, pressure, humidity and so potential temperature above a place to be plotted graphically as **environmental lapse rates** (ELR). Fig. 2.9 illustrates a simplified example of the sort of chart used with examples of typical environmental lapse rates, although the Meteorological Office uses a more complete thermodynamic dia-

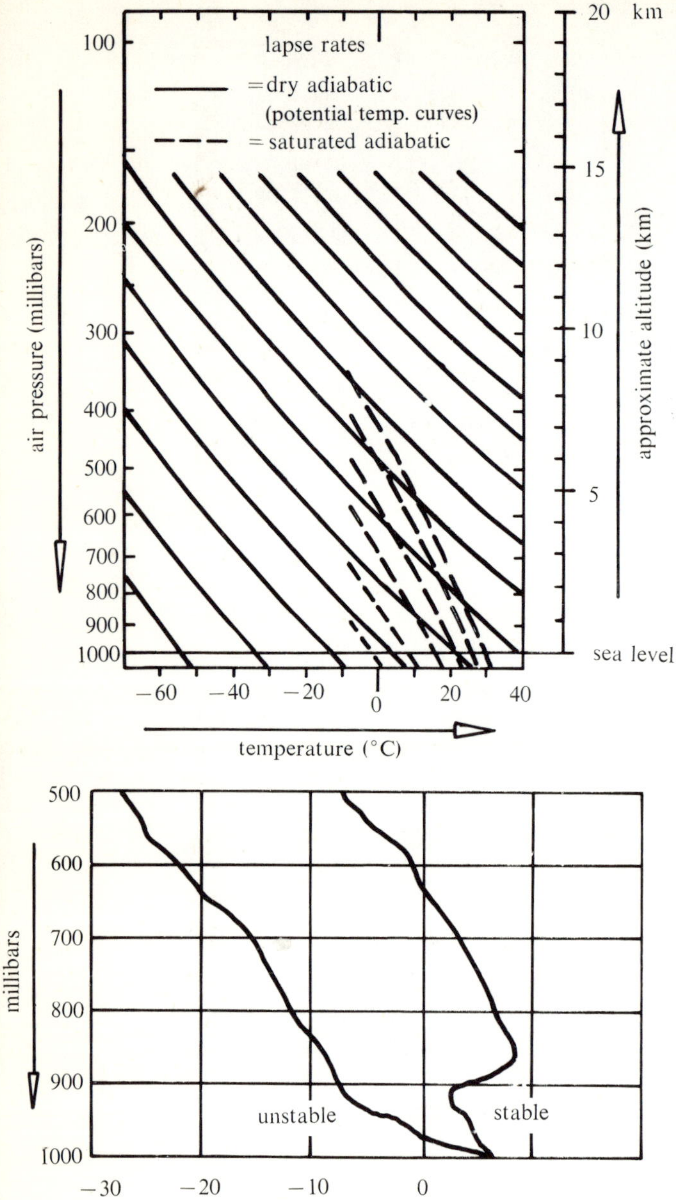

Fig. 2.9 Environmental lapse rates

gram called the **tephigram**. Many other schematic versions are used in the literature. With reference to the examples in fig. 2.9, and remembering the dictum that the *probability* for convection increases as the ELR increases, the reader should seek out a qualitative relationship between the *rate* of convection and potential temperature gradients in the air. This can be accompanied by an investigation into the various states of air mass stability and instability described in most meteorological texts.

CLASSWORK

Lapse rates and air mass stability

1 Consult Taylor and Yates: *British Weather in Maps* (Macmillan 1967 pp. 291–304).
2 Collect Meteorological Office Daily Weather Reports and Aerological Charts for a short period, such as a week, and prepare lapse rates for the different air masses crossing the country.
3 Keep a careful record of the weather during the period in question, then attempt to explain the patterns observed in terms of the stability or otherwise of the air masses involved.

The authors used such techniques for an investigation into the weather systems which caused severe storms and floods across the West Country during 10 July 1968. Fig. 2.10a shows the surface pressure and frontal systems we are accustomed to viewing on television, whilst fig. 2.10b shows the vertical thickness between the 1000 and 500 mb levels during the period in question. The latter shows the 'dome' of humid and unstable air which advanced northward with the depression system. These maps should be compared with the 'contour maps' for the 300 and 500 mb pressure surfaces shown on figs. 2.11a and b respectively. Together with lapse rate information, such maps give a three-dimensional picture of the weather patterns which were responsible for concentrating and steering the humid air across Britain on this occasion. Surface charts only give a small indication of the complete weather picture at any time. As in this unusual case, it is now realised that the behaviour of high speed westerly air currents in the stratosphere (called Jet Streams) is crucial to understanding large scale circulations. Their study has been greatly

a) surface pressure systems at 00 hrs 8-12 July 1968

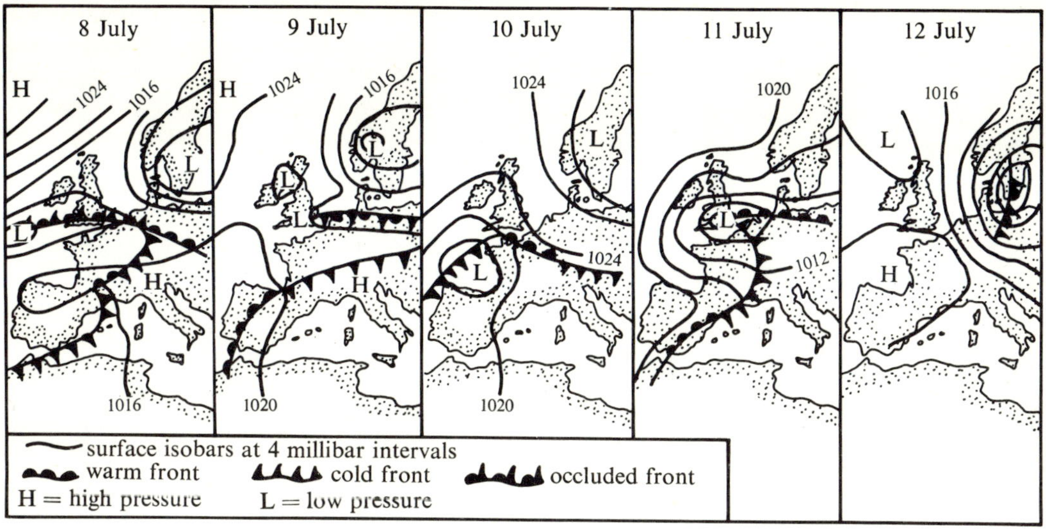

Fig. 2.10 Altitudinal variations of air pressure 8–12 July 1968 over Western Europe. Adapted from Daily Weather Reports and Aerological Records by permission of the Director-General of the Meteorological Office, London

b) thickness between 1000—500 mb levels at 00 hrs 8-12 July 1968

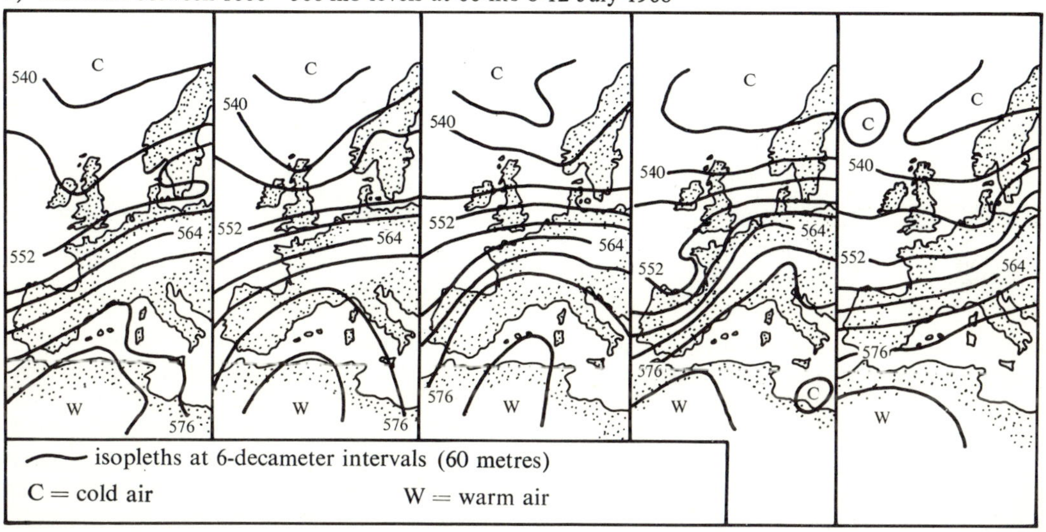

aided by satellite photography of cloud patterns.

We have mentioned this particular example for two reasons; firstly, to illustrate the importance of various isopleth mapping techniques in meteorology and, secondly, to underline that important relationships exist in the increasing scale of convective processes with height and the upper air circulations. Here we see that all the small scale ground level exchanges are 'embedded' within the broad global circulations. Once again it is a question of dealing with systems within systems. However,

a) 300 mb surface, showing Jet Streams

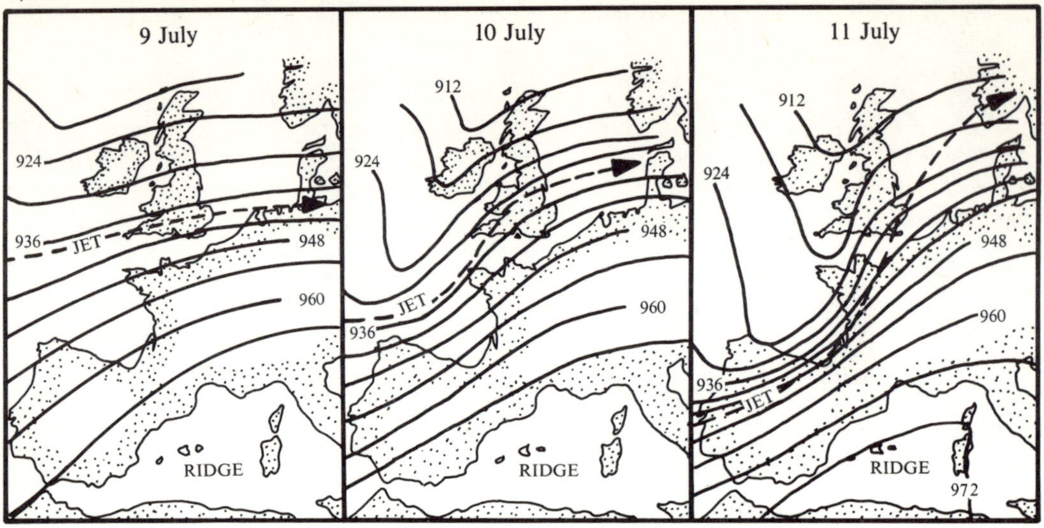

Fig. 2.11 Contours of the 300 mb and 500 mb surfaces at 00 hrs 9–11 July 1968 in decameters above sea level. Adapted from Aerological Records by permission of the Director-General of the Meteorological Office, London

b) 500 mb surface

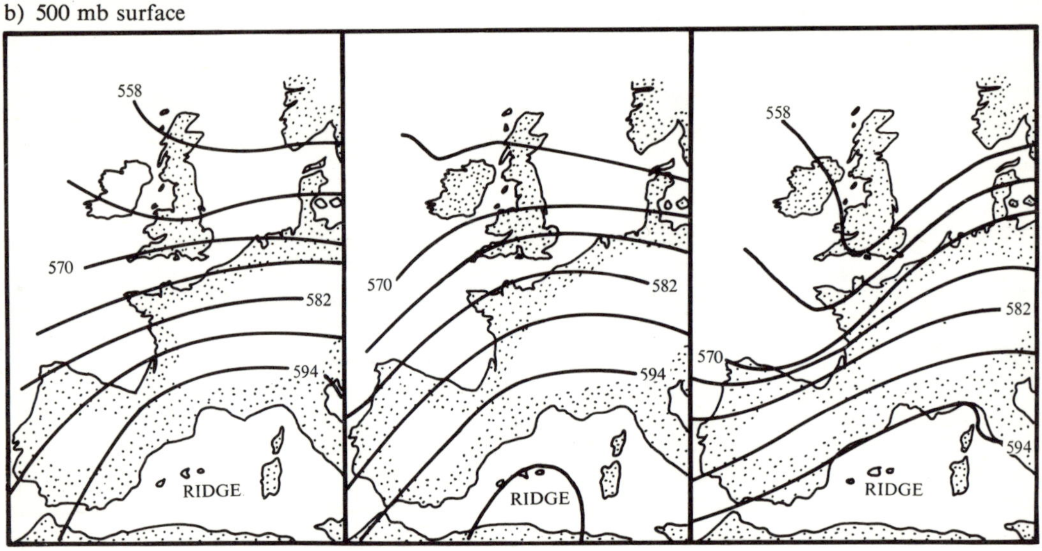

since we have deliberately 'closed' our techniques in favour of the small scale processes which can be observed on the ground, we must come back to earth! We do so in the full knowledge that we are neglecting such time-honoured topics as precipitation processes, although the chapter on hydrology is partly concerned with these. Whilst the British people's 'pre-occupation' with rainfall is more than justified regarding its influence upon our lives, and as probably the most 'visual' of all meteorological processes, we prefer to look upon the more neglected and problematical 'sunny-side' by examining evapotranspiration. After all, there is much meteorological sense in the adage that 'What goes up must come down'!

2B 3 Evapotranspiration

Evapotranspiration is the net upward transfer of water, with its accompanying loss of heat energy, across the ground to the surface of air interface. It does not matter if the former is soil, water, organisms, buildings or the spectators at a Wembley cup final! Winds and convection take over once the water vapour is in the air. In effect, it is a complex *supply and demand* situation. A challenging exercise is to estimate the percentage of water circulating in the atmosphere as vapour and clouds relative to the total water available, given that water covers about two thirds of the earth's surface and that both hydrosphere and troposphere are negligibly thick compared with the earth's radius. A sheet of thin polythene placed against the surface of a classroom globe is a good measure of the relative scale of the tropospheric envelope.

We are all familiar with everyday experiences of heat loss through evapotranspiration whether it be a cup of tea, the bath water or when perspiring. The principle is put to practical use in the porous clay jars which keep water cool in hot weather. As the water percolates through the sides of the jar, evaporation occurs off the dampened outer surface. Just as when we wet a finger and feel it chill on 'drying-out', so the evaporation off the jar withdraws heat energy to keep the contents cool. Such jars are in common use in tropical and Mediterranean countries where refrigerators are still a luxury. From our own point of view, it is interesting to devise various ways of keeping cool on hot days and to isolate the factors involved, e.g. temperature, wind and humidity. These factors demonstrate that the physics of evapotranspiration involves many variables. Complex amalgams of all the processes already outlined are operative at any time. The heat energy lost through evapotranspiration from any moist surface is termed the **latent heat of evaporation** as mentioned earlier. Losses will continue until the vapour in contact with the surface become 'saturated' with the water vapour removed. Thus, any further evapotranspiration from the surface depends upon the saturated vapour being bodily taken away by convection currents or winds and its replacement by more air and unsaturated vapours from elsewhere. Clearly then, measurements of humidity (see page 48) and turbulence immediately above any surface are useful *indirect* techniques of examining the magnitude of latent heat energy losses. Every housewife knows that warm windy weather makes a good 'drying day' for airing washed clothes outside.

Since the latent heat of evapotranspiration remains 'hidden' in the molecules of water vapour removed, one of the most direct ways of determining the amounts of energy involved is to cool a sample of the air independently until condensation causes its 'liberation'. It is then a matter of measuring this liberated heat energy. Physicists refer to such techniques as latent heat calorimetry and, under controlled laboratory conditions, they have many different devices to perform such measurements. In the field, however, it is impossible to use such techniques, and the meteorologist is faced with applying theoretical physical equations to compute both vapour and heat energy losses. Because of the complexity of the processes, however, even mathematical approaches cannot do justice to the study of evapotranspiration. The most acceptable equations are based upon flow concepts. For example, if the surface temperature of the evaporating body is to be maintained, then it requires independent heat energy 'replacements' which can only be supplied by the absorption of radiant energy and its transfer by conduction to the evaporating surface. Thus, we are dealing with an input-output situation in which the heat energies

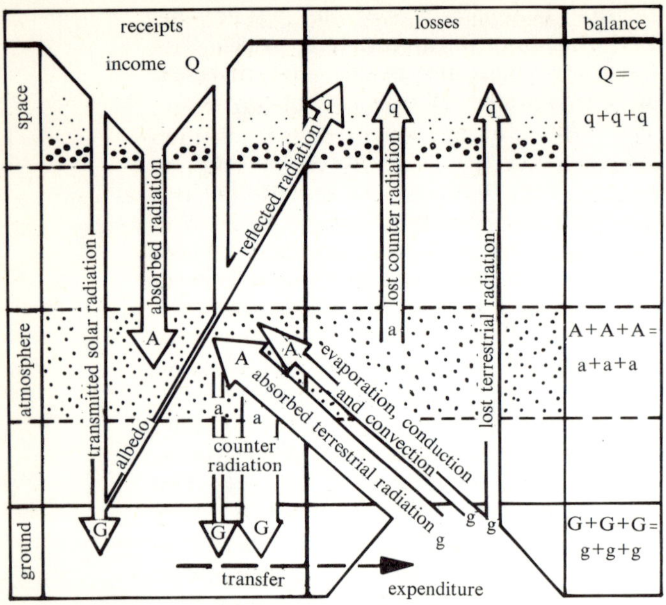

Q, q = energy gains and losses at fringes of space

A,a = energy gains and losses within atmosphere

G,g = energy gains and losses in ground (and water)

⇨ = main energy transfers (thicknesses approximate to mean annual 'transaction')

Fig. 2.12 Heat energy budgets

involved must balance. The basic equation, therefore, is:

R_g	=	H	+
(net radiation heat energy balance at ground surface)		(heat energy gained by air from surface by conduction and convection)	
E	+	S	
(latent heat energy gained by air through evapotranspiration)		(any heat energy gained by air through that *stored* by organisms)	

$$\text{Hence, } E = R_g - H - S$$

We have already dealt with the processes and measurements of R_g and H. It is with the value of S that almost insuperable problems arise, however, particularly in field situations. This is largely because some stored energy is converted to chemical energy by plants, especially through photosynthesis (see chapter 4). Also, we have that stored by animals. Thus, we cannot divorce ecological matters from the equation and the evaluation of S becomes manifestly difficult. Nevertheless, some authorities neglect S for the sake of simplicity. Short of laying ourselves open to the charge of oversimplification, we must acknowledge that most direct practical and theoretical approaches to quantify heat energy and mass losses through evapotranspiration are so involved and controversial as to go well beyond the scope of this book. However, the reader need not feel cheated because, with such a key process, it is inevitable that the topic recurs frequently in later chapters.

The best estimates indicate that, of the total annual heat loss from the earth, some 5 per cent is accounted for by conduction and convection and 20 per cent by the transfer of latent heat by evapotranspiration. Thus, their contribution to the energy which drives the atmospheric 'heat engine' is appreciable. Furthermore, it is very clear that the *physical* geographer must understand the basic *physics* of these processes if he is to equip himself with appropriate techniques to measure the spatial and temporal variations that exist.

2C Heat balances and budgets

In all the processes discussed, *self-regulating* and *equilibrium* concepts are evident. In meteorology, as in other physical sciences, these concepts can be examined geographically in terms of balances. A positive balance at one place in a system will be countered by negative balances elsewhere. Meteorologists use the term **budget** because they are *accounting* for so many factors; a sort of 'income and expenditure' exercise. Indeed, the economic analogy can be elaborated; radiation is the mint, heat energy the currency and the various elements of the atmosphere the central banks and shareholders. Energy surpluses are transferred to make up the deficiences of another area. Unfortunately, the study of these surpluses, deficiences and international exchanges are equally as complicated in meteorology as in economics. Even more obvious relationships are to be made when we reflect upon the money spent to keep warm, to cultivate crops in unsuited environments and to deal with the growing problems of atmospheric pollution.

Fig. 2.12 shows the basis of preparing a balance sheet at a place in terms of measuring heat exchanges. These can be drawn-up diurnally, monthly, seasonally, annually and for longer periods. Whilst annual and 35-year balances are important in quantifying **gross values** and statistical **averages** the smaller time intervals provide equally useful information. In Los Angeles, for example, significant variations in local heat energy balances have been recorded over little more than an hour. In very clear weather, direct solar radiation to polluted industrial and commercial areas was found to be a good 10 per cent less than that to residential districts not more than a kilometre away. The corresponding value for net radiation was 13 per cent down. However, because of the greater thermal conductivity and capacity of the larger industrial buildings significant 'heat island' effects were detectable there, especially at night. We deal with some simple techniques for examining urban climates at the end of the next chapter.

Anyone who has walked far along a road on a very hot summer's day, and then taken a rest on the grassy verge, will appreciate that heat energy exchanges over different surfaces can be very marked in a short distance. Considerable differences can be recorded for cropped and fallow land over short periods, particularly in summer. The following figures are typical of a playing field in southern England during the summer months. We can assume that such fields are fairly large, open and flat. Taking the transmitted fraction of direct and diffuse radiation to the surface as 100% in this case, 75% is absorbed by the ground after reflection off the clipped grass. Since only about 1% of the energy absorbed is stored and used for the growth of the grass and by other organisms, the remainder is returned to the atmosphere in the following proportions: evapotranspiration 42%, conduction and convection 19%, and terrestrial radiation losses 13%. The reader should prepare a simple flow diagram of these exchanges based upon fig. 2.12. By comparing the results with the classwork exercise on heat budgets completed on pages 23–24, important differences will be noted. Discuss the factors likely to explain these differences and suggest how the values given above are likely to vary during the winter months.

Budgets of this nature are very important in climatological studies. However, the figures must not be interpreted too strictly for it is clear that a major difficulty exists because of the influx of heat energy and water vapour from elsewhere by essentially *horizontal* exchanges called **advection.** Herein is the nub of global studies in dynamic meteorology and one of the chief objectives of the current World Weather Watch programme. Since most meteorological texts dwell upon models of the general circulations within the atmosphere, we commend readers to the 'classics' listed on page 221; and, not wishing to stray too far from our brief, we highlight only one aspect concerning advection.

2C1 Advection

If we imagine a vertical column into the atmosphere above *a place*, then it is clear that the air within is subject to both horizontal and vertical motions. The horizontal motions will vary at different levels above the surface. In the lower troposphere we find that the *type* and *nature* of the surface is important whereas, in the upper troposphere and stratosphere, the motions are influenced more by such factors as the earth's *rotation* and *attitude* with respect to insolation. Reciprocal processes occur at the 'boundary' levels; for example, the jet stream activity already mentioned. Therefore, as was found with vertical motions, the *scale* of 'action' becomes *smaller* nearer the ground. This is a response to the infinite variety of surfaces in space and time. Since our lives and experiences are enacted at such scales, we devote the next chapter on local climatology to techniques by which surface air interactions can be studied. Here we outline the basic 'driving force' behind the atmospheric 'heat engine' on a global scale.

Previous studies in physical geography should enable the reader to demonstrate on diagrams of the earth in relation to the sun at different times of the year that differential heating on a global scale is primarily a function of *latitude*. From such diagrams it is possible to derive quantitative relationships between latitude and the lengths of day and night annually. In turn, these will give some measure of the *intensity* and *duration* of radiation inputs and outputs globally. Many useful graphs can be prepared to study such information; for example, angles of incidence of the noon sun at different times of year, times of sunrise and sunset, hence hours of daylight or darkness and so on. Techniques for calculating and illustrating such information for any place are considered in more detail in the next chapter (see page 60).

Fig. 2.13 shows mean annual input and output curves of radiation against latitude. It should be noted that the abscissa is divided according to the sine of the angle of latitude as in a cylindrical orthographic or equal area map projection. The method of constructing this scale is shown in fig. 2.14, where it can be seen that $x = R \sin \theta$ (R being the radius of the earth to whatever scale is appropriate and θ the angle of latitude). Thus, the appropriate divisions of x are a function of the respective areas A_1, A_2, A_3 and so on of each latitudinal or zonal belt. Consequently, the areas of the zones between the two curves on fig. 2.13 are a measure of the heat *stored* by the air at different latitudes. Using these areas it is possible to prepare an analogous graph with the ordinate divided to indicate the maximum possible budget for any zonal belt; surpluses are evident at low latitudes and deficits at high latitudes. Where is the point of balance? Clearly, given potential solar and terrestrial radiation exchanges, similar procedures can be followed to indicate global budgets at different times of year. The results can be transposed as isoline

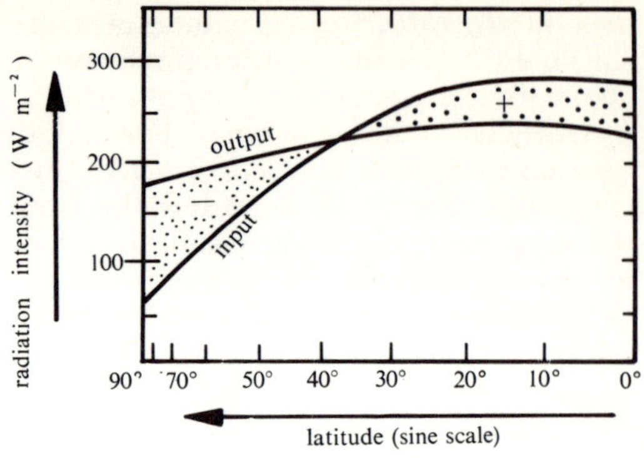

Fig. 2.13 Radiation exchanges with latitude at the roof of the atmosphere

Fig. 2.14 Projecting a latitude sine scale

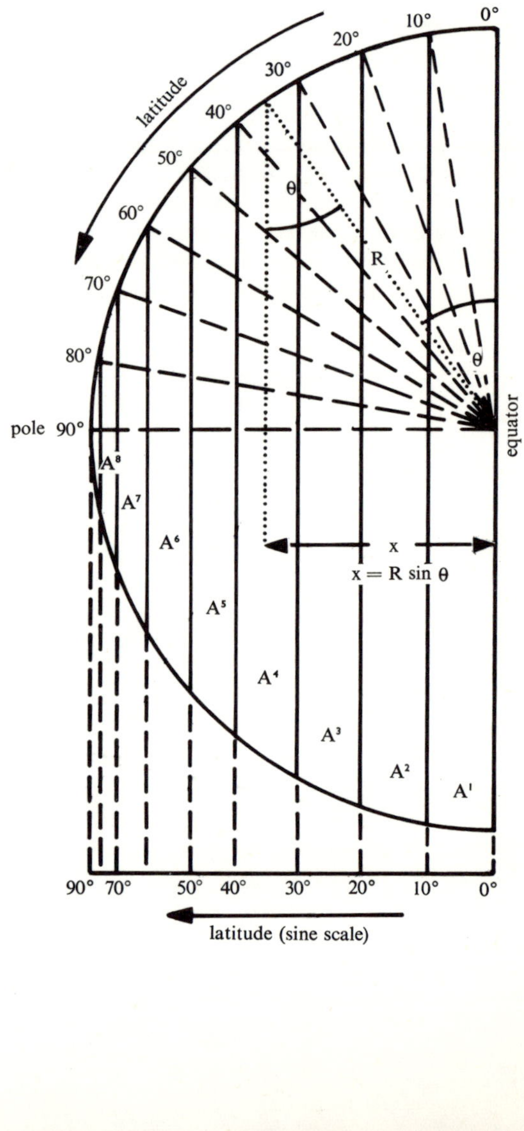

maps on equal area projections of the world. Whilst such exercises ignore important factors like the distribution of land and sea, nevertheless, they do illustrate that the atmospheric 'power house' lies within the tropics and that higher latitudes rely upon influxes of energy by advection to maintain thermal equilibrium. The major circulatory motions of the atmosphere affect these movements; otherwise, tropical regions would gradually become hotter and hotter and the polar regions colder and colder. The great variety and scale of 'weather systems' are manifestations of the global energy transportations at work. Before we dismiss them as being beyond the scope of this chapter, however, it is important to see them in perspective. If we consider the total solar energy received by the earth on any day to be 2^0 (or one unit on a binary scale), then it is possible to 'scale off' typical fractions used by different weather systems; for example, an Atlantic depression system 2^{-10}, a vigorous cold front 2^{-20}, moderate warm sector rainfall 2^{-30}, a local summer thunderstorm 2^{-40}, light winds in a valley 2^{-50} and the convection currents off a tennis court during a game about 2^{-60}. The latter are the orders of magnitude applicable to local and micro-scale studies. We have used an approximate logarithmic binary scale so that readers can relate the weather systems given to the exercise based upon fig. i.ii in the Introduction.

2C 2 The heat budget equation

Finally, it follows that the heat budget immediately above *a place* on the ground embodies net radiation (R), losses by conduction plus convection (H) and through evapotranspiration (E), and gains or losses by advection (A). In addition, we must not forget the heat stored in the soil, water, vegetation and the like (S). Thus, the complete heat budget equation is:

$$R+H+E+S\pm A=0$$

With the proviso that it is very difficult to isolate the stored and advective fractions, each of these factors can be broken down to *measurable* elements as outlined.

2D Measuring the elements

Here we are concerned with direct measurement techniques using 'home-made' and manufactured instruments reasonably available to school and college students. At the same time, we have deliberately introduced more advanced topics to encourage field investigations which will go well beyond the scope of any 'core' syllabus in physical geography. Many of these are presented as open-ended exercises. Because individual and group field studies are an important part of the techniques used by physical geographers, it is essential that experience is gained in the planning and administrative prerequisites to such ventures. It is not a matter of being led by the hand and doing a set series of exercises. Thus we have considered it important to tempt the reader with the vast opportunities which exist to develop personal research programmes, however modest or difficult. Alternatively, or in addition, the increasing emphasis upon course work, greater contacts with related disciplines and the growth of more broadly based school and college societies, all lend themselves to projects more ambitious than possible in the classroom. We wish to lend support to these trends, not just in meteorology, but throughout the fruitful field of physical geography.

In fieldwork our primary objective is to be able to *compare* measurements at one place with those at other places. Thus, it follows that both instruments and methods of observation must be *comparable*. Only in this way can we expect to discover the variations of meteorological parameters spatially and temporally and, therefore, their characteristics in localities rather than specific places. So, we do not advise or support schemes which aim solely at routine observations of the sort undertaken by synoptic and climatological *stations*. Many 'fixed' instruments cannot be justified in cost-benefit terms and student usage, particularly as professionally manned stations are never far away.

Advantage must be taken of the numbers of students usually available to cover any chosen locality with a close network of observations when required. Thus, we concentrate upon *portable* instruments as opposed to the conventional ones housed in Stevenson Screens and at meteorological stations. Immediately, the techniques guaranteeing comparability are even more important. Great thought must be given to precisely what element is being measured and the factors likely to give anomalous readings. Accuracy is essential in observations on a *small scale*. We confine ourselves to measurements of the main parameters previously mentioned in this chapter.

2D 1 Radiation

The main difficulty lies in distinguishing the fractions of direct, diffuse and terrestrial radiation. The devices used can be divided into those of a thermal or **calorimetric** type and those using **photometric** principles.

With the former, the heating beneath different surfaces is measured. A useful technique is to compare a white surface that is highly reflective to incident solar radiation, yet absorbs most diffuse sky radiation, with a matt-black one which absorbs all radiation. Thus, if a good conductor has equal black and white surface areas exposed, the differential temperature increases that occur after a **short** time provide a measure of the proportion of diffuse sky radiation to net radiation. The difference will represent the intensity of incident solar radiation over a short period at a place. Relative orders of magnitude can be determined using two flat trays similarly covered with soot and flour. These can be protected by reflective foil, taken to a site and then exposed to the sun. Thermometers inserted into the soot and flour will show different rates of increase for a short period. These two rates can be measured and recorded as an indication of the effects of net radiation and diffuse counter-radiation respectively. If the experiment is repeated several times throughout a warm sunny day, the results can be plotted graphically to show the diurnal variations of the two fractions of radiation. It is also useful to measure the angles of incidence of the direct beam of solar radiation when each measurement is made and to record air temperatures near the surface. The effects of clouds can also be noted.

Since differential heating of a single conducting material generates an electromotive force in proportion, it is possible to measure this electrically to determine a scale of radiation *intensities*. In order to calibrate such equipment one needs an independent sensor that is not influenced at all by radiation or has a known radiation output. By experimenting with such devices, a great deal can be learnt about the nature of radiative processes as well as their measurement. Should the black and white surfaces be flat and placed normal to the sun's azimuth, or horizontal? What advantages are there in having hemispherical or spherical sensors? Can diffuse sky radiation be measured by screening out the beam of direct solar radiation, or terrestrial radiation by inverting the apparatus? Much information can be gained from the operational manuals of the numerous manufactured instruments to measure radiation using these principles; for example, pyrheliometers, actinometers, pyranometers and solarimeters.

Photometric techniques have the advantage of being able to select specific wavebands by using filters, and are not subject to radiation losses in the same way that calorimetric techniques are. The 'light' values received can be transposed into appropriate radiation energy units. Those who are keen colour photographers will be aware that filters are available to minimise ultraviolet light reception by a film, and so on. Infra-red sensitive films have obvious applications in the study of diffuse sky and terrestrial radiation for, since the intensity of radiation can be translated into colours, it is possible to display local variations in emission on so-called false-colour or 'thermographic' photographs. The latter techniques require very expensive cameras and processing. However, ordinary infra-red black and white film can 'sense' differences in long wave radiation from various surfaces, particularly at night. Photographs of buildings and different types of vegetation can display useful 'patterns' related to the absorption and storage of radiant and heat energy.

Sensitive exposure meters of the photovoltaic type measure *visible* light intensities and so give some indication of solar radiation. They are particularly useful in measuring the albedo of a surface. By taking a reading of light reflected off a white card or board placed on the ground and then on the natural surface, an approximate albedo for the latter can be stated. Again, there are control problems regarding the orientation of the card and meter relative to the sun. Traverses through woodlands (see page 90) can give useful results regarding the 'filtering' of light through different types and amounts of foliage. However, it must be remembered that *light* is not synonymous with *insolation*. Sunshine recorders, which are often to be found at seaside resorts in connection with the local tourist agency, measure the *duration* of *bright* sunshine only. 'Home-made' versions based upon the principles of a pinhole camera can be constructed. All provide useful data which can be plotted graphically and correlated statistically with related information like air temperature, cloudiness and so on.

2D 2 Heat and temperature

There are very many pitfalls associated with the taking of temperatures in the soil or the air. Since many of these arise because of the lax way in which the terms 'heat' and 'temperature' are used, we remind the reader of the strict distinction between the two concepts made earlier (see page 25).

Mercury-in-glass thermometers provide the most familiar means of measuring temperature. However, it should be remembered that such thermometers only measure their *own* temperature, which is why they must be left to achieve thermal equilibrium with the medium whose temperature we require.

To develop this point, we must consider the conduction of heat from the sensor, radiation balances determined by the size and exposure of the instrument and other heat transfers caused by evaporation, wind speed and turbulence. The nearer we are to the ground, the worse these problems become. With so many variables which do not have uniform responses, it is clear that the measurement of temperature is fraught with control difficulties in the field particularly when using mercury-in-glass thermometers. It is of little value comparing temperature readings unless the techniques used can be justified. Most meteorological texts pay special attention to the 'sheltering' of thermometers in Stevenson Screens, but here we deal with temperature measurements to investigate conduction, convection and evapotranspiration processes in the field. Before the reader gets the impression that we are summarily dismissing such instrumentation, however, we emphasise the great value of short term continuous or autographic 'screen' records; for example, the use of bimetallic thermographs, aneroid barographs and thermohygrographs to provide charts of temperature, pressure and humidity fluctuations respectively.

Although designed for evaluating humidity, aspiration **psychrometers** are highly recommended for consistent air temperature measurements. They work on the principle of passing a regular flow of air over well-screened mercury thermometers (for the moment we are concerned with the dry-bulb only). Clockwork driven Assmann psychrometers are commended for their portability and robustness. However, care must be taken to avoid sucking in 'anomalous' air; for instance, near the ground surface the instrument itself can be moved to supply an undisturbed air current. Nevertheless, running whilst crouched or bent double and carrying a heavy instrument is not so easy as it might seem. To avoid accidents, albeit usually comic ones, we commend 'rehearsals' over flat ground before succumbing to the temptation of charging across a rough hillside.

Plate 2.2 Recording wet and dry bulb temperatures with a Whirling Psychrometer

Much less expensive, but prone to dramatic breakages if used carelessly, is the sling or whirling psychrometer (see plate 2.2). It is manually operated like a football rattle and is very easily transported. Inconsistencies and breakage risks can be minimised by coupling them to clockwork or electric motors in the manner of food mixers. All psychrometers are ideal for the sort of traverse work described in the next chapter.

Invariably the most accurate, and certainly the cheapest, techniques of measuring temperatures involve electrical methods. They have the advantage of being equally applicable to observations in air, liquids or solids with little disturbance of either. Most mercury thermometers on the other hand must alter their surroundings, particularly when inserted in soils. Two main electrical instruments are available known as **thermistors** and **thermocouples.** The former have a small capsule containing a substance whose electrical resistance falls rapidly with increasing temperature. Thus, by measuring the electrical resistance, a direct and very sensitive temperature relationship can be obtained. Since the capsules can be miniaturised as 'bead thermistors' they can be used for highly accurate pinpoint readings as sensitive as 10^{-4} °C. However, these can be broken by thoughtless handling in the field.

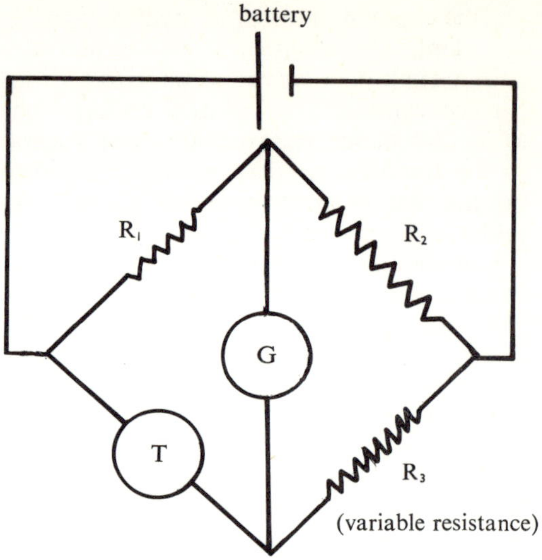

G = galvanometer (used as a zero or null indicator)
R = resistance coils
T = thermistor

Fig. 2.15 Circuit diagram for a thermistor

Fig. 2.15 shows the basic Wheatstone Bridge-type circuit used to measure the resistance of the thermistor. R_1 and R_2 are the so-called ratio or balance arms because they comprise of fixed known resistances, e.g. 100 ohms. Meanwhile, the third resistance R_3 is an adjustable coil resistance, say from zero to 10 000 ohms. Because the thermistor is effectively the fourth and unknown resistance in the circuit, its value (R_n) can be obtained by adjusting R_3 so that no current flows through the galvanometer. In doing this, we are *comparing* the four resistances in the circuit; hence,

$$R_n = R_3 \times \frac{R_1}{R_2}$$

This expression shows that, if R_1 and R_2 are identically balanced then the relationship is simplified to $R_n = R_3$. Therefore R_3 can be calibrated so that it registers the same resistance/temperature relationship as the thermistor. A big problem is maintaining the balance between R_1 and R_2 under different field conditions; so, whilst a home-made laboratory 'set up' is useful for demonstration and calibration purposes, it is best to acquire a professionally-made compact unit for use out-of-doors. Thermistors can be calibrated using heated water baths and a sensitive mercury-in-glass thermometer.

Theremocouples are less accurate but superior in most other respects. If two lengths of unlike metal wires like copper and iron are joined into a circuit and then the two junctions maintained at different temperatures, a small electromotive force is generated from the copper to the iron at the 'hot' junction and *vice versa* at the 'cold' junction. The total electromotive force (e.m.f.) generated is a function of the two materials used and the difference between the temperatures at the two junctions: principles known as the Seebeck and Peltier Effects. Clearly if a sensitive galvanometer is connected into the copper wire section, the e.m.f. can be measured. Furthermore, by keeping the cold junction at a constant temperature (ideally 0 °C in melting ice), it is plain that the hot junction can be used as a sensor and that the e.m.f. values are now directly related to the temperatures recorded.

Again, such a device can be calibrated in heated water baths against a sensitive mercury-in-glass thermometer. Fig. 2.16a shows the complete calibration curve for copper and iron. It has a parabolic form for readings up to 600 °C and the e.m.f.s within the effective temperature range in meteorology are exceptionally small. By using copper and constantan (a copper-nickel alloy) wires, both scales are vastly enlarged (see fig. 2.16b) and the temperature/e.m.f. relationship is virtually linear for all meteorological purposes. Fig. 2.17 shows the basic arrangement of a simple copper-constantan thermocouple circuit.

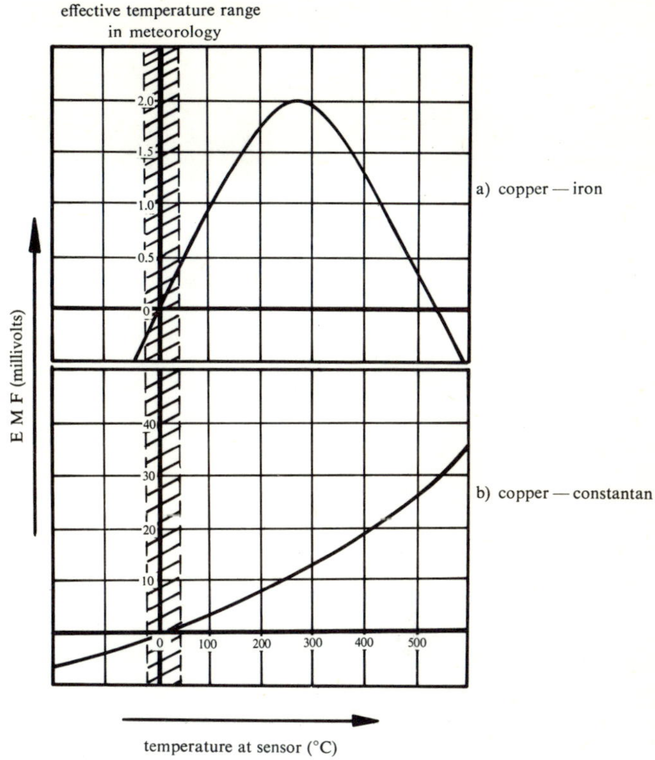

Fig. 2.16 Calibration of thermocouples. Note the different e.m.f. scales

Fig. 2.17 Circuit diagram for a thermocouple

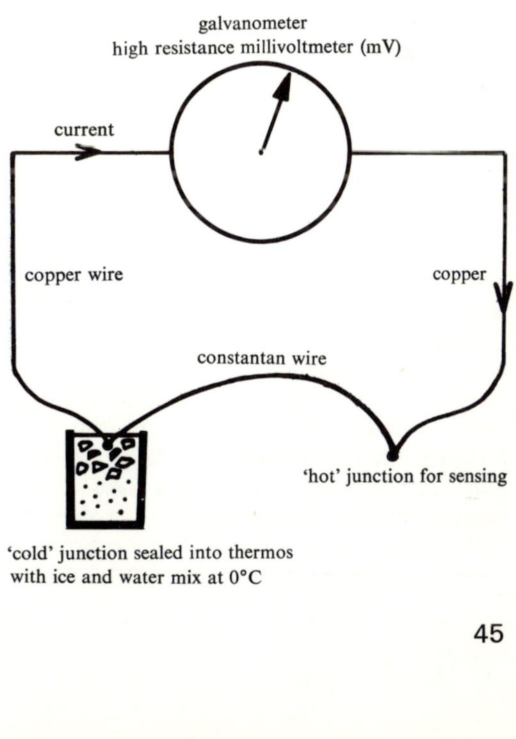

Plate 2.3 Recording leaf temperatures with thermocouples made from disposable hyperdermic needles. Although the cold junctions in the thermos are in view, the observer is well away from interfering with the natural temperatures

Out-of-doors, direct readings on a thermocouple are certain to go awry, particularly if the length of the wires is altered for any reason. This practical difficulty can be overcome by introducing a potentiometer into the circuit. This is a versatile electrical device which will balance and check the known e.m.f. output from a standard cell against the unknown e.m.f. generated by the thermocouple. We encourage those conversant with such equipment to set up a working demonstration of this circuit. For those who do not wish to become thus involved with the vagaries of electricity, we can assure them that compact and robust units are available to do the whole operation quite simply. It is only necessary to 'wire in' the thermocouple circuit as indicated on fig. 2.17.

Given good insulation and carefully prepared junctions, thermocouples satisfy most conditions for accurate temperature measurements in meteorology. Pinpoint positioning is such that it is easy to take temperature differences in skin tissues and similar cellular structures. 'Do-it-yourself' enthusiasts might devise miniature junctions for such studies; a particularly useful one being the insertion of very fine gauge insulated copper wire down the hollow tube of a hypodermic syringe (the disposable types used by veterinary doctors are suitable). Plate 2.3 shows temperature measurements of this type being taken in leaves. Although in meteorology the control or 'cold' junction is usually placed in an ice and water mix at 0 °C in a thermos flask, in some situations it is valid enough to probe the junction into the ground well beyond the depth where significant temperature fluctuations occur. Clearly, the temperature here must be determined independently.

Apart from cheapness and versatility, the chief advantage of such accurate pinpoint measurements is that they can reveal precise temperature gradients and lapse rates in and above the ground for studying conduction and convection processes. Vertical stands or horizontal layouts of thermocouples can be read quite rapidly. Figs. 2.18 and 2.19 suggest simple circuit arrangements for temperature **stands** and **quadrants** respectively. The former will facilitate observations of lapse rates or temperature gradients one metre above the surface, and the latter can be used to **integrate mean temperatures** per square metre. Both provide consistent positioning of the sensors for comparative observations at several sites. Although neither permit simultaneous readings, it is possible to improve the circuits and speed up the recordings by using appropriate switching units. The stand arrangement can also be adapted to obtain soil temperature profiles. Here, one of the great advantages of thermocouples is that they are inexpensive enough to be left in the ground so that the soil need not be disturbed once they have been inserted where required. Such observations are basic to calculating heat energy flows in and immediately above ground surfaces using the equations given earlier in this chapter. Also, much of the climatological fieldwork suggested in the next chapter relies upon such apparatus. The improving availability of cheap electronic equipment affords much scope for constructing such apparatus of considerable value in field measurements. Numerous other electrical resistance techniques are possible, and the advent of semi-conductors and solid state circuits provide much food for thought in making inexpensive and portable meteorological instruments.

The reader is reminded that the other major factors influencing temperatures in and above the ground are the *relative density* of the soil, *specific heat capacity* of the solids, air and moisture present, and the concentrations of *dust*, *carbon dioxide* and *water vapour* in the lower air layers. Since we have already mentioned techniques for the first two, and deal with dust and CO_2 later (see page 68), we can turn to the closely-related topic of evapotranspiration and water vapour.

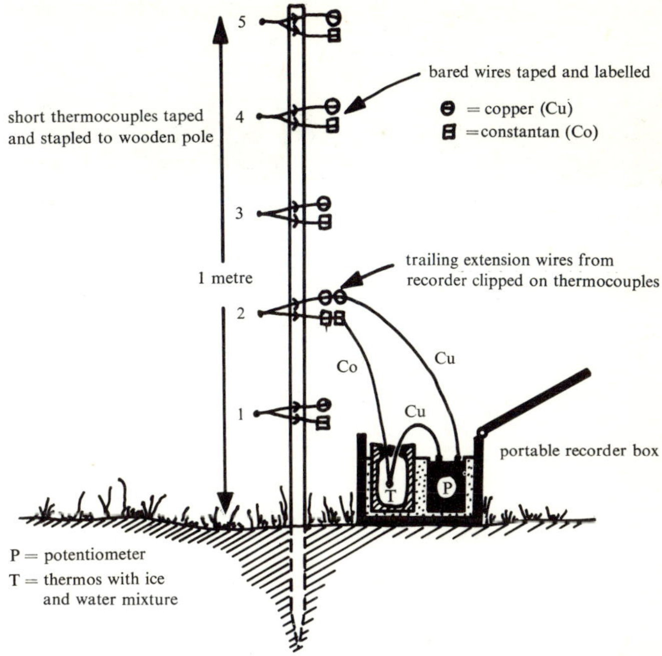

Fig. 2.18 Stand of thermocouples to measure lapse rates near the ground surface

Fig. 2.19 Quadrat for surface temperature observations

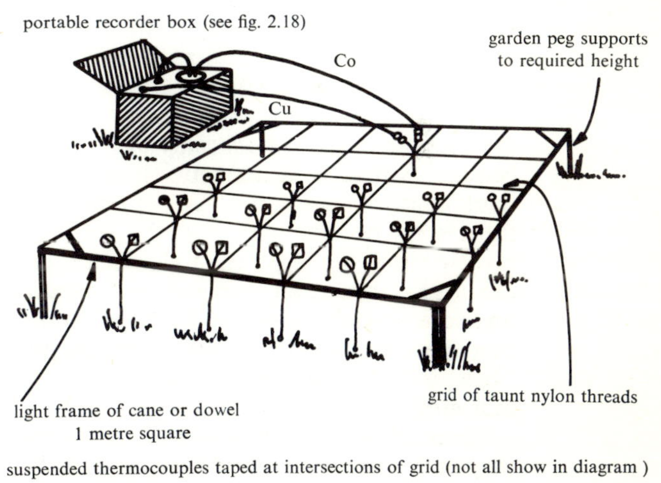

2D 3 Evapotranspiration, condensation and air motions

Studies of evaporation from free-water surfaces are comparatively easy since falling levels or the decreasing weight of a water-filled receptacle give useful *relative* measurements. Similarly, the rates at which damp cloths dry out, condensation clears from window panes and even the curliness of our hair are all related to evapotranspiration. Also, you will probably have heard older people complaining more of their ailments in response to humidity variations. However, these indicators are all indirect and suspect. Much the same reservations apply to thc use of evaporimeters and atmometers in micro-scale studies. These are various types of reservoir and bottles from which water losses can be measured more precisely. Since most rely upon capillary tubes and porous 'candles', the results must not be taken as indicative of what losses occur off *natural surfaces*. Thus, their use is limited to *comparative* measurements by identical instruments. Biologists find such instruments of value in controlled green-house experiments on plant growth and so, for the same reasons that we have encouraged the reader to venture into techniques traditionally the domain of physicists, we also encourage closer contacts with practical biology. As will be found in chapter 4, the geographer and biologist share much common ground here, and have a lot to learn from each other regarding experimental techniques in the field. The measurement of transpiration rates is an immediate example.

Anhydrous cobalt chloride is deep blue but changes to pink as the salt crystallises and takes up water vapour from the surrounding air. Thus, filter paper soaked in a solution of cobalt chloride reverses this colour change on drying out. If we now clip a postage stamp-sized piece of this dried filter paper between two glass slides, 'sandwiching' a leaf as well, then the time taken for the paper to turn back to pink will be a function of how fast the leaf gives up water vapour. By using identically dried paper on different leaves simultaneously, we can *compare* their transpiration rates. Good quality filter paper is essential and accuracy can be improved by checking the colour change against independent standard colours. Strict procedures must be followed to make a success of such experiments, particularly when used in the field. Potometers designed to measure the water uptake by plants in response to transpiration are frequently used as a control with the cobalt chloride techniques. We advise consultations with colleagues studying biology. This is not suggested because we wish to delegate a tedious technique, but because we are anxious to encourage the use of a variety of approaches to the evapotranspiration problem. The physical *geographer* must have a wide repertoire to cope with such a difficult topic. A single technique is most unwise, especially in the field.

Measurements of *absolute* evapotranspiration from soils and vegetation pose considerable difficulties. Large weighing lysimeters (see chapter 5) are the only field instruments which give reliable results. Indirect field techniques which evaluate evapotranspiration *requirements* of air are based mainly upon observations of humidity and air motions. Good measurements of absolute and relative humidity, and so vapour pressure, are based upon wet and dry-bulb temperatures. Psychrometers and suitably dampened thermistors or thermocouples give accurate readings on traverses. An explanation of the former will suffice to outline the principles involved, particularly as their use is commended for several climatological exercises mentioned in the next chapter.

Psychrometers operate upon the fact that evaporation from a dampened surface causes it to cool. Thus, if we surround the bulb of a mercury-in-glass thermometer with muslin and allow it to soak up distilled water through a trailing wick, the temperature recorded will be less than that of an ordinarily exposed thermometer nearby. The amount the *wet bulb* reading is *depressed* relative to the *dry bulb* value must be a measure of the rate of cooling induced by evaporation off the damp muslin into the surrounding air. Differences between the wet and dry bulb temperatures, therefore, give a measure of the *relative humidity* and *vapour pressure* of the air as mentioned earlier; that is, how much more vapour the air can hold. Clearly, if both readings are identical, then the vapour is already 'saturated' and the relative humidity is 100 per cent.

The standard wet and dry bulb thermometers housed in Stevenson Screens respond to the comparatively still air slowly circulating through the screen. This air flow is important regarding rates of cooling. Fig. 2.20 shows that, as air speed increases, cooling is fairly rapid at first but then decreases after about $2 m s^{-1}$, i.e. a brisk walking pace. Thus, steady readings are best achieved if air is circulated past the wet bulb at about this speed. Instruments like the whirling psychrometer referred to earlier rely upon such 'forced' ventilation to acquire consistent readings of humidity. It follows that the temptation to whirl too fast must be resisted or, that to read the depression of the wet bulb after only a few half-hearted swings is also unsatisfactory. It is as well to carry out several test runs indoors before placing too much reliance on one's technique. Assmann psychrometers are designed to overcome such inconsistencies by drawing in a regular flow of air over both bulbs at the optimum speed. In the field the muslin can be dampened by using an eye-dropper filled with distilled water, and it is important that it should remain free of grease and dirt. They should be changed and gently washed regularly.

Hygrometry is plainly a topic which requires caution. This is particularly the

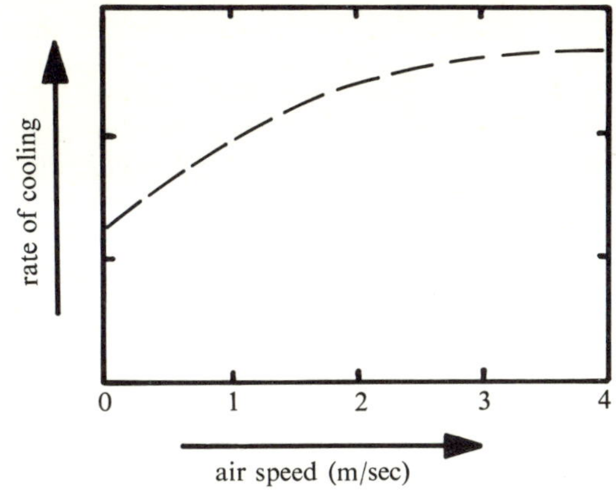

Fig. 2.20 Rate of cooling on an aspiration psychrometer

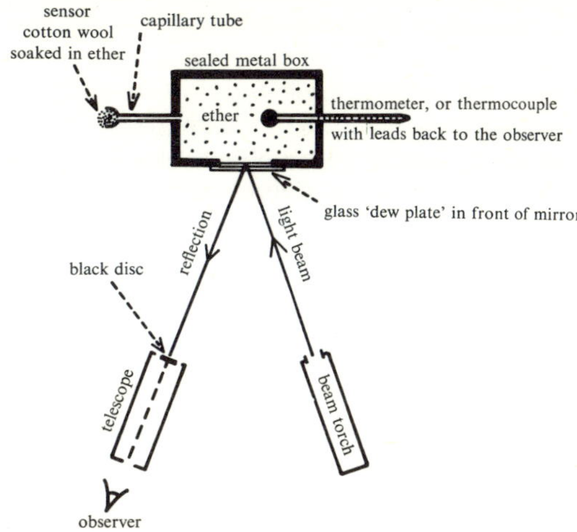

Fig. 2.21 Recording the onset of dew

case when dealing with the onset of saturation and condensation at or near the ground using so-called **dew-point hygrometers.** Given time and a modicum of patience, we can all record the times at which we see dew form on different surfaces like lawns and paths; yet, there is a great danger that we can hopelessly destroy what we are attempting to observe by the proximity of our own breath and body heat. We meet this situation when 'demisting' car windscreens. In the field, it is necessary to 'hide' behind transparent screens or, even better, to use binoculars from a safe distance however ludicrous the spectacle may appear to uninformed onlookers. Various techniques have been devised, and the optical version developed by Moss deserves mention. Fig. 2.21 shows the basic arrangement of an optical dew recorder. The cotton wool sensor is exposed and draws up ether as evaporation continues. The rate of cooling can be measured by a thermometer or thermocouple. Meanwhile, the reflected light beam is 'stopped' from being seen by the observer by a black disc partially covering the lens of the telescope. When condensation sets in on the dew plate, the reflected light will be scattered so that some is transmitted past the stop to the observer. Thus, its precise onset is accompanied by a 'lighting up' of the telescope objective. A photoelectric cell can be used to facilitate readings at a distance, even indoors! Obviously, there are numerous setting up and control problems with such apparatus, not the least being the use of an anaesthetic and highly inflammable liquid like ether! However, such devices make valuable project topics in meteorology, and much will be learnt from discussing the problems involved. Alternative approaches are to use sensitive thermocouples attached to leaves, grass, or whatever surface is being investigated, in order to monitor the liberation of latent heat at the onset of condensation. The same techniques can be applied to the formation of hoar frost.

Air motions, particularly those at a small scale, pose many problems too. Whilst the various mechanical or cup anemometers and vanes are valuable well above the ground surface in giving visually convincing results, they do not respond well to light air currents and gusts at levels where evapotranspiration occurs. The reasons for this are well worth studying in themselves. As already mentioned, plotting the motions of smoke-puffs, colour flares, French chalk and even soap bubbles are the best techniques for examining turbulent currents. Poles tagged with light streamers give useful indications of wind speed variations vertically and horizontally. Controlled runs in a simply constructed wind tunnel using electric fans or hair driers will give a rough measure as to how such devices will fare out-of-doors.

Thermal anemometers are highly sensitive and the only instruments capable of measuring air currents actually among thick vegetation and crops. Most operate on the rate of cooling of an electrically heated wire exposed to the air. By maintaining the wire at a constant temperature and resistance, any fluctuations of electrical current in the circuit will be a measure of the velocity of the air motions. Other versions called katathermometers use the rate of depression of standard alcohol thermometers after being heated to a certain temperature. The reader can derive an empirical formula to relate wind speed with the time taken for the alcohol to fall from say 40 °C to 35 °C on an ordinary minimum-type thermometer. A useful trick in the field is to put the thermometer under the tongue, to bring it to body temperature (effectively 37 °C), and then to remove it quickly into the air so that the time taken to 'chill' to say 30 °C can be measured. We can assume that anyone fit enough to undertake fieldwork will have a 'normal' body temperature! Such techniques are of great interest concerning the *cooling rates* of draughts and so are used in studies of human comfort levels, ergonomics and so on. Physiologists use such techniques in studying the *exposure* risks to people stranded in hostile environments such as high mountains in bad winter storms. Perhaps a study of such hazards in the safety of 'mock ups' in a laboratory (a butcher's or supermarket's cold store is ideal for a start, or even a large domestic deep-freeze) will enlighten those who might otherwise go into the countryside ill-equipped and uninformed. Physical geography is about real problems! Indeed, every process outlined in this chapter affects all of us in some way.

In confining ourselves to processes and measurements at a *place* we have largely ignored concepts and techniques applicable to the more usual global treatments of meteorology. We justify this by re-emphasizing that concepts and techniques are best understood through *experiencing* their validity and usage. Should the reader have jibbed at the large body of theory preceding the use of most techniques in meteorology, then such qualms should have been laid by appreciating the futility of rash approaches to measurements in the field. It will become increasingly apparent throughout this book that one must not be lured into field investigations without considerable forethought about the problems involved. Otherwise so much time and energy will have been wasted. This is very true of the precision needed in *micro-scale* studies. No apologies are necessary, therefore, for our frequent excursions into the realms of physics; after all, we are dealing with *physical* geography and many topics which are common to work in geophysics. It will also be apparent that fewer exercises have been included as we became more involved with basic measurement techniques. This is because instruments in themselves naturally lead to field tests and trials, and also that much of the work suggested in the following chapter on climatology presupposes familiarity with the more basic equipment mentioned.

Clearly, it is through field observations and tests that their scope in physical geography will be *realised*; any place will do whether urban or rural. We rest this particular brief in meteorology with a firm belief in Joseph Conrad's remark that,

> Any fool can carry on, but only the wise man knows how to shorten sail.

3 Local Climatology

Since life on earth could not exist without solar energy and the energy exchanges dealt with in the previous chapter, it is clear that the meteorologist is in the front line of studies in physical geography. This is why we have put its study first in this book. It is also why the reader will discover that meteorological processes can be made the main thread which links all the *physical systems* examined in later chapters, whether *present* or *past*. The immediate question, therefore, is where does the *climatological system* fit in to this pattern? Indeed, how do techniques used by climatologists differ from those used by meteorologists, and how are they applied? Without wishing to beg such awkward queries, we draw an analogy with military affairs; meteorologists are akin to those studying the mobilisation and manoeuvres of the forces, whilst climatologists are like those examining the effects of terrain on less tangible strategies and strengths of the combatants. Too often the generals of climatology are content to remain well behind the lines dealing solely with lists of supplies and casualties. We urge more campaign experience by way of the *battlefield* rather than the *communiqué*! Thus, our emphasis on *local climatology*. At this smaller scale of study the interface between ground level meteorology and climatology is clearly of greater importance.

From the time that man donned clothes and sought shelter he has responded to the physical environment on a *small* scale. Although human endeavour has seen the harnessing of energy at unprecedented levels, we are very far from competing with commonplace events like thunderstorms let alone 'managing' them usefully. Whilst some have set their sights upon large-scale enterprises like reclaiming deserts or modifying permafrost wastes, we must not neglect to assess the total impact of society's interventions at a local level. The quest for distant horizons often blinds us to what is going on around us; yet, the *aggregate* of these small scale processes is probably more far-reaching. This is particularly true in the study of climates; for example, sources of atmospheric pollution, local irrigation needs, and so on. So we feel justified in concentrating upon the techniques which give greater insight into the climates that we *experience* and can *examine*. Although the value of climatic tables and statistics cannot be disputed, we dwell upon field investigations as being more instructive. In the words of S. W. Wooldridge (1956):

> few things are more sterile than the generalisations of climatology . . . The realistic physical geography must be sought not in books but shod with a pair of stout boots.

Ask several people from a locality to describe its climate independently and then attempt to isolate their common conclusions. Such an empirical exercise is a useful guide to our perception of what is meant by a *climate* as well as the difficulties which arise in rational descriptions of its characteristics. Usually, our 'integrated experiences of "weather"', to use Manley's definition, are strongly weighted towards recalling extreme events rather than generalisations about day-to-day and seasonal patterns. Inevitably, we are very subjective and selective when it comes to 'appreciating' climates; ask different people *when* and *where* they prefer to take their holidays, and to give their reasons. What suits some will not attract others. Thus, in studying climates, we must find ways of analysing the 'extremes' as well as the 'averages'.

To a large extent, all definitions of climate have changed as the techniques which enable us to study climates have improved. More accurate weather records and increasingly sophisticated methods of statistical analysis have wrested us from the tedium of representing climates solely by a plethora of averages. Climatology is no longer confined to *static* values. Systems theory enables our approach to involve more explanatory or *didactic* treatments. We argue here that the original Greek concept that climate was largely influenced by the *slope of the ground* is far more valid as a basis than the later stress upon 'a clime' being a broad *latitudinal zone*. After all, when we look about us we see a landscape comprising of slopes, not a sphere with lines of latitude and longitude. Like the present-day human geographer, we see greater relevance in *locational* and *ecological* treatments of climatology.

A slope provides us with a 'given area' whose characteristics can be defined. Inputs and outputs of energy (e.g. radiation and heat) and of mass (e.g. precipitation and evapotranspiration) within such systems can be examined on two counts; first, how the system itself affects the inputs and outputs and, second, how certain inputs and outputs can transform the system. The question of *feedback* is a recurrent theme in all branches of physical geography (see page 178); in climatology the vast majority of feedback mechanisms are *negative* or self-regulating, although for short periods *positive* or self-reinforcing feedback can occur to alter the slope system. Whether or not the effects are lasting, as might be argued for climatic change, depends largely upon what we consider to be 'change' and the *scale* of study, especially that of *time*. Clearly, it is wiser to concentrate upon negative feedback situations at this stage. Climatic change and its effects are dealt with in chapter 8. In explaining the processes involved, *patterns* emerge which can be examined in space and time.

The 'mosaic' of slopes in a locality, therefore, gives a sounder basis for such studies than the traditional *station* network. In most instances, the only accessible information about a station consists of three coordinates; namely, its latitude, longitude and altitude. Thus, it is inevitable that much teaching in climatology is centred around these three factors, whereas our experiences indicate that variables like aspect, shelter, shade, type of surface and so on, are of more immediate significance. Consider, for example, one's climatic priorities in seeking a picnic spot, the best seat in a cold lecture room, a good place to get a suntan and how we 'orientate' ourselves with respect to the sun and slope we are lying on. Ask a gardener his reasons for the layout of various plants or the *aficionado* of the bullfight for the prices of seats around the plaza.

3A Observations on Slopes

In climatology, the main elements of a slope are its *aspect* and *angle*. Contours on large-scale topographical maps can give some indication of both but it is usually necessary to go into the field for accurate measurements. Good compasses and clinometers are essential. Morphometric mapping techniques are applicable, particularly in delimiting breaks and changes of slope (see page 149). Whilst stressing the geomorphological pitfalls in interpreting and visualising landscapes in terms of slope 'facets' as fundamental units, it is clear that their geometry at least provides a *basis* for the climatologist. Only when we attempt to refine the techniques to include vegetation and soil factors does the approach become unduly complex. For the moment we confine ourselves to the geometrical variables as forming discrete subsystems within the drainage basin as the fundamental morphological unit (see page 116).

Slope aspect and angle can be determined in much the same way that geologists measure the direction and angle of dip of sedimentary rock strata. Methods of finding the *maximum* inclination in three-dimensional terms are given in most introductory geological texts. Clearly, a ball rolling down an inclined plane follows the steepest path possible. Simple experiments with a clinometer on sloping surfaces indoors will clarify the problem; indeed, the reader can variously tilt this book and find the directions and angles of the maximum inclinations to the horizontal.

Fig. 3.1 illustrates an example of maximum slope mapping in the field using

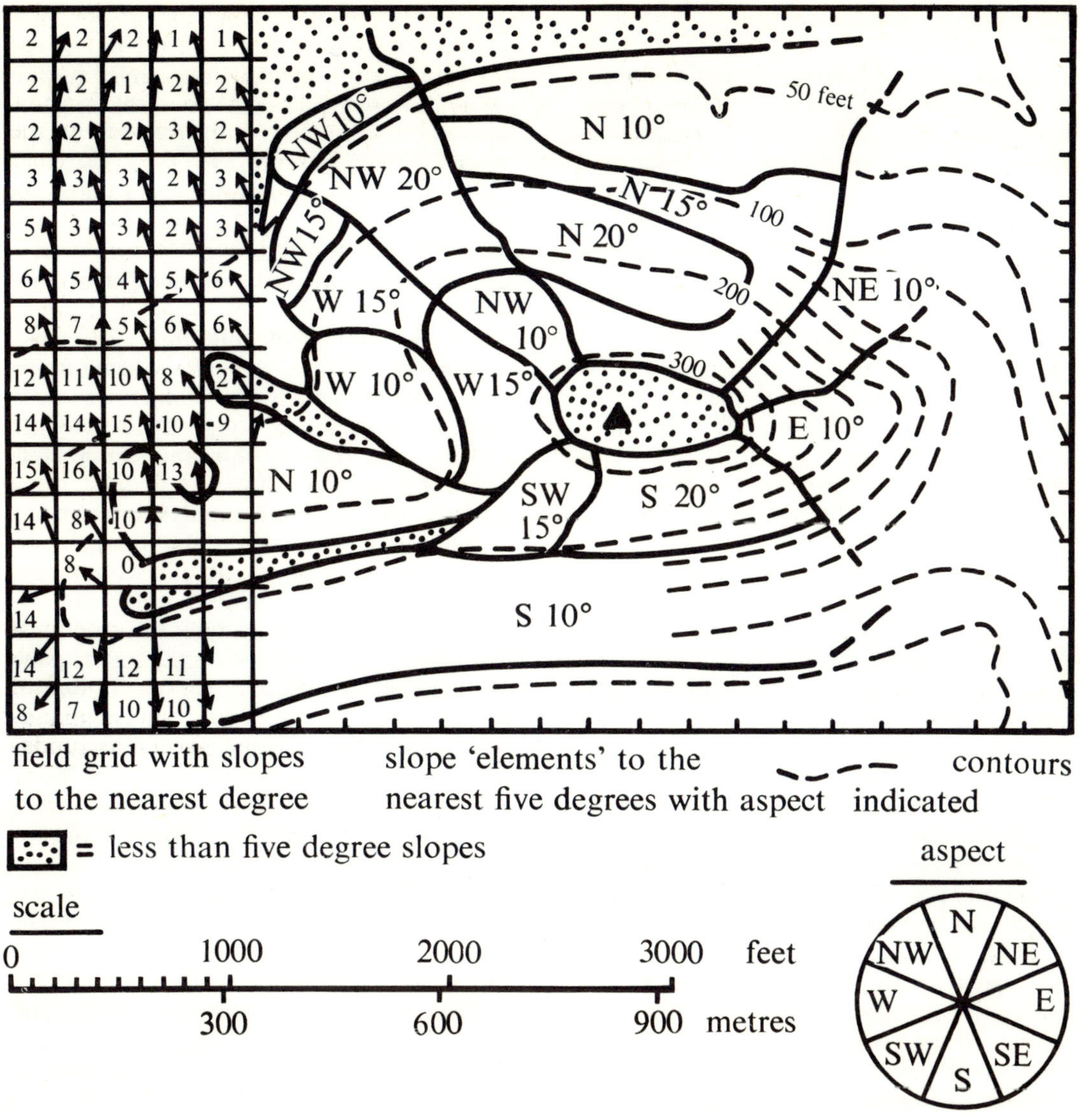

Fig. 3.1 A slope survey for climatological fieldwork on Banwell Hill near Weston-super-Mare, Somerset

large-scale base plans. The hill shown has been arbitrarily divided into slope elements: angles of maximum inclination are shown to the nearest 5 degrees with their corresponding aspects for eight 45-degree sectors relative to true North. Alternative field techniques for plotting the slopes are indicated at the western and eastern margins of the map. The former illustrates a superimposed grid with the direction and angle of maximum slope measured at each intersection. Such a co-ordinate approach minimises the risk of subjective sampling, whilst the significance of the grid dimensions is discussed in a later chapter (see page 100). When making the measurements it is invariably more accurate to take readings up the slope. Even then difficulties arise on gentle slopes, with obstructed views and in 'smoothing-out' minor relief irregularities nearby. The eastern margin of the map shows 'close' contours prepared by conventional step-levelling techniques in the field. A rapid way of doing this is to start from a known datum point like a benchmark, stand upright taking a horizontal sight through a clinometer or level to an identifiable feature *on the ground upslope*, and then take long 'metre' paces to this point. Clearly, this point is whatever height it is from the ground to eye level when you are upright. Thus, taking your paces as the hypotenuse and your 'eye height' as the rise, the slope angle will be the *sine* of the latter divided by the former. The rise alone enables contours above the datum to be located. Accuracy can be improved, of course, by having a friend to act as 'stooge' and using a tape.

Since directions of maximum inclination are at right angles to contours, and the horizontal spacing between successive contours is a function of the map scale and gradient, it is possible to determine various slope elements. The reader should redraw and complete the map shown, with a view to applying similar techniques to a familiar locality. So-called unit area methods of land classification described in Garnier's book on practical geography (see page 218), can also be adapted to such work, and form-line maps, prepared from viewing stereoscopic pairs of vertical air photographs in conjunction with topographical maps are an alternative for those who cannot afford the luxury of time-consuming levelling in the field. Also, many readers will have experimented with building relief models from contour maps. Such models, if carefully built true to scale with no vertical exaggeration, can be lit by spotlights to examine light and shade relationships of different slope elements. These will give the climatologist a useful insight into the effect of slopes upon the receipt of direct insolation; a topic we will develop shortly. Nowadays, it is possible to reduce the tedium and improve the accuracy of model studies of this sort by images of terrains being generated by computers. These can be 'viewed' from any angle desired, and such work is being facilitated by the Ordnance Survey storing all topographical information in a digital form on magnetic tapes. However, it will probably be a very long time before geographers will be able to trade in their plan chests for a computer linked to a tape library in Southampton!

Vertical cliffs and walls are equally as important as inclined slopes, of course. Built-up areas are obvious cases where upright and horizontal surfaces invariably dominate the landscape. Aspect and *height* are key factors with all buildings. The former is easily determined from large-scale maps and building plans, or by compasses in the field, and the latter by simple measurements on the ground with tape and clinometer. The techniques of horizontal and three-dimensional building analyses used by Toyne and Newby may be adapted to portray aspect as well. In densely-housed areas, the heights of buildings can be 'contoured' relative to the relief of the ground, whilst 'blocked' buildings can be treated as isolated units. Schools often fall in the latter category, and superimposed building elevations give useful data of the *total area* of wall directly exposed to given aspects. These can be prepared from architect's drawings or from suitable photographs of facades. Such techniques are important in building design concerning lighting, window space, radiation exchanges and heat transfers. We shall refer to such topics in more detail at the end of this chapter. Schools lucky enough to have scale models of their buildings will find it instructive to train spotlights upon them in a darkened room to simulate incident sunlight at various times. Surfaces in the light and shade can be examined at leisure and mapped. Those who do not have models might well make their own with the co-operation of interested departments. Apart from their instructive value when extensions are proposed, they are versatile teaching aids and pleasant to display.

3A 1 Sampling and station networks

On any given slope element we are faced initially with the problem of *where* to take observations which will be valid for the *whole*. A definite sampling plan must be adopted which avoids any bias. Since we cannot assume that regular climatological patterns exist across a slope, it is permissible to use grid or co-ordinate and traverse line techniques. However, where soils, vegetation and so on are uniform, random point or line sampling procedures can be used to locate a *representative* site. The statistical reasons for adopting such techniques are explained by Toyne and Newby, and will be found in many of the texts we recommend in the final chapter of this book. Considerable thought must be given to the sampling plan if time is not to be wasted. It is a good test to *correlate* the results of two groups working on a slope, one of which takes say temperature readings with a regular plan at the *same* time as one using random sampling techniques.

The *number* of sites taken is the second issue concerning the *confidence* which can be placed in the results as being representative of the whole. In climatology, as in most statistical studies, this involves analysing the *variability* and *deviation* of the results. The different kinds of variability concerning monthly, daily and hourly measurements in climatology are given in Conrad and Pollak's standard work on *Methods in Climatology*. Many of the statistical procedures for dealing with a large amount of climatic statistics are just the same as those used by the human geographer, e.g. quartile ranges, variance, standard deviation and coefficients of correlation and variation. Indeed, most statistics texts examine the computational methods required. If there is likely to be a wide scatter of values it is essential to have a large sample, whereas if the slope is regular far fewer sites will probably suffice. For the slopes illustrated on fig. 3.1 the most number of sites to be representative of any slope was 25. Again, it is instructive to compare the results of two field parties: using random sampling plans, one group should take temperatures at 100 sites and the other at 20. Although we indicate how many samples to take in similar exercises throughout the book, such as in sediment analysis, you may well adopt

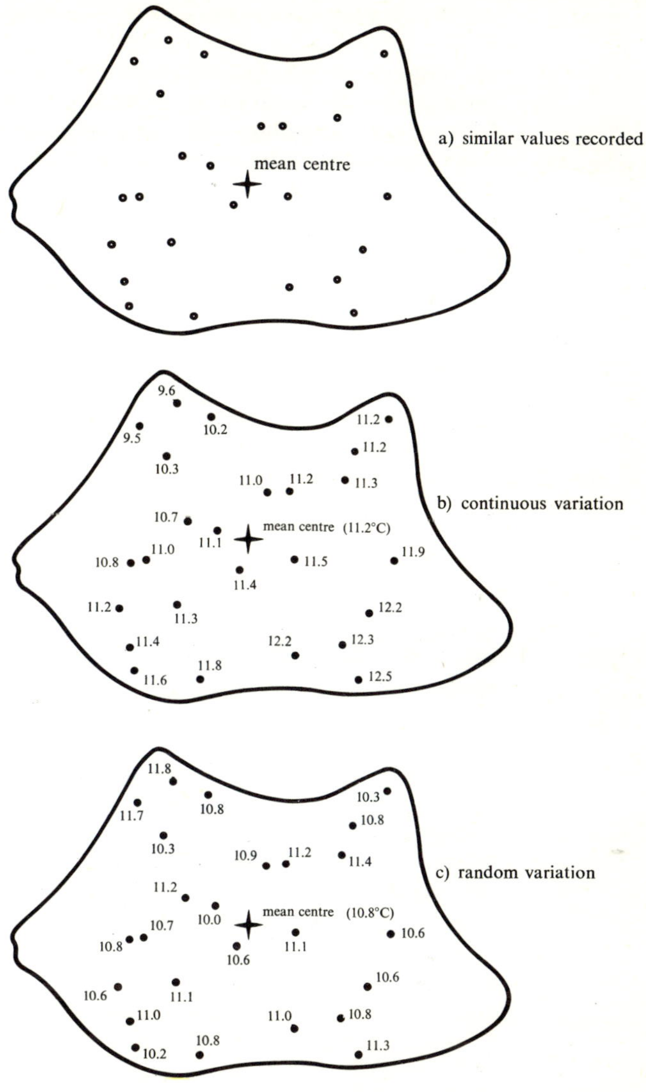

Fig. 3.2 Temperature observations at randomly distributed sites on a slope

the above technique to find out exactly how many samples are applicable to the particular work you undertake. Assuming that the observational methods are strictly comparable, such as all temperatures being recorded by whirling psychrometers at a comfortable height above the surface like a metre, it should be possible to ascertain the optimum number of sites that need to be considered.

If the results of any survey over a uniform slope element are plotted as a distribution of point values on a map, we can locate their **mean centre.** In spatial terms, this is the position around which similar values are distributed or at which the most representative value occurs; in effect, the 'centre of gravity' and arithmetic mean respectively. Three situations can be postulated regarding climatological observations: firstly, where all values recorded over a slope are the same: secondly, where they vary continuously, and finally, where random variation exists. Fig. 3.2 shows these situations for temperature readings over a south-facing slope, at one metre above the ground surface.

With similar values the mean centre can be found by cutting out the shape of the slope in card. This is then suspended from one edge by a drawing pin and allowed to swing freely until it stops. By using a weighted thread attached to the pin a vertical line can be drawn down the card. If the process is repeated using different points of suspension, the vertical lines will be found to intersect at the centre of gravity or mean centre of the card. The reader should test this regarding fig. 3.2a. Clearly, the same results can be determined mathematically for those who prefer pens to pins.

In the situation where continuous variation occurs, the mean centre is best found by drawing isolines at suitable intervals. We pointed out the importance of such mapping techniques in the first chapter regarding weather maps, and are sure the reader is well aware that the contour line is the most common example of an isoline. In this case we are dealing with the **isotherm**; however, the techniques for plotting these from point values (analogous to the spot heights of a contour map) are the same. In fact, all **isopleth** maps are prepared in similar fashion. Many will have drawn such maps 'by eye' in the past, but inaccuracies are unavoidable when intermediate positions are merely guessed. We commend the reader to the Thiessen polygon technique outlined on page 119 concerning lines of equal rainfall amounts or **isohyets.** Only by such methods will the reader solve the problem of locating an isoline in an indeterminate area.

Having grasped these techniques, the reader should prepare isotherms on a copy of fig. 3.2b. The midpoint of the mean isotherm will give the mean centre of the continuous variation of the temperature readings plotted. Since we can argue that uniform slopes will exhibit comparatively regular variations of this nature, the mean centre will closely approximate to the centre of gravity of the area concerned. This point can be demonstrated by plotting alternative sets of uniformly-spaced isotherms across the slope in question.

Where the readings taken are found to be random as in fig. 3.2c, we have a similar situation to that in fig. 3.2a. Thus, the mean centre will be once more at the centre of gravity of the slope element. The statistical reasons for these **isomorphic** situations are well worth examining.

In the three instances discussed, therefore, we find that the mean centre of a uniform slope element is an ideal place at which to take representative observations. Accordingly, it is an important concept in choosing sites for permanent instrumentation, or for regular readings over given periods. In fact, on this basis it is possible to choose station sites for all the slope elements shown on fig. 3.1 to produce a comprehensive network for local climatological observations in the area. The reader should attempt this for his home locality. Careful observations of the melting of uniform snow covers are often good guides to the extent of ground representative of any station; a technique used in siting certain permanent meteorological stations. It is instructive to examine why this should be so in terms of the radiation processes and heat balances over snow already mentioned in the previous chapter.

In practice, of course, the range of validity of such 'core' area measurements will not cover the exact bounds of the slope. Here we face one of the central problems in statistical methods applied to geography; namely, of defining **confidence limits** in a two-dimensional situation and then *comparing* the 'fit' of the boundaries. The problem may be visualised in terms of the

Venn diagrams and set theory 'vocabulary' increasingly used by mathematicians and many others studying natural and number systems. To geographers, these techniques have become powerful aids in studying the relationships between the distributions of different *elements* or *members* characteristic of particular areas or regions. These problems are discussed with regard to 'interaction fields' at the end of Toyne and Newby's book. Needless to say, had we ventured into the controversial realm of delimiting large scale 'climatic regions', it would have been necessary to undertake an exhaustive examination of such techniques in this book too. In local climatology, however, the demands for such treatments are not so pressing. Fig. 3.3 gives a simple indication of various correlations between the boundary of a slope and specified confidence limits enclosing a really-grouped data. The reader can extend these concepts in terms of causal relationships between sets of climatological importance like urban areas and air pollution, drainage basins and storm 'cells', relief and rainfall areas and so on. Superimposed isopleth maps and transparent 'overlays' are useful ways of gaining some visual insight into correlating different distributions; for example, the obvious relationship between relief and rainfall in the British Isles.

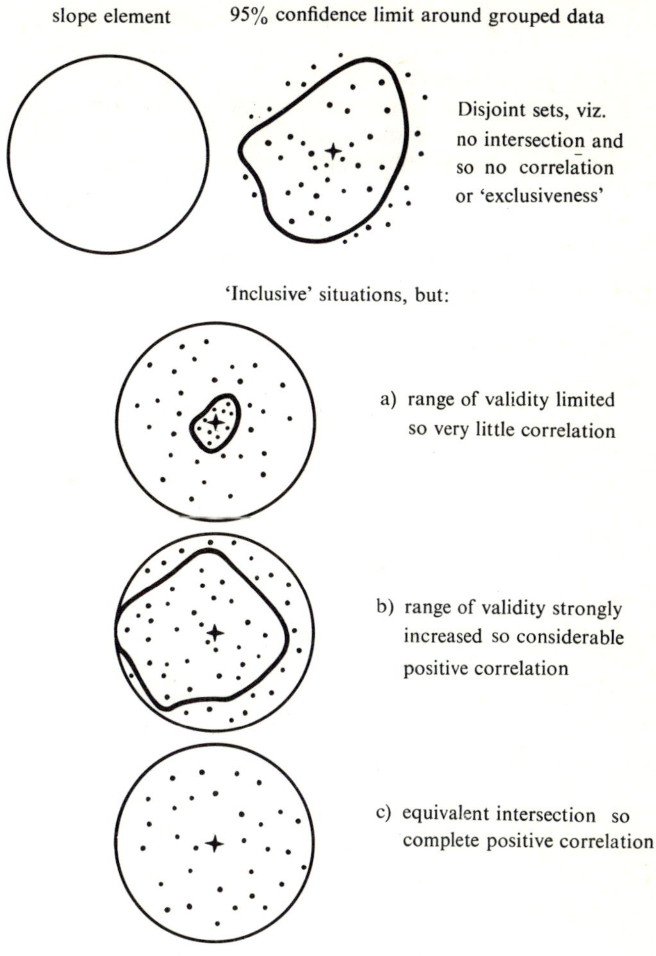

Fig. 3.3 Venn diagrams illustrating degrees of correlation

FIELDWORK

Grid surveys around a site

1 Using pegs, knitting needles or skewers, grid an open uniform surface surrounding a garden shrub, bush or clump of flowers. Clipped lawns or flat patios are best.

2 On a calm sunny day:

a Using suitably screened thermometers or thermocouples, take *surface* temperatures every hour at *each* peg between sunrise and sunset.

b Note the pegs in shadow and direct sunlight every hour.

3 Using the field data and a scale plan of the grid:

a Plot *isotherm* and shadow maps of each hourly set of readings on tracing paper.

b Assuming uniform increases and decreases of temperature, prepare graphs of the rise and fall of surface temperatures at each peg.

c Calculate the time of day when a suitable threshold temperature was reached at each peg and plot an *isochrone* (isoline of equal time) map to illustrate the results.

d Prepare an *isohel* (isoline of equal direct sunshine) map.

4 Attempt explanations of the isopleth maps drawn.

a Is surface warming related to the duration of direct radiation?

- **b** What effect does the shrub, or whatever, have in restricting direct radiation?
- **c** Is aspect of importance regarding the warmest surrounds?
- **d** Is the filtering of sunlight through the foliage significant, and does this vary with the angle of the sun?
- **e** What effect does the shrub, or whatever, have in restricting outgoing radiation?

5 What other isopleth maps would help in this enquiry?

6 Repeat the experiment on a calm cloudy day, when the ground is dry, when it is wet and at different times of year with varying amounts of foliage present. Compare the results. After a frosty winter night, using temperatures from grass minimum thermometers, prepare and analyse *isoryme* (isoline of equal frost) maps of the site.

3A 2 Some slope characteristics and their effects

It is possible to illustrate the *exposure* climate of any slope in terms of aspect and angle of inclination. Fig. 3.4 shows the effects of uniform unobstructed south- and north-facing slopes upon radiation exchanges. Direct solar radiation is shown to both slopes, and it is seen that the same amount of energy (Q) is spread over a larger area of the north-facing slope. This means that absorption will be more at site S_1 than at site S_2, all other things being equal, so that ground temperatures S_1 and S_2 will respond accordingly. Since it has been shown that about 80 per cent of terrestrial radiation from any site is directed towards a 'dome' of sky 60 degrees either side of the zenith, Stefan-Boltzmann's law

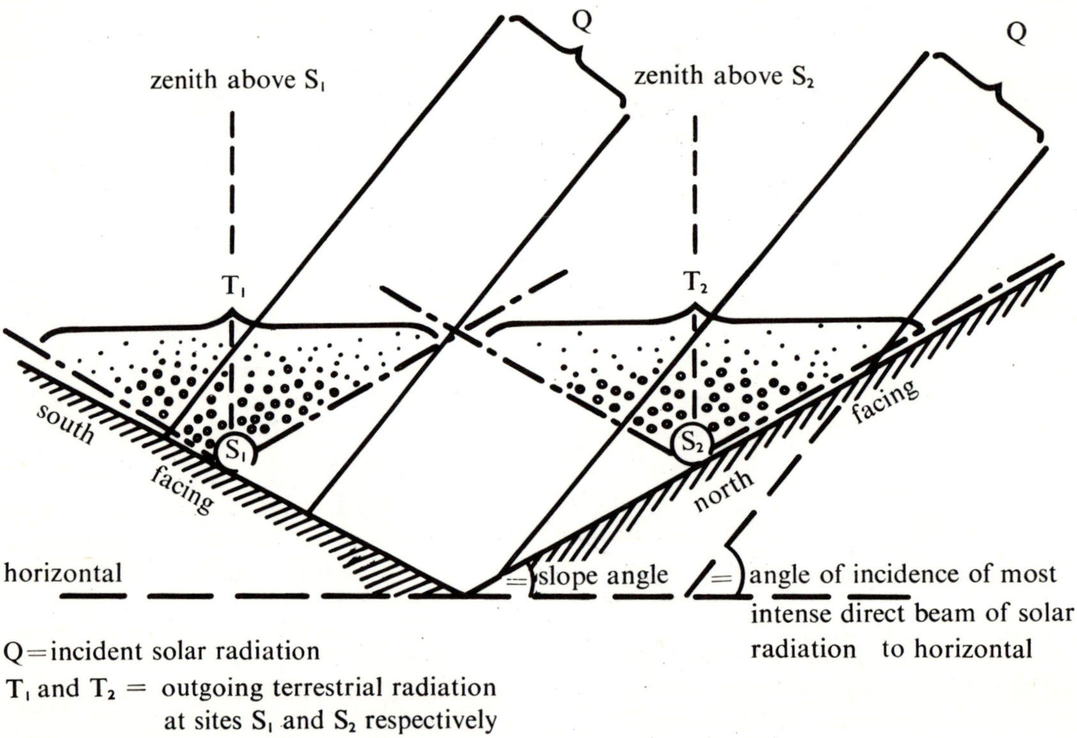

Fig. 3.4 Radiation exchanges on the south and north facing slopes of a valley

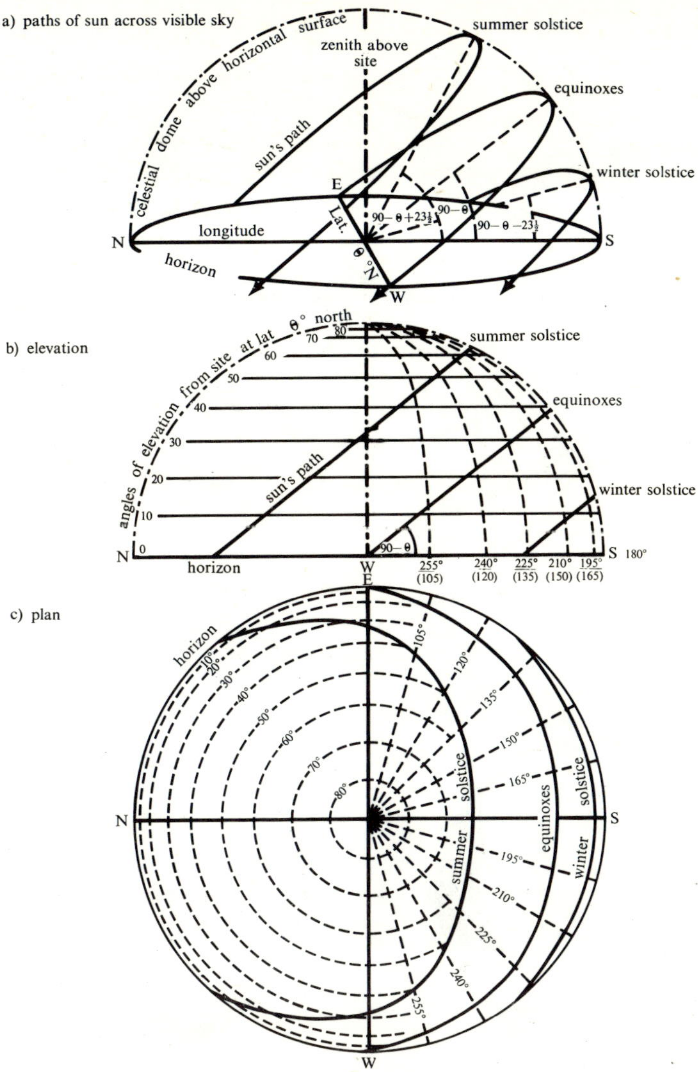

Fig. 3.5 The sun's paths above a horizontal surface at Latitude θ° North

$\sigma\ T_n^4$ gives a measure of the long wave energy returned to this part of the atmosphere. This will be correspondingly greater from S_1 by virtue of its higher surface temperature. As terrestrial radiation is partially absorbed above each site, mostly within the lowest 100 metres of the air, temperatures immediately above the south-facing slope will be greater too. Counter-radiation returned to the ground from the 100-metre layer is less directional so that its receipt on adjacent slopes of a valley will not differ significantly. In most valley situations this will be aided by air motions and 'mixing'. With the 'outward' facing slopes of hills, however counter-radiation also favours the slope with a southerly aspect. Most terrestrial radiation is lost in the direction of the zenith whilst counter-radiation is least from this area of the sky. This situation can be visualised by redrawing fig. 3.4 as a hill whilst retaining all other characteristics given. Further drawings to show what happens in deeply-dissected

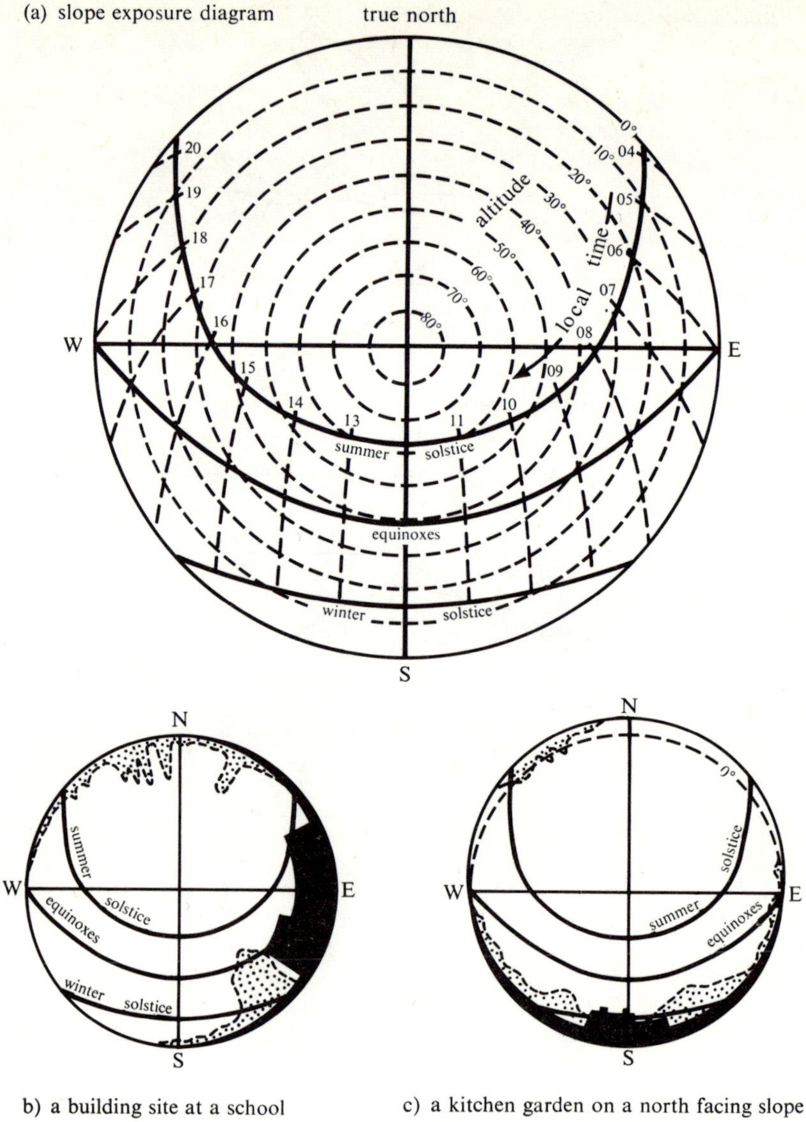

Fig. 3.6 Slope exposure diagrams for sites at Latitude 51°N

valleys, including suitably aligned streets with different building heights, will illustrate that it is not until slopes exceed 30 degrees that appreciable interference with radiation exchanges occur. We meet such situations in woodland clearings, pathways and the canyon-like streets of many large cities.

In rural areas with strong relief, or urban areas with tall buildings, most sites are screened and shaded at certain times. The paths of the sun across the sky are critical. Slope aspects and angles must be studied in relation to their latitude, time of day and the date in the year. Fig. 3.5 shows the relationships of the latter factors, and readers not familiar with the basic principles and terms involved will find them explained in many of the books listed in the final chapter. Hardware model building enthusiasts will find much scope for devising working demonstrations of the factors involved by reading the appropriate sections of Balchin and Richards' book on practical and experimental geography. Fig. 3.5a is a perspective viewpoint, showing paths of the sun across the visible sky above a flat surface at latitude $\theta°$ North, whilst the two diagrams below give elevation and plan views respectively. The projection techniques by which the latter have been prepared should be examined carefully in so far as they demonstrate many essential features in climatology, and geography in general. Although latitude 51°N has been chosen here, it is clear that such drawings can be prepared for any site.

Fig. 3.6a shows a more pleasing and functional development of the plan projection in the previous diagram. Apart from simple reorientation, the concentric circles depicting angles of elevation above the true horizon are linearly spaced as opposed to the awkward cosine scale of the projection. In addition, local time curves are indicated across the main sun paths. The derivation of these, in terms of the durations of night and day at specified latitudes, is another basic exercise to be mastered. Here it must be remembered that the times given are *local*, and so, in order to relate them to a *standard* time like Greenwich Mean Time (GMT) or British Summer Time (BST), the longitude of the site must be found. Familiarity with the data published in *Whitaker's Almanack* is essential in determining any values needed for particular occasions at a place. This is an annual publication which contains calendar tables giving an exhaustive range of astronomical data; such as, the times of sunrise and sunset at any place, sun angles and, for those who wish to plan ahead for an overnight vigil to record nocturnal radiation losses or temperature fluctuations, the phases of the full moon! A useful exercise is to plot on a graph the times of sunrise and sunset throughout the year at your school or home. The information from *Whitaker's Almanack* will enable you to do this without getting up at the 'crack' of dawn for 365 consecutive days.

The importance of this basic slope exposure diagram is that features above the true horizon from a site can be mapped-in. By setting up and orientating an appropriate base model on a plane table at the mean centre of a slope, it is possible to survey the surrounding skyline with a clinometer as in the examples shown. Fig. 3.6b shows details for a prepared and levelled building site at a school. Existing buildings shown to the east in silhouette block out the early morning sun for most of the year. The stippled areas indicate trees to the southeast through which mid-morning sunlight penetrates. Only in the summer months will direct morning sunlight reach the proposed building at ground level. Fig. 3.6c is for a kitchen garden generally sloping at an angle of 10 degrees northwards from the dwelling house. From the site chosen it can be seen that the sun at the winter solstice will be barely visible above the skyline. Both house and surrounding trees will cast shade over a large proportion of the garden. However, from early spring to late autumn, crops grown will get the maximum benefit possible on this comparatively steep north facing slope. The reader should study these examples and their implications regarding the receipt of solar and diffuse sky radiation.

Similar exposure diagrams can be prepared for any local climatological study. They are particularly valuable in conjunction with measurements of radiation exchanges, temperature profiles in and above the ground and soil moisture readings. Also they can be combined graphically with vector diagrams showing wind directions and velocities over a slope. Small gardens, backyards, sheltered tennis courts, streets, parks and woodland clearings are usually ideal for such investigations. After preparing an exposure model at the mean centre of the area in question, the whole site can be pegged out for an observational grid network. This will enable techniques similar to those given in the previous field exercise to be used; and, if sufficient observers and instruments are available, the comprehensiveness of the investigation can be vastly extended by taking soil temperature profiles, lapse rates up to a metre above the surface, relative humidities, and so on. In this way, a large enough body of readings will be accumulated on one calm, sunny day to make the effort of hiring or loaning a class set of instruments extremely worthwhile. The subsequent analysis, mapping and interpretation of the data in the manner suggested will provide considerable instruction and food for thought.

We have emphasised the choice of a calm sunny day because it is clear that the mixing and the advection effects of winds upset both the horizontal and vertical distributions which are the main purposes of the above investigations. The following techniques are commended regarding an examination of the effects of *shelter* upon *winds* at similar sites. Once again, we base our observations upon the site exposure model and an appropriate grid.

The standard graphical presentation of wind *direction* is to plot the frequency at which winds are recorded *from* a given compass direction upon a **rose or vector diagram.** Depending upon the degree of accuracy required, these directions are read to the nearest 32-, 16- or 8-point compass scale. Experience shows that 16-point readings are the most practical with the more turbulent eddies experienced near the ground at the sort of sites suggested above. Clearly, reference compass points can be marked out on the ground and observers, using puffs of French chalk or talc (see plate 3.1) at 2 metres above each station (that is, well over head height for most of us), record wind directions at appropriate intervals; 5 minutes, 10 minutes, or whatever. With less experienced observers, more accurate results can be achieved if observations are made on a 32-point scale and then reduced to a 16-point or even 8-point scale. The Meteorological Office uses this **reduction summary** technique when they are dealing with the so-

Plate 3.1 Talc puffs from squeezy bottles to observe light air currents above a water surface

called *free wind* at least 10 metres above the ground surface; that is, observations from a vane at that height.

Fig. 3.7 shows a typical wind rose prepared at a site during an hour with readings taken 2 metres above the surface at 5-minute intervals. The table shows the *actual* and *percentage frequencies* of wind directions recorded out of a possible number of twelve in the hour. The former have been plotted on to the rose diagram, and it is clear that the same 'petal pattern' would have appeared had the percentage figures been used instead. It should be noted that, as the readings were taken in the kitchen garden illustrated in fig. 3.6c, it has been possible to reproduce the 'skyline' of the site. Since the free wind was from the south at a steady 12 to 15 knots (i.e. a 'moderate breeze', Force 4 on the Beaufort Scale), it is possible to assess the influence of the buildings and trees under such conditions. It is seen, for example, that the main flow is significantly around the west side of the house. This can be attributed to the fact that, although the trees to the east of the house are taller, they are less dense and so *permeable*. Those to the west are thick evergreens which are effectively *impermeable* to all winds. Fig. 3.8 shows the basic airflow patterns over such 'barriers', the pecked line showing the upper limit of the *sheltered* areas. The great amount of research carried out in this field, which is usefully summarised in Caborn's study of shelterbelts (see page 221), shows that there is a more efficient reduction in wind velocity to leeward of the permeable barrier because of its filtering effect. The reader can observe this when the smoke from bonfires penetrates through hedges, but swirls with strong eddies over walls. There is a strong relationship between the *permeability* (expressed as a percentage of void to total volume), *height, width* and *profile* of a barrier and the extent of the area sheltered to leeward. In this instance the relatively low pressure created downwind of the buildings and evergreens and the 'drag' effect of the deciduous trees, causes the main airflow to be diverted to the west. Hence, the reason for the higher frequency

	N	NNE	NE	ENE	E	ESE	SE	SSE	S	SSW	SW	WSW	W	WNW	NW	NNW
frequency	–	–	–	–	–	–	–	–	1	1	3	5	2	–	–	–
percentage frequency	–	–	–	–	–	–	–	–	8	8	25	42	17	–	–	–

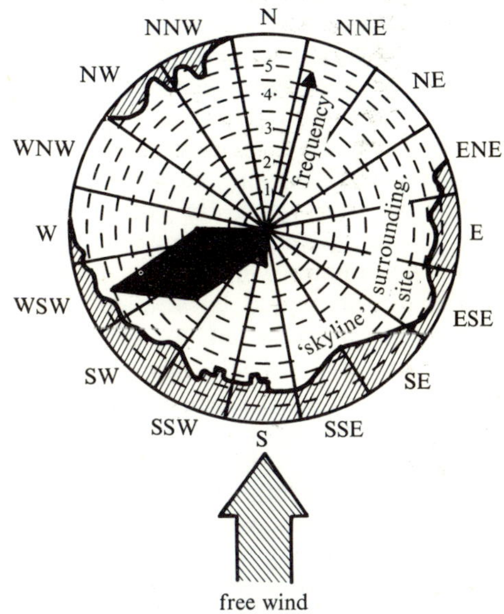

Fig. 3.7 Wind directions and frequency on a wind rose. (c.f. with fig. 8.15 regarding till fabrics)

Fig. 3.8 Air flow over different trees

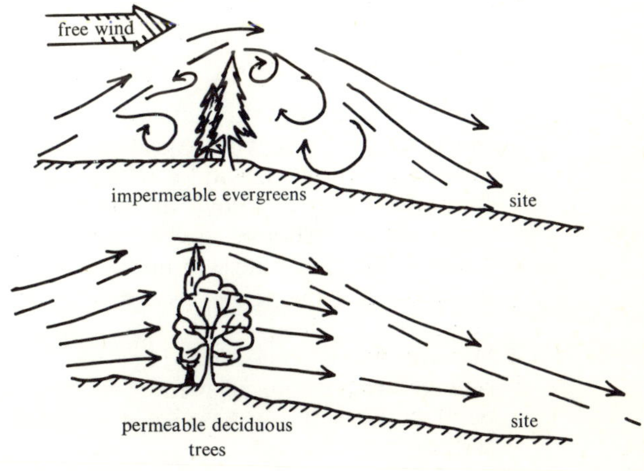

of winds from this direction at the kitchen garden site when the free wind is southerly.

The measurement of wind *speed* near the ground is fraught with difficulties. Whilst you might experiment with some of the methods described in the last chapter (see page 50), it is likely to prove wiser to use the approach adopted by Sir Napier Shaw who grouped wind velocity into three classes: light (Force 1–3, up to 10 knots), moderate (Force 4–7, 11–23 knots) and gales (Force 8–10, 24–55 knots). A handkerchief will just be lifted by a 10-knot wind, loose clothes will billow vigorously in 23-knot winds, and you would be safer staying indoors with gusts exceeding 50 knots! With this comparatively simple classification the reader should be able to analyse the relationships between wind speed and wind direction at any site. To what extent do *slope* and *shelter* influence the direction and frequency of light and moderate winds?

On pegged-out and gridded plots it is clear that, given certain free winds over the whole site, observers can log the direction, frequency and strength of the 2-metre air currents. These results can be examined on suitable site plans and exposure models. Given a good sample of observations, it will then prove possible to prepare airflow or **streamlines** over the plot as flow diagrams. These can be accompanied by isopleth maps of **wind direction reliability**; that is, the percentage frequency of the 'prevailing' streamlines across the plot.

FIELDWORK

Grid surveys on a slope

1 Select a suitable plot, such as a tennis court, mark it out with a grid and then take observations of wind directions and strengths 2 metres above the surface at each station. Choose appropriate time intervals like 5 minutes, 15 minutes or 30 minutes. Repeat the experiment with different free winds.

2 Prepare wind roses showing the percentage frequencies of winds on a 16-point or 8-point compass scale for each station.

3 On a site plan, draw flow diagrams to illustrate prevailing streamlines and isolines for wind direction reliability.

4 With reference to your maps and appropriate exposure models from the mean centre of the chosen site:

- **a** Explain the effects of slope and shelter on air flow over the plot.
- **b** Relate particular airflow patterns to soil moisture and the relative rates at which certain parts of the plot 'dry-out'.
- **c** Attempt to distinguish between the effects of winds and of the receipt of direct radiation regarding 'drying-out'.

5 Based upon the information discovered by the above exercises, select contrasting stations on the plot to record soil temperature profiles and lapse rates in the lowest metre of air. Assess the effects of winds and radiation upon heat energy losses at each station.

Similar tactics, of course, can be adopted to study the directions, strengths and frequencies of free winds on a *larger* scale, always assuming that enough 10-metre poles or masts are available to sample a district effectively. The effects of local relief become more important at this level of investigation; for example, valleys affording wind lanes, sheltered leeward slopes, exposed hilltops and so on. We can also introduce new techniques to examine the **deviations** and **percentage probabilities** that particular weather will be associated with certain wind directions and speeds. The easiest elements to record are the *visual* ones like cloudiness, precipitation, mists, fogs and frosts. However, to enable valid statistical analyses of the relationships, it will be necessary to accumulate a large body of observations over a long period. This is why large-scale studies require teamwork and a continuity of records. Those who start such a project may not have the satisfaction of seeing their hard work come to fruition. We are only too aware of many long-term 'weather' projects which have foundered when its initiators have moved on or acquired other interests. So, we prefer the smaller scale short-term project with the benefits of rapid results. The mobility of the student population is a factor which cannot be ignored when planning fieldwork in physical geography.

Apart from **celestial, topographical** and the **vectorial** aspects of slopes and the shelter they afford, yet another use for the exposure model is for **cyclical** data. Quite simply, the circular base can be used as a clock-dial or calendar to show **harmonic** patterns. For whatever element is being recorded, the centre of the circle can be zero and the circumference a suitable upper limit. Alternatively, to make comparisons between separate models easier, the information plotted can be in calculated percentages between zero and 100 as in the case of illustrating the percentage frequency of winds. Since most climatic elements can be recorded on a cyclical basis, a variety of such information can be superimposed or overlaid on a single model.

FIELDWORK

Energy inputs to slopes

1 Take surface temperatures every hour throughout a clear sunny day at the mean centre of a slope partially screened by buildings or trees. Walled gardens and most school playgrounds are suitable. A thermograph will facilitate such observations.

2 Divide the exposure model of the station into a 24-hour clock dial, adapt the concentric elevation circles as a suitable temperature scale and plot the hourly temperatures.

3 What degree of correlation is there between shading at various times and surface temperatures?

4 Repeat the exercise at different sites; either superimpose the results on a single model or use tracing overlays to compare the sites. Also, examine the possibility of introducing additional climatic data on the lines suggested in the preparation of figs. 3.21 and 3.22 (see page 79).

3B Relating observations on several slopes

Maps and diagrams afford the obvious means of relating climatic observations visually, whether in space or time. As climate is an abstract concept, however, we cannot treat it as for thematic maps of topography, geology, soils and so on. As seen above, scale is the most important factor; for example, the majority of climatic maps encountered in atlases and text books are based upon a broad network of stations which sample characteristic elements. Given records for long enough periods, it is possible to make generalisations which will hold for a much wider surrounding area than the station site itself. Nevertheless, these remain generalisations until systematic observations are undertaken at intervening sites. We are then faced with the problem of reconciling observations from a dense network of temporary stations with those from the wider mesh of permanent stations. It is clear that a dense network of observations locally, even if only for short periods, can 'pick up' information which coarser networks will 'lose'. Since the former involves much more consideration of the local features likely to influence the observations, as much attention must be given to the *terrain* itself as to the climatic observations *per se*. Indeed, the denser the network, the more we can rely upon terrain types as climatic 'indicators'. The term **topoclimatology** is particularly apt in this case.

In small areas, variations of latitude and longitude are insufficient to alter significantly the *duration* of potential solar radiation from place to place. A single exposure model base will suffice for studying the more important *topographical* variations on several slopes. Counter-radiation and advective processes pose problems, however, because of the sort of 'interference' already mentioned (see page 59). It is difficult to isolate those processes *internal* to a slope system from those which are *external*. We can only reaffirm that, in

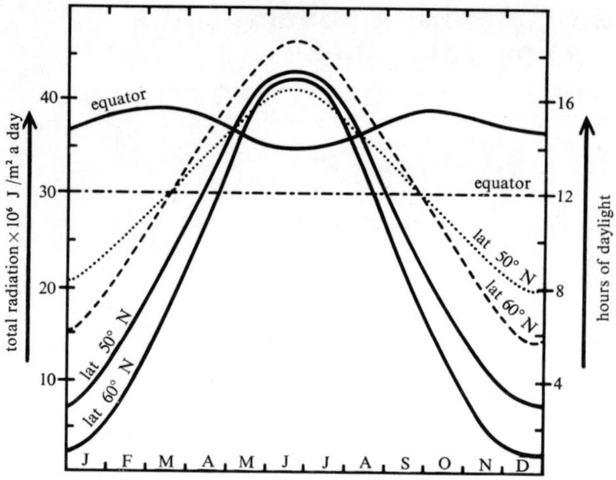

Fig. 3.9 Total daily radiation to the atmosphere and hours of daylight at the equator and latitudes 50° and 60° north

relatively well-defined valleys, the receipt of counter-radiation will be similar. Thus, there is a strong case for confining local climatological investigations to slopes within a valley. Here we may draw a parallel with geomorphological studies in selecting the *drainage basin* as the fundamental unit. In urban areas, the street provides a useful counterpart. By choosing small-scale drainage basins or streets, therefore, we can minimise the need to consider counter-radiation and rely upon *direct* inputs and outputs of energy and mass.

3B 1 Topoclimatic data and mapping

Having mapped the slope elements within a selected drainage basin and prepared representative exposure models for each, we can relate their 'performance' regarding direct energy exchanges without recourse to elaborate and lengthy field observations. Data on the following basic topographical factors is needed: the unobstructed or *free* sky to the direct solar beam, the area of visible sky either side of the zenith, a representative albedo and the aspect and angle of each slope. Surface temperature, vapour pressure, carbon dioxide and dust concentrations are the primary climatological observations required. We consider each one in more detail below.

Whilst charts for computing radiation fluxes have been in common use since being devised by W. M. Elsasser in 1942, empirical formulae are just as useful. Most of the latter stem from theoretical calculations based upon the solar constant (Q_s) to the roof of the atmosphere and the seasonal declination of the sun. Fig. 3.9 shows the *total daily* solar radiation (Q_d) to the atmosphere and potential hours of daylight (D) at representative latitudes. The curves can be used to evaluate how much direct solar radiation $\tau(1-A)$ is absorbed by a unit surface area of slope on a particular day, viz.

$$\tau(1-A) = Q_d(1-A)(0{\cdot}3+0{\cdot}5\frac{d}{D})$$

In small areas, Q_d and D are constants easily determined from diagrams like fig. 3.9. The variables for any slope will be the albedos (A) and the *actual* duration of sunshine (d). Albedos can be measured by the technique explained on page 43, whilst the latter is effectively the number of sunlight hours above the horizon less periods when clouds intervene. The last value is usually common to a small area too, being determined by the mean daily cloud amount. For example, if the sky remains clear, then d will be the number of hours the sun shines on sites during any day as calculated from exposure models. Mean cloud amounts given in oktas, or eighths of the visible sky covered by cloud, can be used on overcast days, although even with complete cloud covers some 20 per cent of solar radiation can penetrate to the ground. The reader should attempt to compute how much solar radiation arrives at the slopes of a drainage basin on selected days, or at selected times of day for that matter.

Having 'arrived', the actual input of radiation to any slope will be a function of aspect and angle with respect to the incidence of the direct beam. This is easily demonstrated by shining a powerful torchlight on to different sloping surfaces. The

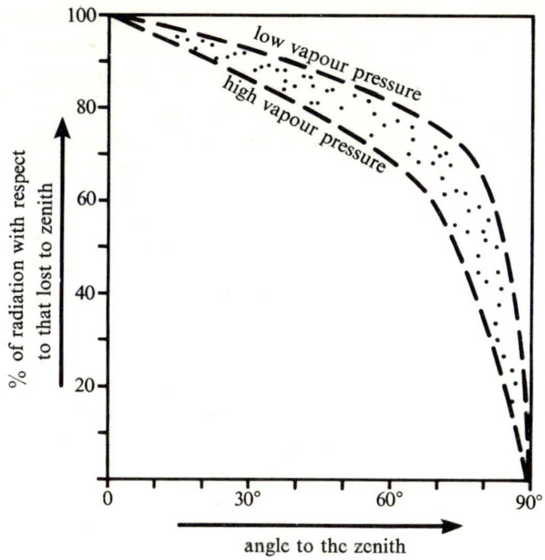

Fig. 3.10 Terrestrial radiation losses with respect to the zenith of a site and vapour pressure of the air layers near the ground

geometry of such situations should be examined carefully regarding the differential heating of slopes, and so their ability to emit radiation accordingly. Measurements of radiation, as explained in the last chapter, and of surface and soil temperatures will verify general conclusions as to 'favoured' and 'unfavoured' slopes in the basin. If further proof is required, we suggest that a relief model of the area be made so that the problem can be studied at leisure indoors.

The free sky at any site can also influence radiation outputs, especially when it is remembered that the majority is lost to a 'dome' restricted to 60 degrees either side of the zenith. Fig. 3.10 illustrates the approximate relationships involved, the stippled area being a zone within which specified curves for different water vapour pressure (e) fall. Since vapour pressure is a function of temperature, and so humidity, fig. 3.10 is significant regarding night cooling and diurnal fluctuations of terrestrial radiation. What will be the effects on say north- and south-facing slopes, of relief, trees or buildings screening a site, or by a nearby source of water vapour such as a river or lake? How will mists, fogs and low level clouds affect radiation losses? Such queries are useful discussion points, and the conclusions might be verified by field observations; for example, temperature stands and quadrats to evaluate heat energy losses from the soil at selected sites (see page 47).

FIELDWORK

Energy outputs from slopes

1 By adapting the observations and analytical techniques used in previous exercises, determine the factors affecting heat energy losses at night from selected sites.
2 What controls do exposure and the proportions of free sky impose on the processes assisting heat energy losses?
3 Where are the coldest and warmest parts of your garden or the school grounds?
4 Using wet and dry bulb thermometer readings at a comfortable height above the surface, such as 1 metre, examine the effects of humidity and vapour pressure above the warmest and coldest sites identified above. What other factors are involved, and how would you set about measuring their effects?

As we have seen already, carbon dioxide concentrations also influence the fraction of terrestrial radiation absorbed in the lower air layers. As critical increases in CO_2 are consistently reported in industrialised countries, it is worthwhile mentioning methods of measuring its concentrations. Trapping gas samples out-of-doors is not so simple as it might seem, however, and many of the methods require elaborate equipment and lengthy laboratory analyses. Yet, we feel that the reader should have some knowledge of the techniques commonly used, if only to reinforce our plea that the physical geographer must be prepared to understand and use complex instruments in difficult field situations. There is hardly a more worthy cause than accurate measurements and monitoring of atmospheric pollution!

Devices for sampling known volumes of air are required which use vacuum or pumping principles. The CO_2 concentration of a sample can be determined in a number of ways. Infra-red absorption instruments are the most sensitive, but techniques which involve chemical reactions can provide useful results. If the sample is brought into contact with a caustic soda solution, CO_2 is removed from the air and an appropriate drop in pressure occurs. This can be measured by sensitive aneroid barometers in an enclosed chamber. Alternatively a sodium chloride solution with some phenolphthalein will react in such a way that a known electrical current passed through the solution will generate an electrolytic action from which the weight of CO_2 in the sample volume of gas can be calculated. Another method is to relate the percentage of CO_2 to the amount of calcium carbonate ($CaCO_3$) that water in equilibrium with the air will dissolve. The experimental controls and chemistry of these processes are well worth examining in so far as they explain much about carbonation or solution processes of erosion in geomorphology as well as the measurement of CO_2 for climatological purposes.

One important use of such techniques is to measure the variations of CO_2 above contrasting surfaces at different times of day at different seasons. For example, the concentrations usually increase at night and during winter in response to plant respiration, temperature gradients and air circulations. You should study the affects of CO_2 variations regarding the carbon cycle illustrated in the next chapter (see page 104). Where the air is relatively rich in CO_2, the absorption of outgoing radiation will be greater and air temperatures correspondingly higher. The reader might reflect upon this concerning the temperature increases in a poorly ventilated, centrally heated, room in which people are exhaling CO_2. For what reasons are potted plants commended as beneficial indoors, and can you think of any scientific basis for the 'old wives tale' that such plants should not be kept in a bedroom?

Dust is another important feature of the lower air layers. It influences the scattering of radiation and, since many of the particles are hygroscopic, having an affinity for water, it assists in the absorption and condensation of water vapour. Indeed, if you can examine water droplets on a glass slide which has been exposed to a mist or fog under a very high-powered microscope, you will find minute dust particles in them. The 'dirt' left on windows after a rainstorm is an indication of the amount of dust in the atmosphere. The reader should collect some of this and study it under a microscope. Dried samples can be stirred into a beaker of distilled water and the variations of pH measured (see page 95). One way of getting dust directly from the air is to expose a Buchner funnel set into a flask from which the air can be withdrawn by means of a filter pump. Damp filter paper set into the funnel will trap dust particles as air is drawn through it. This can then be dried gently in an oven ready for analysis. If attempts are made to count the number of particles on a rafter slide (see page 103) a measure of the concentrations in given volumes of air will be possible. In his classic study of London's urban climate, Chandler drew air through filter paper up to periods of 24 hours and then measured the darkness of the stain as a means of assessing daily concentrations. We are sure that the results will impress the reader with the importance of dust in the lower air layers, and will probably horrify those who take their samples in heavily industrialised areas! An obvious effect of dust in the air is the improvement of visibility after a rainstorm has 'washed out' the majority of particles. You can check this effect by sampling dust concentrations before and after a summer thunder shower.

Because of its hygroscopic properties, dust-laden air is often associated with 'premature' condensation before complete saturation. The particles serve as *nuclei* for the absorption and concentration of water vapour molecules. Relative humidity readings taken in polluted atmospheres will verify this point. Great concern exists about the likely medical and climatic hazards of increasing atmospheric pollution in urban areas, and so the analysis of airborne dust and its variations in time and space is of considerable importance. Thus, under the Clean Air Act of 1956 all local authorities regularly measure the degree of pollution and the results are published in the Medical Officer of Health's *Annual Report*. Sufferers from hayfever will be only too well aware of the significance of the so-called pollen counts published in some national newspapers! Such data can be plotted graphically and related to other climatic characteristics.

More exciting and graphic methods of measuring dust concentrations have been developed recently using the properties of scattered laser light and three-dimensional images called holograms. So-called pulse echo techniques with a ruby laser can be used in the field and particles as small as 1μm can be analysed in laboratory situations. Since colleagues studying physics may well be conducting simple laser experiments, it is a good idea to join forces with them to analyse the effects of a laser beam through different air samples. Also, lasers can illustrate much about air turbulence for, by using stroboscope techniques, the movements of dust or tracer particles can be examined visually or photographically. In addition there are numerous field methods which can be adapted to measure the size of water droplets in mists and fogs without interfering with the air itself. However, such techniques require very sophisticated equipment.

Relative inputs and outputs of radiation, or any other areal measurements for that matter, on a given set of slopes in a drainage basin can be mapped using the chorochromatic and choropleth techniques, outlined by Toyne and Newby. Such maps are used in all branches of geography and the methods of their preparation are important to master. These can be used in conjunction with isopleth maps prepared from the point-sampling of selected elements in a district. Examples of such maps abound in most textbooks and, when combined with statistical techniques like principal component and factor analyses, such maps are powerful aids to climatological research at all scales. Whilst the latter techniques go beyond the scope of this book, the reader will do well to explore the scope for both choropleth and isopleth mapping in studying the relationships of sets of field data. After all, we have already mentioned that conventional topographical maps already display contours as isolines, viz. isohypses. Their comparison with overlays or sieve maps of isotherms, isohyets, isohels, isorads and so on is basic to studies of local or topoclimates.

FIELDWORK

Network surveys

1 With reference to a local district, such as the area surrounding your home, school or the nearest small drainage basin, take systematic temperature, rainfall and sunshine observations at the mean centres of its slope elements. Relate these observations to the longer term records available from River Authorities and other organisations maintaining climatological stations in the area. Make sure that the observations can be reasonably compared in terms of the instruments used and the methods of taking the readings.

2 Plot suitable point data on a base map of the station network, e.g. mean daily temperatures, temperature ranges, monthly or annual rainfall totals, rainfall variabilities, hours of bright sunshine, etc. Using Thiessen polygon techniques, prepare appropriate isopleth maps of each set of observations on tracing paper.

3 Superimpose or overlay the isopleth maps on a relief or slope map of the district.

- **a** What degree of correlation is there between the maps?
- **b** What are the effects of local relief and the exposure of different slope sites?
- **c** Is there a case for nominally classifying the slope elements on a 0–9 scale range of climatic 'favourability' for different purposes like horticulture, agriculture or residential housing?

4 What other data would you require to make a more complete climatic survey of the chosen locality? How would you gather the necessary information?

3B 2 Traverses and transects

So far we have concentrated upon relating observations between *places* and *patches* of land; the location of the observations being determined by sets of slopes within chosen drainage basins or perhaps streets. In such work a major difficulty lies in getting simultaneous readings. With a veritable army of students, all equipped with the necessary instruments, such problems can be overcome in small areas, of course. Although good spatial sampling procedures can minimise these requirements, there remains the possibility that variations determined more by the *timing* of the observations will confuse the analysis of the results. Since it is important to be able to differentiate between the effects of both, traverses across representative slopes can help solve the time factor and prove more economical for smaller field groups where fewer instruments are available. Some of the techniques described below can be carried out successfully by a single enthusiast.

Traverse designs can take several forms, the more usual being illustrated in fig. 3.11. Linear traverses within a drainage basin can be chosen to include different slope elements; those across the basin are termed **transverse** and those along the valley floor **longitudinal**. These correspond to the valley cross-sections and long profiles (sometimes called thalwegs) commonly used in fluvial geomorphology. Radial traverses about a chosen centre are obviously ideal in school and college situations with the regular movement of students to and from their residences. In practice, of course, the traverse need not be a straight *line*, and might be a belt or *zone* rather than a line in its strict sense. The distinction is important in certain studies, particularly those of an ecological nature (see page 99). Since all linear traverses are methods of systematic sampling, it is important to appreciate the pitfalls of choosing the 'route' in terms of the *locus* and *number* of sampling points along any line. For the most part, each line

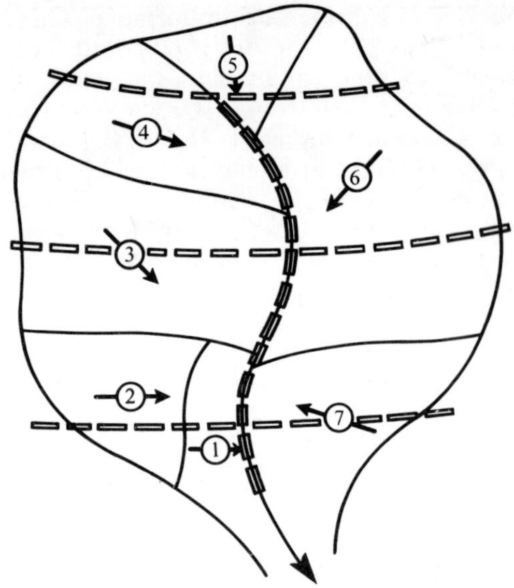

a) linear traverse lines in a drainage basin

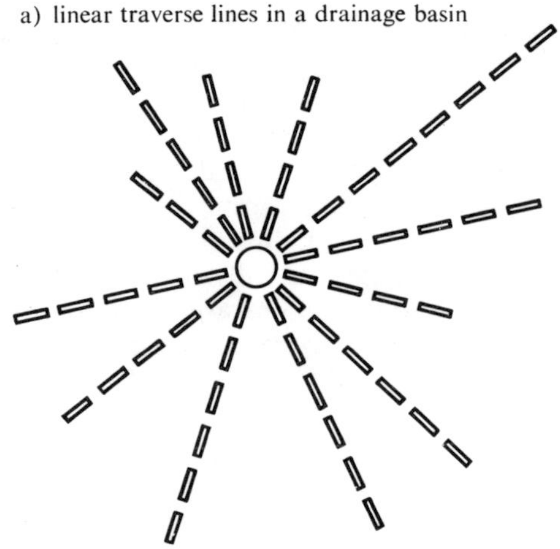

b) Radial traverse lines from a centre

Fig. 3.11 Simple traverse designs

should be randomly spaced, even if parallel, and the stations along them ought to be randomly located too. As explained earlier, this will minimise the risk of sites being chosen subjectively and so of bias in the results. On uniform plateaux and lowland surfaces, the locus of the traverse lines can be a random 'criss-cross' pattern to avoid missing climatic features which might be regularly distributed. It is essential to prepare alternative traverse designs on large-scale topographical maps *before* observations are made, and it is plain that a

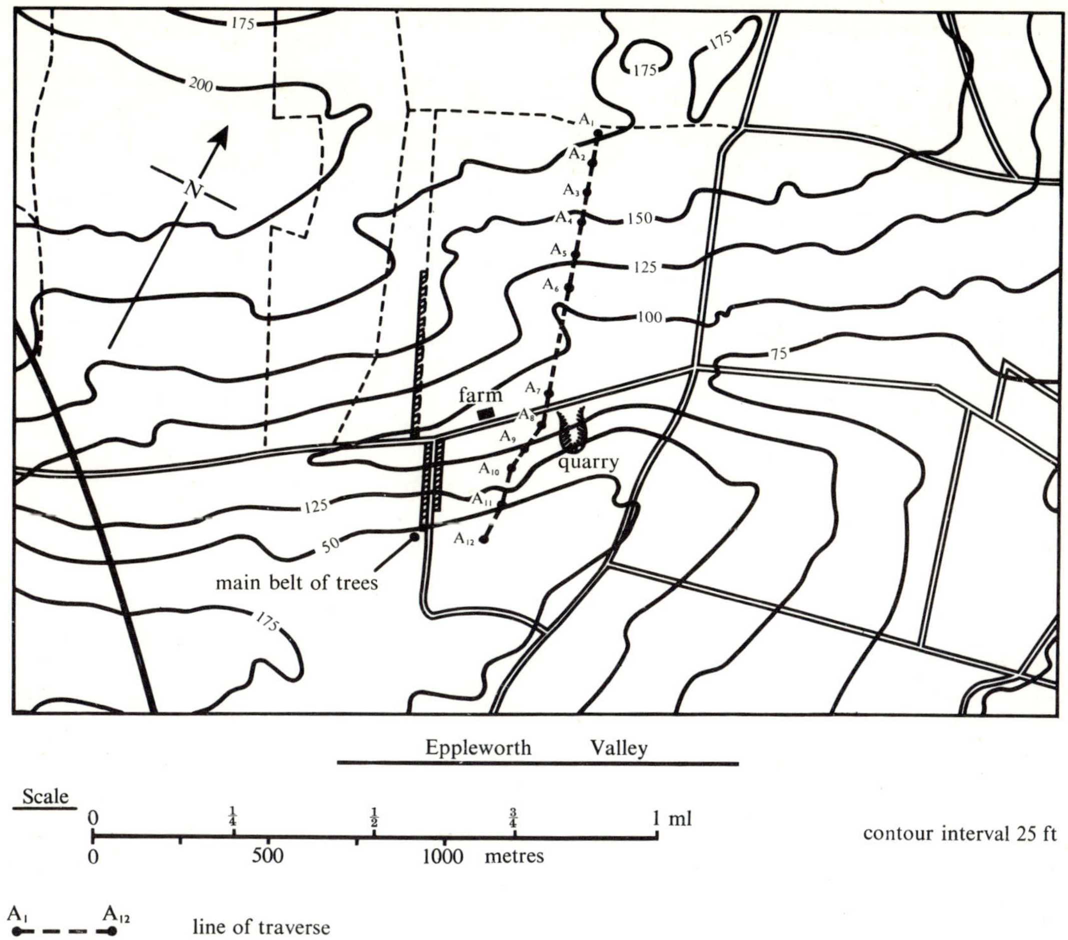

Fig. 3.12 The Eppleworth Valley, near Hull, showing a transverse traverse line with twelve observation sites

table of random numbers should be to hand.

The results of the following traverses near Hull in Yorkshire and across the Sahara Desert will give some idea of the scope for such investigations in contrasting areas at different scales. In both instances, a carefully calibrated whirling psychrometer was used at arm's length about one metre above the ground surface.

Fig. 3.12 shows twelve stations along a traverse across the Eppleworth Valley on the outskirts of Hull. This is a dry valley on the chalk dip slope of the East Yorkshire Wolds to the northwest of the City of Hull. At the time of the survey in the autumn of 1956, the fields involved were either short, permanent grass or thick stubble. Stations were located by triangulation using a prismatic compass read to the nearest degree. The first set of readings were taken on October 14/15 when a stable anticyclone lay across the British Isles giving clear skies and calm conditions. Five separate traverses were made across the valley between 1500 hrs GMT on the 14th to 1100 hrs GMT on the following day. The duration of each traverse varied largely because of the greater difficulty in reading the instruments at night. The exact times of taking dry bulb temperatures at each site were carefully recorded in the field. Subse-

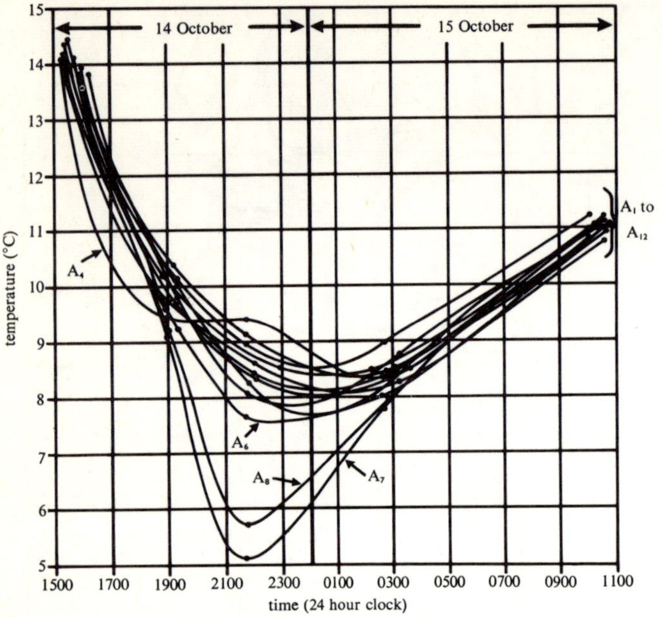

Fig. 3.13 Overnight temperatures at twelve observation sites across the Eppleworth Valley, Hull, October 1956

Fig. 3.14 Temperature profile across the Eppleworth Valley, Hull, 14–15 October 1956

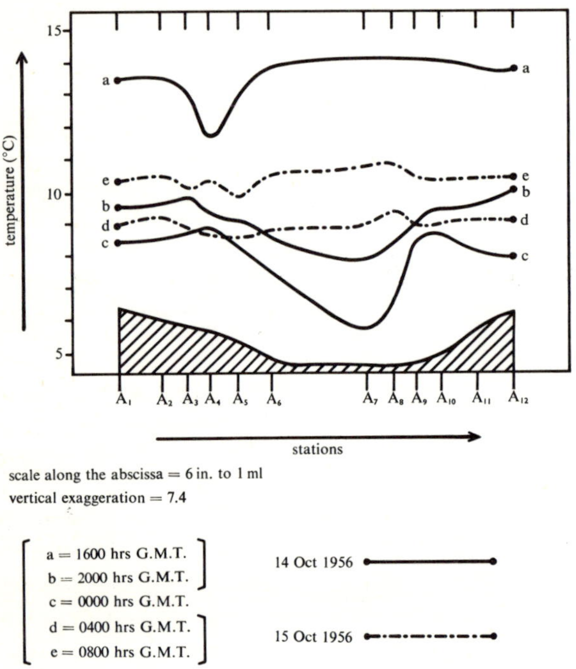

quently, the dry bulb readings for individual stations A1 to A12 were plotted assuming consistent temperature variations between the times when measurements were made. Although this need not have been the case, this procedure at least enabled temperatures at certain *standard times* to be interpolated to prepare profiles across the valley at four-hourly intervals.

Fig. 3.13 shows temperature curves for individual stations and fig. 3.14 valley profiles at selected times. The former generally illustrate that slightly warmer air lay over the south-facing slopes, that the rates of initial cooling were more rapid at sites in the valley floor which 'lost' the sun earlier, and that minimum temperatures were significantly lower here too as a result of cold air draining off the valley sides. The temperature profiles on fig. 3.14 confirm the occurrence of a temperature inversion prior to the cold air drainage. Smoke released near the ground at the time 'flowed' downslope at some 2 knots (*c.* 1 metre per second). Overnight fog gradually built-up in the valley and ultimately 'overspilled' on to the surrounding hills. This accounts for the general increase of temperatures apparent, partially through the release of latent heat upon condensation and the blanketing effect of the fog upon further nocturnal radiation losses from the ground.

Another survey was undertaken in similar anticyclonic conditions on 6 November. Instead of separate traverses, a continuous 'to and fro' technique was used to reveal minor fluctuations of temperature not evident from the longer intervals between readings on the mid-October traverse. Fig. 3.15 shows the results of individual stations from 3 p.m. to 7 p.m. Where observations were insufficiently close, pecked lines have been used. Eventually, in the preparation of the temperature profiles, times were selected to avoid those zones as much as possible (see fig. 3.16). Again we notice the more rapid initial cooling in the valley floor. The smaller time scale of study, however, shows an early temperature discontinuity which some authorities attribute to the release of latent heat as dew forms.

Interestingly, this was more evident over the grass surfaces than the stubble. More certain, however, is the cyclic nature of cold air drainage into the valley and the corresponding 'drawing in' of warm air to the shoulders of the slope. Smoke from a belated November 5th bonfire verified this, and plumes of smoke moved steadily downvalley at about 5 knots (*c.* 2·5 m/sec.). It was possible to 'trace' these katabatic movements by wet and dry bulb readings on longitudinal traverses along the valley floor. Ultimately, fog moved downvalley as on the previous survey.

Only two observers carried out both surveys, one taking the readings and the other booking the data. It is not difficult for one person to do both, however. Simple techniques like this can reveal much about radiation processes, fogs, dew formation, temperature inversions and so on in relation to topography. Similar approaches can be applied to urban areas, particularly as many towns are located in valleys. We shall elaborate upon such techniques shortly.

FIELDWORK

Traverse surveys

1 Select a well-defined valley and choose suitable traverse lines across representative slopes. Establish observation stations at intervals along each traverse line. Under calm and clear conditions, preferably stable anticyclonic weather in winter, take wet and dry bulb temperatures at each station at a standard height like one metre. Traverse to and fro along the line of stations recording the times at which temperatures were taken and the types of surface at each.

2 Plot the information as indicated on figs. 3.13 and 3.14. Analyse how the relief of the valley, including slope aspects, exposures and types of surface, influences temperatures and humidities along the traverse.

- **a** Is there any evidence for temperature inversions and cold air drainage?
- **b** Where are the frost and fog-prone areas of the valley?

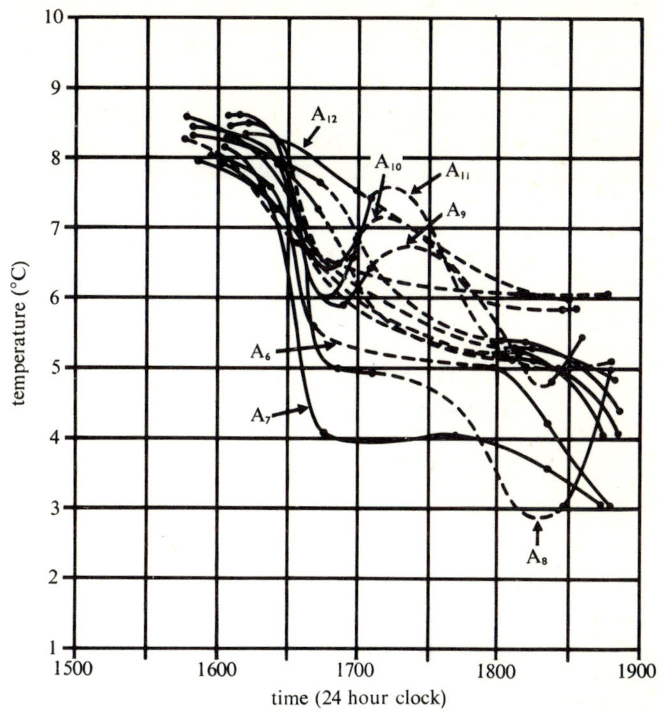

Fig. 3.15 Afternoon and evening temperatures at twelve observation sites across the Eppleworth Valley, Hull, 6 November 1956

Fig. 3.16 Temperature profiles across the Eppleworth Valley, Hull, 6 November 1956

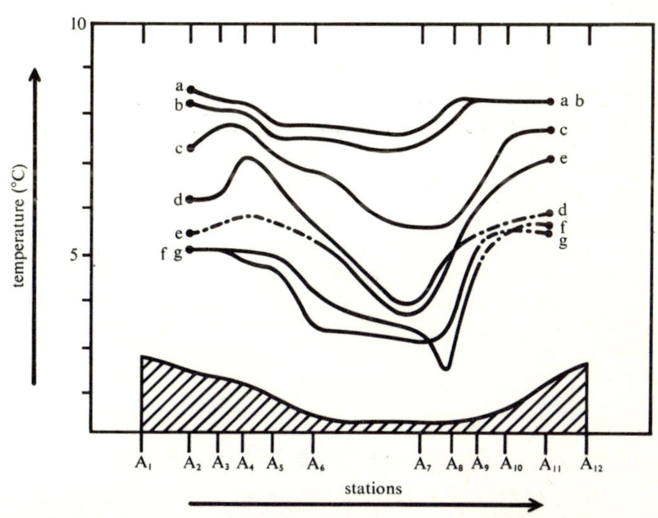

a = 1600 hrs G.M.T.
b = 1615 hrs G.M.T.
c = 1630 hrs G.M.T.
d = 1645 hrs G.M.T.
e = 1700 hrs G.M.T.
f = 1815 hrs G.M.T.
g = 1830 hrs G.M.T.
scale = 6 in. to 1 ml
vertical exaggeration = 7.4

c What are the effects of buildings and vegetation upon the patterns of cold air drainage?

3 Using talc, French chalk or smoke, attempt to measure the speed of down-slope air flows at different times. Is there any evidence for cyclical flow off the shoulders of the valley?

4 What other measurements would be of value to make a more complete climatic survey of the valley? Compare readings taken under different conditions, e.g. a clear summer's day, a cloudy day, a windy day, and so on.

5 Repeat your experiments with profiles along the valley floor.

Similar traverses can be made on much larger scales in time and space on extended field courses. Fig. 3.17 shows simple weather observations recorded on a journey across the Sahara during January 1964. Each day, wet and dry bulb temperatures were recorded on breaking and setting-up camp in the morning and evening respectively. Notes were made on cloud amounts and present weather. This data is shown on fig. 3.17 as on standard meteorological reports (c.f. notations on Meteorological Office Daily Weather Reports). The altitude of camps were established by means of the largest scale topographical maps available and an aneroid barometer, whilst position was fixed by standard navigating techniques. In effect, the relief section illustrated on fig. 3.17 has been topologically transformed with respect to travel times and overnight stops for the particular journey. The reader might check the actual relief on atlas maps of the Sahara. It will be seen that the Hoggar Mountains lie on the Tropic of Cancer in the south of Algeria.

Many significant climatological features and processes are demonstrated on fig. 3.17: diurnal variations of humidity, drier air and

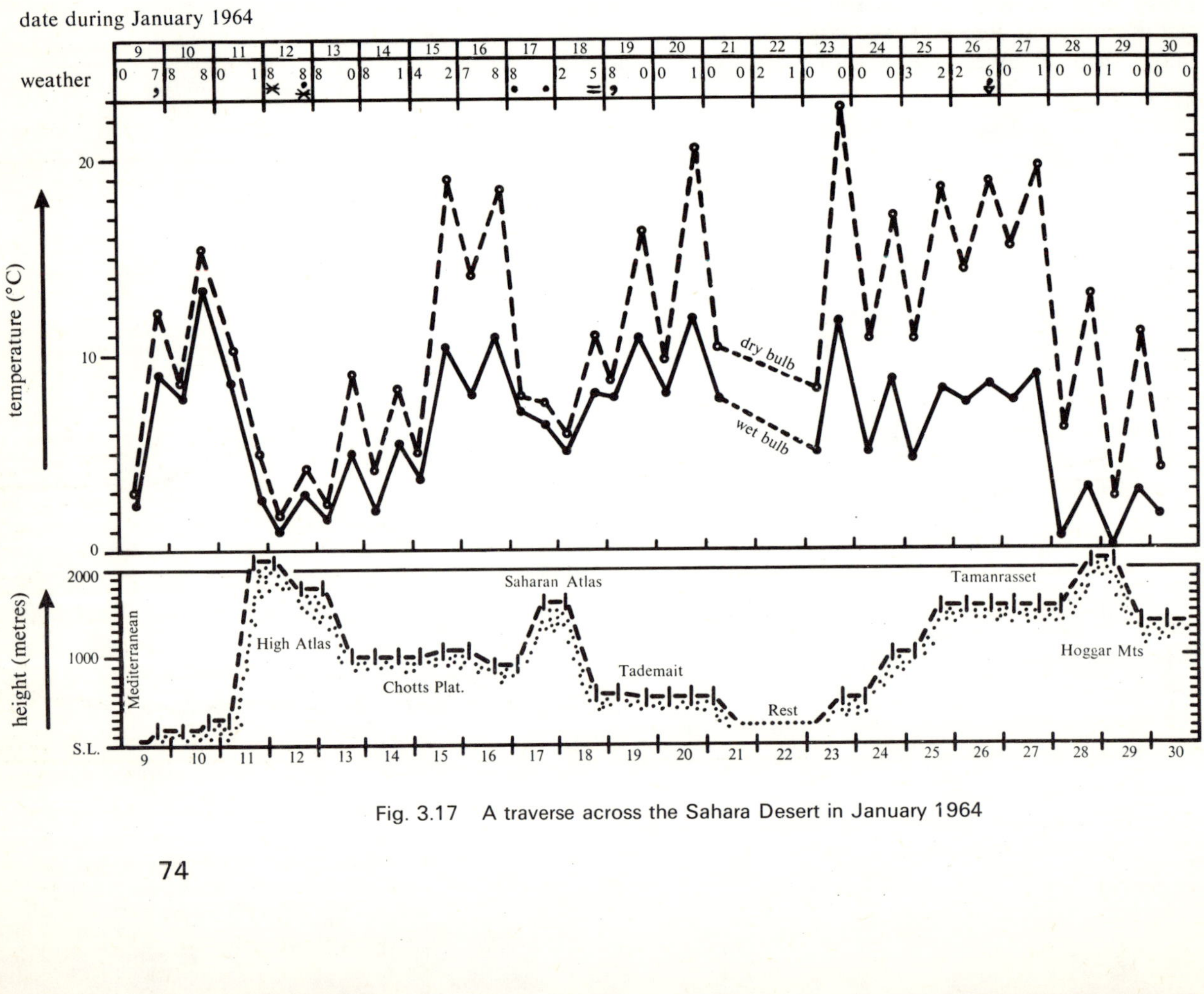

Fig. 3.17 A traverse across the Sahara Desert in January 1964

greater temperature ranges progressively further inland, the close correlation between temperature, humidity and altitude, and so on. However, it is important to realise that the temperatures recorded do not represent actual maximum and minimum values: the latter only being recorded at selected camps and during one overnight period on the tropic (see fig. 3.18). Also, it is important to bear in mind that synoptic situations would have varied throughout the period concerned. Although this does not detract from the value of the data recorded on the trip across the Sahara, this would not be so on less ambitious traverses in this country or across Europe. The small time scale for shorter journeys will help, of course, but it will still be necessary to check that any fluctuations recorded cannot be accounted for by synoptic changes. In itself, this provides a valuable exercise in isolating local influences from macro-scale weather situations. Thus, it is necessary to acquire the Daily Weather Reports, and perhaps even the Aerological Records (see page 34), to make such an investigation thorough and meaningful.

The weather data on fig. 3.17 shows appreciable cloud covers over the Atlas region with associated winter rains and snows. This is in marked contrast to the convective storms evident on the plateau around Tamanrasset on 26 January. The explanation for such differences are fundamental, concerning the distribution of different types of precipitation. Also, it should be noted that heavy dews were recorded at all the mountain and most of the plateau camps. This is a well known source of moisture for desert plants and animals, including the traveller! It should be remembered, of course, that January is the coldest month in the Sahara.

Fig. 3.18 illustrates data recorded on a 227 km journey along the plateau to the north of the Hoggar Massif on 29 January. Wet and dry bulb readings were taken periodically by the simple expedient of whirling the psychrometer when leaning well out of the moving vehicle! Whilst we cannot commend this particular technique on our busy roads, it can obviously be adapted with safety on certain local studies. The reliability of such techniques is often question-

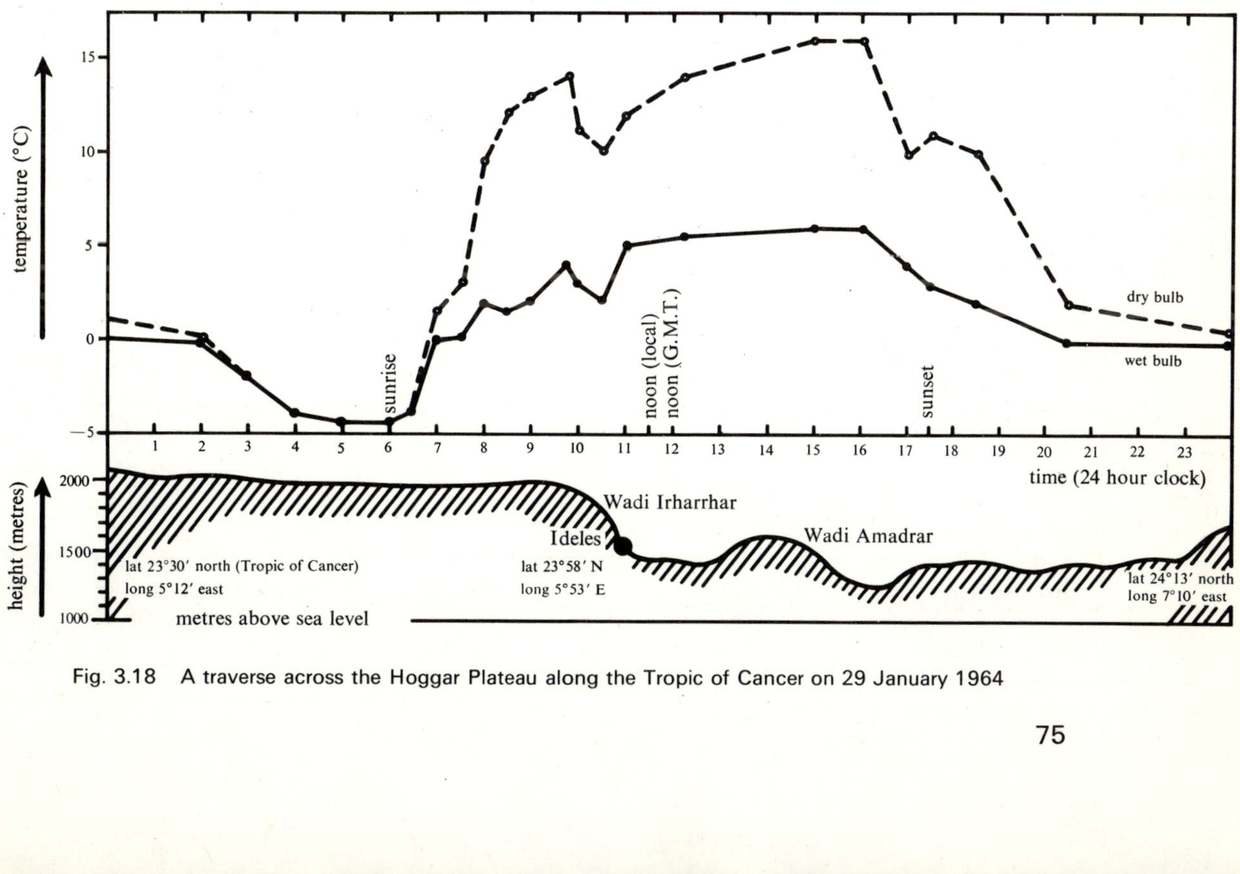

Fig. 3.18 A traverse across the Hoggar Plateau along the Tropic of Cancer on 29 January 1964

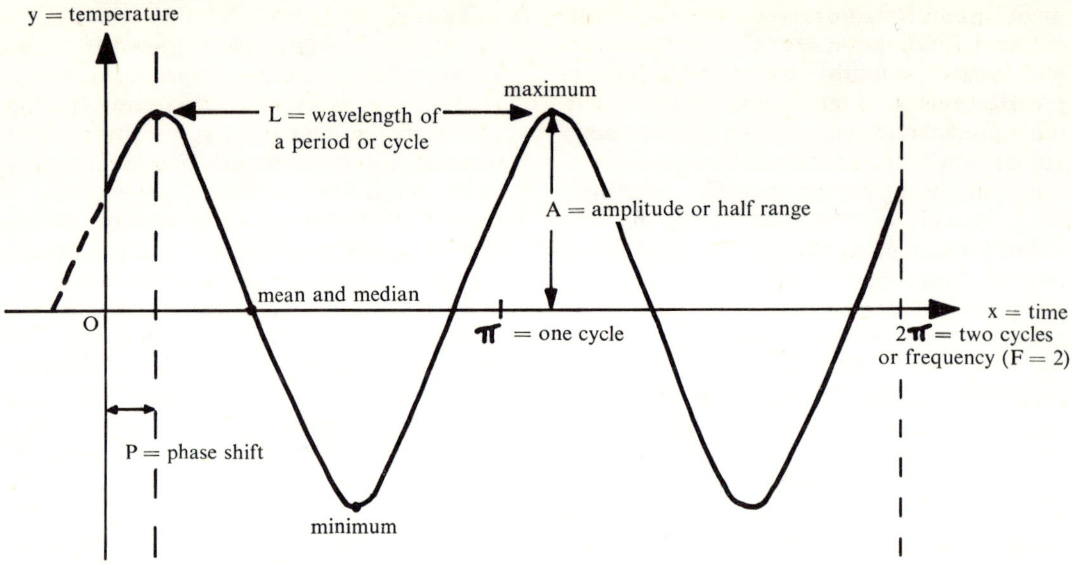

Fig. 3.19 Ideal form of diurnal and seasonal temperature variations

able, however, and the accuracy of the results must be treated in this light. Nevertheless, fig. 3.18 does display useful diurnal patterns, and the fall of temperature at the slope into the Wadi Irharrhar is an interesting reflection of the heat economy on partially-shaded slopes at the season and time of day concerned. By the same token, the increase of the evening dry bulb temperature on the west-facing side of the Wadi Amadrar demonstrates the importance of slope aspect discussed earlier.

Variations of such traverse and transect techniques relying on different time scales can be combined with 'static' measurements spatially. One can be used as a means of sampling and extending the range of validity, the other at selected times or places. There is no lack of published literature in this field; for example, the study in the New Forest by Morris and Barry revealing the significance of aspect and ground cover upon soil temperatures, and that by George who examined the relationship between temperature profiles and fog in the Irfon valley, mid-Wales. The latter, in particular, underlines how useful the most modest techniques are in local climatological studies.

Just as we concluded the previous section by briefly mentioning *cyclical* data, it is clear that one of the best methods of relating observations on different slopes is to compare their effects upon selected elements over appropriate *time scales. Diurnal* and *annual* cycles are the best to choose; especially the former over a period like a week. Fig. 3.19 shows an ideal cyclical pattern of temperature variations over two days for a single slope. Since this is in the form of a *sine curve*, it is possible to express the cycles by the general equation for such 'waves', viz.

$$y = A \sin F\left(\frac{x}{L} + P\right)$$

or, with reference to the curve on fig. 3.19,

$$y = A \sin\left(2x + \frac{\pi}{4}\right)$$

Thus, it is possible to compare the fit of actual temperature curves with such 'standard' cycles. A simple way of doing this is by superimposing the graphs for different slopes and relating their forms by the properties labelled on fig. 3.19. Equations for curves which have complicated fluctuations can be 'built up' from a number of different wave patterns by manipulating the phase shift, wavelength and amplitude of the 'standard' curve. Such techniques are called **Fourier analysis** and obviously involve a lot of elaborate calculations. We mention this to indicate that, however complicated the fluctuations in any data, it is still feasible to quantify them for accurate comparisons to be made. Although we have illustrated the approach using temperature curves, it will be appreciated that all cyclical data can be examined by these methods. Such techniques are used in identifying climatic changes at different time scales, and you will find similar treatments for studying the meanders of rivers in chapter 6.

3C Studies in urban areas

As well over eighty per cent of the population spend the vast majority of their lives in towns and cities, we must elaborate upon some techniques for studying urban climates. Although the larger part of our countryside is hardly more 'natural' than a built-up area, there are now growing indications that urban spheres of influence are as significant in climatology as in human geography. The man-made environment of concrete is no less important than that of the cereal crop. Yet, understandably perhaps, the physical geographer has been much more concerned in the past with 'dust bowl' problems than building designs and their effects. Nowadays, this imbalance has been redressed in response to the increasing scale of the latter, although this does not imply that the problems of the former have been solved! We turn to these questions in the next chapter.

The increasing control of widely distributed energy resources, and their collection and conversion to heat energy in areas of concentration, clearly deserves the attention of geographers. After all, according to Sellers, the annual world use of energy is now only about one hundreth of the total amount of solar energy received daily. This is on a par with that which drives the vast monsoonal circulation! Some fear that the increasing dissipation of this energy as heat and CO_2 into the atmosphere will boost the observed natural increase in temperatures until the amount of water vapour held aloft is such that, given the slightest cooling through some extra-terrestrial cause, condensation would occur and another ice age be 'triggered off' prematurely. Whilst urban areas, on a global scale at least, are still in a minority, it makes sense to examine their effects upon climate to give more substantial information as to whether such fears are justified. Although you will not expect to reach conclusions on such matters through the techniques which follow, at least an insight will have been gained into the best examples of man-made climates currently available to the physical geographer. We should be thankful of the opportunity to look forward with an 'onward' time scale rather than the customary 'backward' one, even if the prospects appear somewhat bleak!

Before detailing specific techniques for investigating urban climates, we remind readers that most of those already given can be carried out with equal success in the city. For instance, traverse techniques which measure radiation, temperatures, humidity and dust concentrations are particularly suitable. Similarly, parks and playing fields (many of which contain a meteorological station anyway) are ideal for most of the techniques based upon the exposure model.

3C 1 Building form and design

Just as grasses, shrubs and trees present irregular surfaces to energy exchanges in the natural landscape, so suburban houses, business premises and tower blocks represent increases in the *scale of roughness* of such surfaces. By comparison, therefore, man effectively lives closer to the soil environment than the tropospheric one. This is particularly true in the densely built-up Central Business Districts of many American cities. Consequently, on a human scale, we can start *indoors* slowly working outwards to the broader scope offered by the city as a whole.

A useful start can be made at home, or in school, by taking temperatures at floor level, head height when sitting (about a metre) and near the ceiling (say 2 metres high) using a grid plan or random points. The data will allow vertical temperature profiles and isotherm maps at different levels to be drawn. However, a more useful technique is to prepare an isometric block diagram of the room, select the mean or median temperature recorded and then construct an **isothermal** surface as shown in fig. 3.20. In this case the 18 °C surface has been chosen, the air above being warmer and that below cooler. The method of construction is similar to that used in drawing block diagrams of relief. The form of the surface in fig. 3.20 clearly shows the drift of warm air towards the window from the heater, and there is some indication that denser cold air 'creeps' beneath the door. Other situations can also be examined; for example, with drawn curtains and draught excluders, with the door and window open and so on. The circulation of air can be checked with French chalk puffs or smoke. You may have seen an inversion set-up in a room when cigarette smoke becomes stratified at about head height.

Observations of airflow in rooms will show you that windows are the chief cause of heat losses. By using the heat flow equation

$$\Phi = Ak\frac{\delta y}{dx}$$

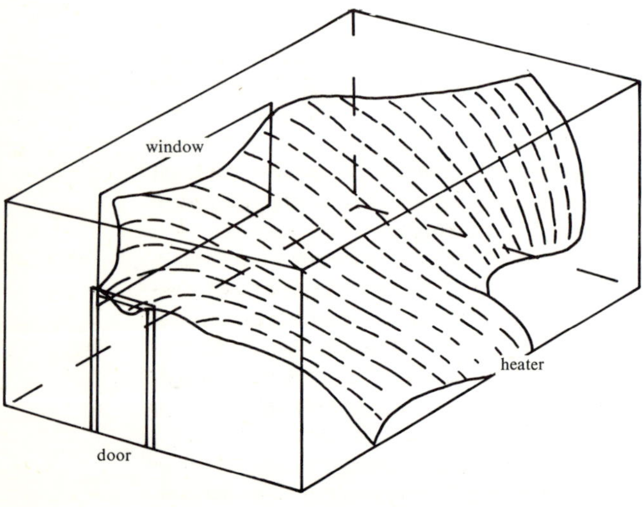

Fig. 3.20 The 18 °C temperature surface in a living room

(see page 26) it is possible to evaluate such losses under given conditions. Temperatures taken near the inner and outer surfaces of the window pane will give the temperature gradient

$$\frac{\delta y}{dx},$$

and the thermal conductivity of glass is about 0·8 W m^{-1} per degree Celsius. Thus, with a room temperature of 18 °C and an outside temperature of 10 °C, the heat flow through a window would be 6·4 watts per square metre. If this is now multiplied by the area of window, say 5 m^{-2}, then the total heat loss is obtained, e.g. 32 watts. In practice, the conduction of heat energy through windows is more complex by virtue of the thin 'films' of still air on either side of the pane. These films are important in reducing heat losses; a principle which is taken advantage of with double glazing. Also, the thermal conductivity of different types of glass does vary. So, those who wish to calculate the heat losses from their rooms, to make a case for better heaters or curtains, are advised to consult the appropriate manufacturers or traders! We feel that we have outlined sufficiently on *room climates* for the reader to be more aware of the problems facing architects and heating engineers. Obviously, *aspect* is important too, and the higher and more exposed the window the greater the heat losses on the whole because of increased turbulence and lower air temperatures. Needless to say, the scale of enquiry can be subsequently increased to study the climate of a *house*. Most of us will know of the warmest and coldest rooms in our homes or schools. Very often, simple remedies will improve living and working conditions so long as we can find the causes of particular problems scientifically.

Venturing outside once again enables us to study the climate surrounding a house. Since the external walls face different directions, we have an ideal situation for studying the effects of *aspect* and *exposure*. Assuming we have four walls facing north, south, east and west, it will be possible to

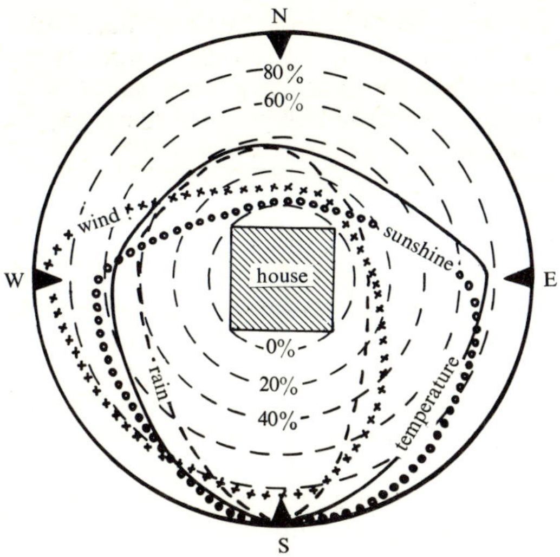

Fig. 3.21 Climate around a house for two weeks in July

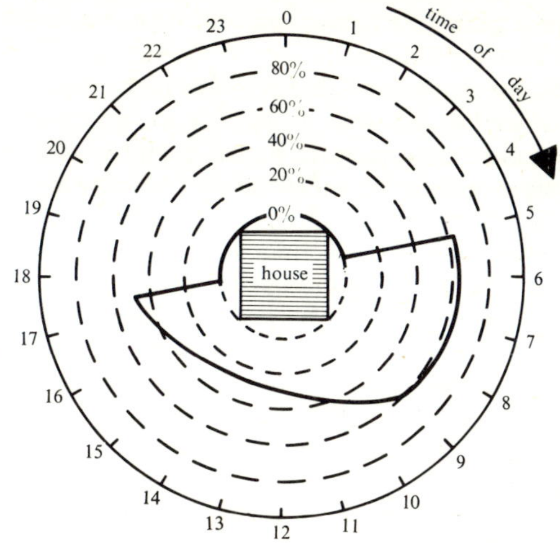

Fig. 3.22 Receipt of bright sunshine to the walls of a house for two weeks in July

keep records of a wide range of observations for each. Obvious examples are: albedos, durations of direct solar radiation (hence evaluations of absorbed radiation), surface temperatures and temperatures at standard distances from the walls, humidities and air currents. Given enough readings it should prove feasible to prepare heat budgets for each wall. Over shorter periods, diagrams like fig. 3.21 and fig. 3.22 can be prepared; for the want of a term, we will call these *precinct* climates.

Fig. 3.21 shows various climatic elements recorded near the four external walls of a house in mid-July. Rainfall was measured (see page 120) each day and the fortnight's totals compared by plotting each as a percentage of the maximum catch recorded near the south-facing wall. Similar procedures were adopted for the incidence of winds in excess of 3 knots (Force 1) every 3 hours, mean daily temperatures and intensity of sunshine measured with a light meter every 3 hours. For the period in question, it is seen that the south-facing side of the house is the warmest and sunniest but also the wettest. The west side is the windiest, which might well explain the lower rainfall recorded there, and the east side the most sheltered. As might have been expected, the north side is coolest and most shaded. However, it is well sheltered from both wind and rain. It is of interest that the south-east part of the house is significantly warmer throughout the year despite its marginally greater window space. This can almost certainly be attributed to early morning warming and turbulent air flow. Thus, it is possible to build up a useful picture of the climate characterising the precincts of a house. It would have been possible, of course, to plot the percentages of each element against the *means* and *totals* rather than the *maximums* as in fig. 3.21. We have chosen the latter because it solves the problem of values in excess of 100 per cent regarding mean, and of smaller variations if totals were used. Also, there is a more obvious comparison with the most and least 'favoured' sides of the house by using the maximum in each case.

In order to check the effects of early morning warming, the receipt of bright sunshine was recorded at each wall for a fortnight in July. Amounts *actually* received each hour were calculated as a percentage of the *potential* which could have been recorded had the sky remained perfectly clear. The former was measured by inexpensive, if not exactly accurate, pinhole camera-type instruments and the latter computed from an exposure model of the site. Fig. 3.22

illustrates the results obtained. It shows that, because of the build up of cloud typical of summer days, the southeast part of the house experiences relatively more bright sunshine. Furthermore, the receipt of radiation on the walls of the house facing southeast would be greater in summer. Numerous authorities have found this to be true of hillsides and even plough furrows! However, there is a tendency for the 'favoured' aspect to swing southwesterly in the winter months because of the early morning mists and fogs which occur at this time of year. This is particularly so in urban areas, of course, with the increased use of fires and thus greater concentrations of atmospheric dust during the colder winter months. Which sides of your home and school are warmest at different times of year? How might the climatic characteristics described be taken advantage of in designing a house?

Those of you who live in houses exposed to the wind will know only too well of the noise created by gusts buffeting the windows and, when it rains, of the incessant drumming of water on the window panes. Usually, the higher the building the worst these effects are. Indeed, the airflow around tall buildings must be taken into account when they are designed, and there is a good case for making them as aerodynamic as possible. Rough angular corners exposed to prevailing winds will cause violent air currents to be forced around the sides of the building, and the pressures exerted upon windows will be considerable. To open a window under these conditions is to run the risk of loose objects being blown about the room. Clearly, there is a strong case for studying the climate of a *site* before large buildings are erected. What factors would need to be taken into account?

3C 2 Urban functions and activities

Your studies in human geography will show that, as urban areas get larger, there is a tendency for sets of functions to become grouped in distinct zones or sectors within the built-up area. Each functional 'district' has an ideal location within the urban structure, and gives rise to characteristic morphological features; for example, towering departmental stores and office blocks in the Central Business District, smaller shops and rambling factory complexes in the suburbs. These locational and morphological features are significant to the physical geographer too, and so your models of urban areas will give many clues regarding local climates within the confines of the city.

Clearly the major difficulty lies in assessing the extent to which man-made structures have modified the 'natural' climates. This demands *comparative* approaches using control data from the surrounding countryside. Once again, isopleth maps derived from *point* data throughout the city and traverses provide the best techniques; for example, temperature, humidity, rainfall, the incidence of fogs, dust concentrations and the duration of bright sunshine. The distribution of students' homes can make the basis for point data and journeys to school or college can be used for traverse information. For many instrumented observations the schools themselves will suffice since they are usually well distributed in any large town. With a little amount of organisation and the cost of several telephone calls, schools might share their observations on chosen days. After all, this is but a modest version of the system used by all meteorological stations. Perhaps there is a case for posting or announcing appropriate climatic data with the sports fixture results at morning assembly!

Having mapped such information, it is possible to isolate the *anomolies* created by the presence of certain functions in a district and by the city as a whole. Whilst establishments near the main power station might well be at the bottom of the 'sunshine league', they will undoubtedly fare better in terms of temperatures! Given sufficient readings on a single day, or throughout a chosen week, it is possible to determine whether they are **positively** or **negatively correlated** with the limits of the functional districts revealed by the sort of surveys suggested by Toyne and Newby. How do the *degree* and *type* of functions influence the amount and extent of certain climatic anomalies? Does the time of day and year play any part in the distribution patterns evident?

Chandler's study of London's urban climate graphically illustrates many of the above queries. We may single out smoke concentrations as showing the most marked regional variations. As might be expected, thicker concentrations were recorded around factories and major power stations, especially where both surface relief and that of the 'townscape' precluded the dispersal of discharged pollutants. In other situations, the isolines were clearly drawn out downwind of the source areas. This in turn can be related to the incidence of smogs, bright sunshine, maximum and minimum temperatures, humidity and so on. What is of interest here is that, in his early work, Chandler gained useful results while commuting in a car between the outskirts of north London and Liverpool Street. In many cases, only one day was sufficient to reveal the heat island effect of central London. Confirmation was possible by comparing the records of suitably located climatological stations over longer periods on superimposed graphs. Most large cities have such stations in central parks and the airport well outside the suburbs. Additional information is published by Health Departments and in many newspapers; these often show significant improvements in atmospheric pollution since the Clean Air Act of 1956, the advent of smokeless fuels and alternative forms of domestic heating to the traditional fireplace. Such data can be statistically correlated in space and time with the incidence of deaths and diseases attributable to respiratory and chest ailments. Even more morbid comparisons may be made by examining the occurrences and locations of multiple road crashes on climatically hazardous sections of motorways. Why are certain sections of motorways fog prone?

Chandler's pioneer work in this field established the existence of urban heat islands and thermal winds generated by the temperature gradients both horizontally and vertically over cities. This has led to much research into the possibility of increased precipitation of a conventional type associated with large built up areas, and recent findings show this to be the case over London and many other cities. Clearly, stone and concrete buildings of irregular heights also assist turbulence as well as altering the heat and water budgets. Apart from its capacity to absorb and store solar energy, the city also provides its own heat through fires and even people. How much heat energy does your home supply to the surrounding air given certain room and outside temperatures?

In May 1955, Franken enlisted the help of 270 volunteers to map the times that forsythia blossomed throughout districts in Hamburg. It was found that those in central parks were significantly ahead of their neighbours in the suburbs; a fact explained by the additional warmth of the city in winter. Diaries kept of such observations can reveal much about the effects of urban climates. Consider, for instance, the rates at which snow melts on the streets, when trees come into leaf, how fast pavements dry out after a rainstorm and even how we feel walking down the road on a hot summer's day. There is much scope for studying local climates in any town or city, and those who consider physical geography to be the prerogative of the countryman are sadly deluding themselves. Indeed, there has been little throughout this chapter, or the previous one for that matter, which has not been equally as applicable to urban settings as to rural areas! Just as the architect must take climate into account when designing individual buildings, so the town planner needs to consider local climates when conceiving the layout of streets, parks and industrial complexes. Those of you who live downwind of the local gasworks will have reason enough to appreciate the significance of physical geography!

We have dwelt upon the neglected field of local climatology in physical geography because of its greater reality and accessibility. Therefore, our dismissal of the more remote aspects of global climatology is at one and the same time an attempt to draw attention to climatic systems we are accustomed to whilst giving coherence to the problems and techniques studied throughout this particular book. We even suggest that useful climatological principles can be discovered whilst lying in bed or the bath! Although acknowledging that ultimate objectives in climatology go well beyond the scale range of the bath and the drainage basin, we see much geographical sense for the young student in the remark made by the nineteenth-century physicist John Tyndall, that:

> 'Knowledge once gained casts light beyond its own immediate boundaries.'

As a postscript to this chapter, we underline the greater opportunities in local climatological studies for understanding the distinction between the geographer's *objective appreciation* of climates through *measurements* and the layman's *subjective perception* of them through *experience*. Their confusion was one of the main pitfalls of those who saw physical geography as the primary cause determining particular land-use patterns. In dealing with some applications of local climatological studies, we hope it will be realised that attention must be given to how *man* interprets and supposes rather than just what physical geographers measure and map. The subtle relationships between these two aspects remains the great meeting place for human and physical geographers; a 'place' of increasing importance as our knowledge of physical controls and human responses improves.

4 The Plant and Soil Community

Most of the elements studied in physical geography have long inspired mens' imaginations whether they be associated with the sun and slopes, rainfall and relief or trees and topography in general. Many have even been elevated to the position of gods and made the subject of countless great paintings and poems. Not so the soil: far too many people still regard soil as 'simple and lifeless', to use the words of Sir J. Russell. Is it merely dirt or muck? Think of the unfortunate implications of common phrases like 'soiling one's hands' or 'having feet of clay'. At best, soils are associated with the so-called 'virtues' of toiling and winning a living from the earth; yet, paradoxically, they remain our most crucial resource and the basis of civilisation. Indeed, we have telling lessons to learn from former civilisations who have misused their land thereby precipitating their own downfalls.

Anyone who cares to direct strong lights upon suspended soil samples and then collect the myriad of escaping organisms below (see page 103) will appreciate that plants, animals and soils comprise an harmonious whole, albeit largely microscopic. This entity is the *ecosystem,* being essentially a microcosm of the more familiar systems studied by the geographer, including many features applicable to human geography, at different scale levels, of course. Thus, as geographers rather than botanists, biologists or pedologists (soil scientists), we see our terms of reference embodied within A. G. Tansley's simple analogy of an ecosystem with a *place* or 'home'.

Just as different human *communities* can be characterised by the particular nodes and links by which they are constrained, we can approach a study of natural habitats through an examination of how their *structures* are linked by *energy* or *mass* flows. Once again, we encounter the concept of feedback mechanisms in that both structures and flows are interdependent. The eternal 'hen and egg' problem is to solve the extent that structures influence the flows and *vice versa.* It is much the same as attempting to isolate the effects of our homes upon the ways we live as opposed to how our ways of life affect the actual fabric of the house. As we know only too well, such mutual relationships are never static but inevitably in a state of *dynamic equilibrium.* Those occasionally moved to rearrange the furniture and redecorate their rooms will appreciate the point.

The plant and soil subsystems are clearly the most important *regulators* regarding the transfer and storage of energy (heat) and of mass (water). Much as the way we reorganise a room depends upon our *past* reactions and responses to living there, so what we see here and now concerning plants and soils is not necessarily correlated with contemporary processes. Often, we are dealing with delayed responses and the relicts of former adjustments. However, the relatively more rapid *relaxation times* of vegetation and soil variables means that shorter *time scales* can be used in ecological studies than in say geomorphology. For example, the creation and restoration of the American Dust Bowl illustrates rapid adjustments to changes in the way plants and soils regulated the transfer and storage of heat energy and water in particular. Thus, both the *space* and *time* aspects in plant and soil studies are best viewed at a *small scale.*

In adopting the small scale approach, we are immediately confronted with techniques of **point sampling** vertical *columns* of soil and *stands* of vegetation. Thus, we conceive the soil and plant subsystems in terms of horizontal layers across which essentially vertical energy and mass transfers occur (see fig. 4.1); a view not dissimilar to that we adopted regarding the atmospheric system in chapter 2. This is not to say that lateral transfers do not occur, of course, since we

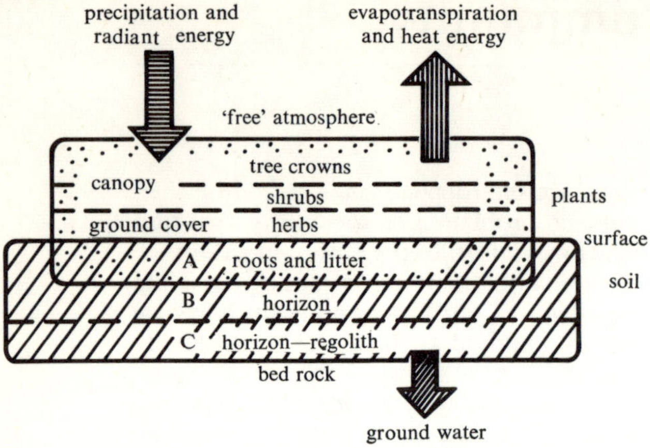

Fig. 4.1 The plant and soil subsystems

shall see later in this chapter and the next one that slopes cause significant soil water flows as well as the advection of heat energy. At this stage we enter the more familiar field of *lines* and *networks* in physical geography. To a large extent, it is symptomatic of the small-scale point sampling technique (used by soil scientists especially), that the coverage of up-to-date Soil Survey maps of Great Britain is limited in comparison with those on say geology or natural vegetation. You should check this coverage, particularly as the locations of published sheets indicate a lot about the establishments where present-day research in the field is being developed. There can be no doubt that these studies have been made more urgent by demands for increased food production and a general concern for the fertility of the land. The Rothamsted Experimental Station is the headquarters of the Soil Survey in this country whilst vegetation is mapped through official bodies like the Natural Environment Research Council or the Institute of Biology and British Ecological Society (see page 219).

Having stressed our particular approach to the problem in space and time, we would be foolish not to mention the alternative and more traditional global and 'developmental' treatments to be found elsewhere. Whilst they have much to commend them, there are also certain pitfalls to trap the unwary. A comparison of the appropriate maps in various atlases, for example, will reveal many significant discrepancies in the delimitation of world natural vegetation and soil types. This is understandable largely because the macro-scale approach is based upon the controversial task of defining major 'climatic regions' and takes both vegetation and soils as manifestations of these distributions. Figs. 4.2(a) and (b) summarise the main links which contribute to the formation of these *zonal* soils dominated by climatic controls and the greater impact of rock type in the development of *intra-zonal* soils. Both imply a lengthy sequence of development: first, a weathered rock *waste* mantle; second, the establishment of organic life to form a *rudimentary soil* and, subsequently, a gradual *succession* of vegetation aiding the *skeletal soil* to develop a distinctive morphology until a stable *mature soil* is attained. The soil *profile* (see page 97) is taken as the indicator of soil maturity, and the concept of climatic *climax groupings* of plants as the end product of the vegetation succession. If

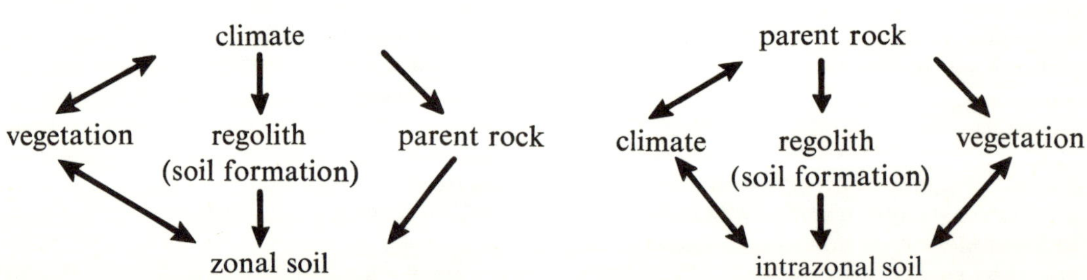

Fig. 4.2 Zonal and intrazonal soil formation

the various *horizons* of the soil profile are poorly differentiated the soil is regarded as immature or *azonal*, and much the same criteria are used concerning the stratification of the vegetation cover. The concept of a stable climax community was put forward by F. E. Clements as long ago as 1916, the climate being taken as the dominant control. Terms like subclimax, disclimax, preclimax and even post-climax are used to define various *degrees* of vegetation development.

The Clementsian 'developmental' philosophy places a great deal of stress upon the **time scale** of study. Here, we can do no better than to quote the words of Kershaw (1964, p. 53):

> 'It is impossible to define stability without reference to a time scale and since these are usually related to the life span of man himself, vegetational changes which occur over a 'geological time scale' are not readily appreciated. If the expectation of life was one minute then a population of groundsel...would be a climax vegetation. Conversely if our life expectation was measured in thousands of years the coniferous forest subclimax of North America would be merely a successional stage which precedes the establishment of deciduous forest.'

We shall meet much the same arguments regarding geomorphology and recall a personal experience to illustrate the point. Once, when excavating a deep shaft on the Mendip Hills in search of a possible cavern in the limestone bedrock, it was decided that a careful record should be made of the soil horizons to determine the events subsequent to the burial of the cave beneath the soil. Well-differentiated soil horizons were exposed near the surface suggesting a time scale of immense age, possibly back to the Pleistocene period (see page 183). Whereupon, much deeper at the soil-limestone interface we were astonished to unearth a beer bottle! Clearly, we had been seriously misled by our interpretation of the profile and, but for digging to the bedrock, might well have remained blissfully ignorant of the fact that the site had been worked earlier. Indeed, so far as most of us are concerned in Britain, and most areas overseas for that matter, it may well be wishful thinking to imagine that we can find undisturbed soil profiles illustrating uninterrupted developmental stages in response to climates. Similarly, with the possible exception of some isolated fells and moorlands, man's 'management' of the countryside in Britain has been so comprehensive that all vestiges of any natural successions must have long since disappeared with the trees. In our fieldwork, we will be dealing with complex adaptations of both soils and vegetation to human activities and we urge great caution when applying theories of soil *formation* and plant *succession*.

An alternative hypothesis to that proposed by Clements was suggested by H. A. Gleason shortly afterwards in 1917. Basically, this is the smaller scale view of the plant *association* or *community* in which both soil and vegetation are continuously responding to changes in a variety of environmental factors which can include man. We shall return to this more satisfactory concept in the section on 'energetics and dynamics'. This is not to condemn the climatic and time-centred approaches as fruitless, because we do not deny that soils contain important 'fossil' features such as pollen grains from former boreal forests. We have left the problems of such *stratigraphical* techniques to chapter 8 and, whilst we look at the unattended parts of our gardens with considerable sympathy for what is meant by a 'pioneer scrub phase', we prefer to deal with the processes causing natural groupings of plant species and the techniques for their detection.

To the ecologist, then, *biotic* and *edaphic* (soil) factors are inseparable. It is pertinent to mention that many of the techniques used in both human and physical geography were largely developed by ecologists. The adaptation of such techniques to problems in human geography, for instance, has been one of the most fruitful and striking features of recent advances in this field. Many human geographers are prepared to adopt the term *human ecology* to describe their approach to the subject, and many ecologists for that matter are happy to call their work *geographical*. Most of the point sampling, quadrat, transect and many distribution mapping techniques used by geographers owe their origins to the quantitative approaches and statistical procedures developed in studying the vast populations and complex associations of ecosystems.

4A Morphology of the community

Early this century, Raunkiaer developed a classification of plants based mainly upon their *height* and Dokuchayev proposed a corresponding classification of soils which, in essence, depended upon their *depth*. Although these concepts have their limitations, there are clear relationships between climate and rock type regarding the thickness of vegetation covers and the depth of the waste mantle or *regolith* supporting the actual soil. Areas with shallow regoliths and thin soils rarely give rise to tall plants in any abundance. Thus, there is a high degree of correlation between the percentage distribution of the different plant species at various levels above the surface and the percentage volume of the major components of the soil at each horizon. An obvious contrast can be made between the communities found on dry sandy heaths and moist peaty marshes. The former are characterised by scrub layers, with few trees because of the high percentages of minerals and air in the soil, whilst the latter support a profuse cover of grasses and sedges, including many alder and willow trees.

Sometimes you will be able to detect similar contrasts in your home district, local park or kitchen garden by simple inspection. By mapping the locations of *dominant* tree, shrub and ground species in relation to soil texture in particular, it may be possible to form general conclusions about the distributions of different communities. The results can be compared with slope exposure and climatic maps of the area. This sort of survey has often been the limit of many physical geography courses in biogeography, but it is plainly not adequate for intensive studies. What are the dominant species and what other soil factors are involved? How can we be sure that the relationships we see are significant? Our conclusions must be founded upon quantitative methods.

4A 1 Measurements and data collection

For the moment we will concentrate upon the plant and soil subsystems separately.

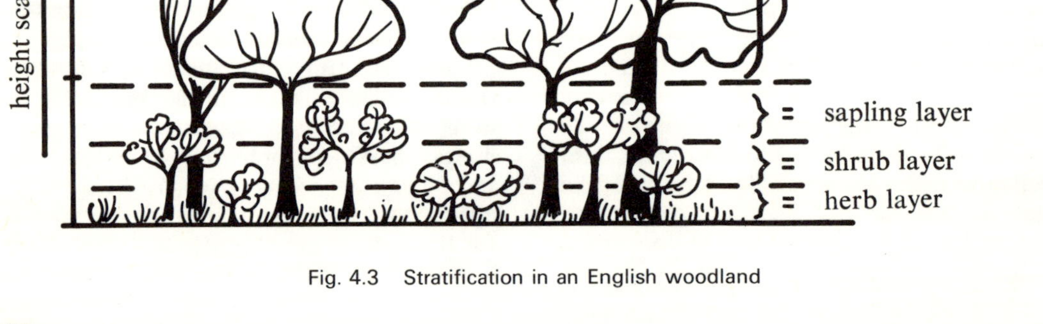

Fig. 4.3 Stratification in an English woodland

From what has been discussed already, it is apparent that stratification, horizontal distributions and the abundance of individual species are the three most important morphological aspects to measure in vegetation surveys.

Fig. 4.3 shows the typical stratification found in a British woodland. The most straightforward method is to measure the heights and widths of the species colonising a particular site. A plot of about 500 square metres can be taped-off and a site plan drawn up to scale. Extended levelling staffs and metre rules with plumb lines are then used to measure the vertical heights to the lowest branches or shoots and to the lowest parts of the crowns or heads. Maximum widths of the latter can be measured by tapes stretched horizontally beneath or beside each tree and shrub. By using a simple notation such as A for trees, B for saplings or shrubs and C for herbs the information can be mapped in on the site plan; for example, $A_3^{10}8$ would refer to a tree whose crown lay between 3 and 10 metres above the ground and was 8 metres wide. Sometimes, of course, it will be necessary to estimate the heights or use a clinometer. Having mapped the plot systematically it will then be possible to determine the *mean* heights and widths of species at each layer so that a **scale profile** can be drawn illustrating the characteristic stratification. The *size* and *shape* of crowns will also become apparent from the drawings, as in fig. 4.3.

In practice, many problems arise with profile surveys. The most common are the density of the herb and shrub layers and several overlapping tree layers. Simple practical techniques will solve the former and statistical techniques the latter. A dense herb and shrub area can be gradually cleared as the measurements are made. This should be done in such a way that the profile is 'actually seen' as the survey progresses across the plot. Some workers even fell the various tree layers in turn, measuring them on the ground. However, we do not commend readers to this rather drastic action since reasonable results are obtainable by the use of 'maypoles' with attached tapes or

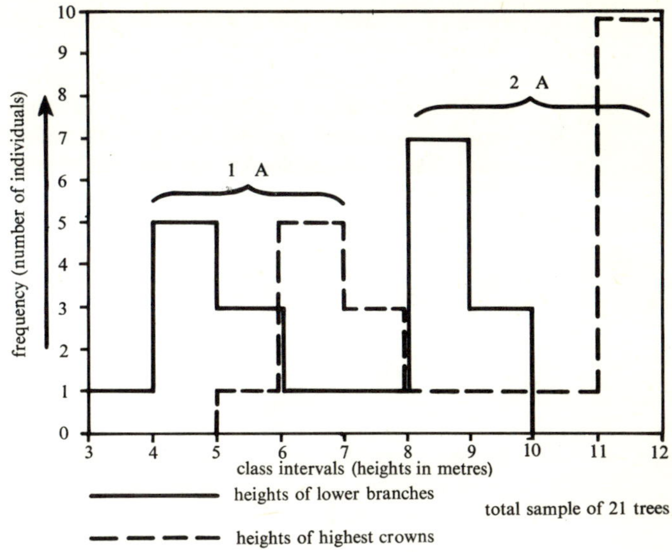

Fig. 4.4 Tree layers in a small coppice

extending ladders. Also, it may prove convenient to use the notation 1A, 2A, 3A and so on for low, medium and tall trees. This introduces a subjective element, of course, and it will prove much more satisfactory to plot frequency histograms of the measurements made in order to determine the number of layers present. In the example illustrated in fig. 4.4 there are clearly two main tree layers: 1A between 4 metres and 7 metres and 2A from 8 to 12 metres. In larger scale surveys it would be necessary to apply far more rigorous statistical procedures to determine the presence of distinct layers. A parallel can be drawn here with the techniques for determining the levels of possible erosion surfaces in geomorphology (see page 185), and the reader should consider the value of a presentation like the altimetric frequency curve for establishing and comparing the stratification of woodlands.

Recording the distribution of individual species on large-scale site plans of woodlands can be very tedious and time-consuming, but highly instructive nevertheless. We recommend that you choose an easily identified shrub or tree to map rather than a herb; the latter will be examined later since they are best surveyed by quadrat sampling techniques (see page 99). The resultant distribution maps can be analysed in much the same way that the human geographer studies settlement patterns. If different surveyors choose particular species, then each distribution plotted on tracing paper will enable useful pictures of the plant communities to be built up. Groups can be assigned to map particular layers so that their overlain maps show the various shrub and tree species respectively.

FIELDWORK

Vegetation mapping

1 Make profiles and plans of the trees and shrubs that occur in your garden, a small wood or the local park.

2 Attempt an explanation of the stratification that exists in terms of the ages of the plants, rates of growth and competition for space.

3 Can you explain the areal distribution of plants? What are the respective influences of climate and management?

4 What is happening in unattended parts of the sites you have chosen?

Some ecologists use a more rapid but highly subjective assessment of *plant abundance* by estimating the **occurrence** of particular species 'within view' or on a sample plot. The usual terms used are 'dominant', 'abundant', 'frequent', 'occasional' and 'rare'. As you will imagine, one has to be very experienced to apply such techniques successfully and even the experts disagree more often than not. A slightly more accurate technique is to describe the **cover** (occurrence) and **grouping** (distribution) of individual species by the so-called **Braun-Blanquet** rating system given in table 4.1 below:

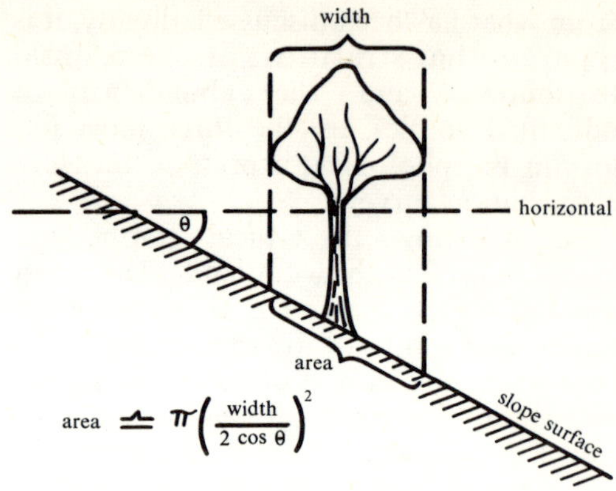

Fig. 4.5 Cover area on steep slopes

Errors with this system are caused mainly by over-estimates of brightly flowered species and large leaved plants because they attract attention. However, it is a useful technique to use in groups on field excursions where time is limited, particularly if individuals in the party make independent estimates and the results are then pooled. Many workers claim greater accuracy by using a modification of the Braun-Blanquet cover scale called the **Domin scale.** In effect, this is a rating of the percentages of an area covered by a particular plant using a ten-point scale.

Having counted, or perhaps estimated, the *numbers* of each species of shrub or tree in a woodland or plot, it is possible to calculate their respective *densities* and *covers*. Density is defined as the numbers of a species per unit area and cover as the percentage of the ground surface overlain by the crowns or canopies of particular plants. With the latter, therefore, it is necessary to sum the areas of foliage perpendicularly projected to the ground. Fig. 4.5 shows this to be a

COVER	GROUPING
+ = sparse, cover small	Soc. 1 = isolated individuals
1 = plentiful, but cover small	Soc. 2 = grouped or tufted
2 = numerous, cover greater or = $\frac{1}{20}$	Soc. 3 = patches or cushions
3 = any number, cover $\frac{1}{4}$–$\frac{1}{2}$	Soc. 4 = colonies or carpets
4 = any number, cover $\frac{1}{2}$–$\frac{3}{4}$	Soc. 5 = pure populations
5 = covering greater than $\frac{3}{4}$ of area	

Table 4.1 Braun-Blanquet rating system

function of slope angles, the formula

$$\pi\left(\frac{\omega}{2\cos\theta}\right)^2$$

being the approximate area of ground covered by an individual plant. Thus, on a uniform slope element of area A, like those mapped in the previous chapter (see fig. 3.1), the **percentage cover** of a particular species will be:

$$\frac{100}{A} \times \left[\pi\left(\frac{\omega_1}{2\cos\theta}\right)^2 + \pi\left(\frac{\omega_2}{2\cos\theta}\right)^2 \ldots + \pi\left(\frac{\omega_n}{2\cos\theta}\right)^2\right]$$

or,

$$\frac{25\pi}{A(\cos\theta)^2}\cdot(\omega_1^2 + \omega_2^2 \ldots + \omega_n^2)$$

where ω_1, ω_2, and so on are the widths of individual plants as indicated earlier in this section. This technique is largely applicable to steep slopes.

For any woodland it is possible to prepare a list of the various species of shrub or tree mapped giving respective density and cover data. This information can be added to that already discussed regarding mean heights and widths in order to compare different woodlands using statistical correlation co-efficients. We shall return to methods of analysis later.

If a sufficiently large woodland is available (most authorities agree that at least 6 hectares or 15 acres is required), it is possible to determine the **relative importance** of specific shrubs or trees to all the others in the community. The most useful technique is to evaluate the **relative density** of particular plants, for example:

$$\text{relative density} = 100 \times \frac{\text{number of individuals of a species}}{\text{total number of individuals of all species}}$$

Similar measures of relative frequency (see page 101) and relative dominance have been devised by some ecologists. Clearly, relative density is the percentage occurrence of a particular plant to those comprising the whole woodland community; high percentages will indicate dominance and low percentages a sparse occurrence. With a very mixed woodland, therefore, it will be possible to list the different trees and shrubs in rank order. Such lists can then be compared with the mean heights of each species to see whether there are any relationships between stratification and densities of distribution. However, a word of warning is necessary because one of the limitations of density values is that they do not give much indication of location and distribution patterns. Although high values must be the result of a widespread and probably uniform distribution, with low values it is by no means certain whether the plants concerned are grouped in 'clumps' or widely dispersed. Much the same applies to cover percentages. Since we are dealing with systematic plot surveys for the moment, however, such problems can be resolved by inspecting the field maps. It is when methods of sampling are used that major difficulties arise (see page 99).

Apart from *edaphic* and *climatic* controls, the distribution and stratification of vegetation are affected by environmental factors created by the plants *themselves*. We can call these *biotic* controls and, although it is realised that the effects of man and innumerable animal organisms are increasingly important biotic factors to be reckoned with, we focus our attention upon some aspects directly attributable to the plants as such. Light and shade are probably the most important features as those of you who have observed plants competing for sunlight will be well aware.

Light affects all aspects of growth and reproduction, the *photosynthesis* process being the most crucial regarding the supplies of energy to the plant. Below certain light intensities the inputs of energy are less than outputs through respiration alone, and so the plant must use-up stored energy in the form of food to maintain its vital functions. Above certain light intensities the plant has an excess of income over expenditure so that growth and storage are possible. Each species has its particular 'point of balance', and for many this is a critical threshold. Others can adapt within limits and may be considered more tolerant. A useful technique, therefore, is to determine what is called the **extinction point** of a species in the shade. Having located the most shaded specimen of a plant in full daylight, leaf-level light readings are recorded (see page 43) and compared with similar readings in the open nearby. The former reading can then be evaluated as a percentage of the latter. Clearly, time of day and seasonal characteristics are important here; the penetration of light through the leaf canopy varies a lot with the angle of the sun, and the filtering characteristics of the canopy at different times. Ideally, therefore, readings should be taken under varying conditions throughout the year and the percentages recorded plotted as mean monthly maxima and minima. The different times that various deciduous trees come into leaf in the spring is often associated with the changes in growth of the ground vegetation.

Light intensity profiles and traverses will provide much useful data concerning the degree of shade cast by different plant layers and the tolerance of those species on the ground. Although trees like beech and oak cast dense shade, their own seedlings are relatively tolerant so that regeneration is not hindered. Many conifers like the Scots pine, however, also cast dense shade but their seedlings will not grow until the parent trees lose their lower branches. This phenomenon can be observed in most Forestry Commission plantations where conifers are grown close together quite deliberately to cause the early death of the lower branches; hence, less knots in the timber cut from the main trunk.

It has now been recognised that the spectral 'quality' of light alters significantly on being filtered through different leaf canopies, and *leaf mosaics* seem to be influenced by this as well as genetic controls. Clearly, leaf orientation (aspect and angle), slope, area and arrangement are all influenced by the interception of light. You will all have seen particular plants which appear to be 'seeking' more light, often at the expense of others, and this must be widespread to a greater or lesser degree. Apart from carefully measuring and keeping records of leaf mosaic data for selected plants, useful evidence of competition and aggression can be gleaned from photographs taken periodically at a particular site. One of the standard techniques for measuring the growth of plants is to uproot them carefully, remove the excess soil and then dry them slowly in an oven. After brushing away all remaining particles of soil the dried plant is then weighed. By taking regular samples, it is possible to plot the *dry weight* of vegetable material against *time* as indicative of rates of growth. The obvious disadvantage with this method is that one removes and destroys a plant each time. However, we can recommend a carefully controlled experiment on an isolated part of your school or college grounds. In the country you may get a landowner's permission to study a small part of a woodland in this way.

FIELDWORK

Effects of sunlight

1 Record the dates when flowering shrubs and trees blossom and relate these times to local climate, especially the duration of early spring sunlight.

2 Plot the flowering times of a single shrub on a polar diagram. Is there any evidence that those receiving more sunlight are earliest? What are the effects of height above the ground surface?

3 Using a simple clinometer and compass, measure the aspect and angle of the leaves on a broad leafed shrub at various times on a sunny day. Is there any evidence that the leaves orientate themselves with respect to incident sunlight to maximise the photosynthesis process? What happens on the shaded parts of the shrub?

4 Persuade your parents or the school gardener to let the lawn grow for about a fortnight between cuttings. Grid out the lawn and weigh the clippings caught in each grid square. Accumulate the totals over a long period and prepare a choropleth map of rates of growth over the lawn. Can you explain the distribution patterns evident?

Since we shall be returning to plants later, it is now appropriate to consider some basic measurements regarding the study of soils in the field. Obviously, the approach has to be quite different because the soil is hidden from view. Therefore, it is impossible to prepare complete maps in the same way as for vegetation. Like the geologist, the soil surveyor is thankful to inspect any exposures whether in quarries, alongside roads and railways or the foundation trenches for buildings. After a little digging and augering we feel sure that you will understand this view!

The community concept enables us to envisage the soil as a 'home' for its plant 'occupants'. Furthermore, much as we study homes or buildings in terms of their building materials and design, so soils may be examined through their *textures* and *structures*. In addition we must consider the basic *mineralogical* and *chemical* make-up of different soils.

Texture refers to the **particle-size distribution** of the inorganic materials in a soil sample. Hand specimens of soil are dried slowly in an oven and then their different particle-size fractions separated by passing the material through a nest of sieves (see fig. 4.6). The mesh sizes are stacked in such a way that each sieve retains different size ranges, viz. pebbles (over 2 mm in diameter), coarse sand (2·0–0·2 mm), fine sand (0·2–0·05 mm) and the remaining silt, clay and fine organic materials are caught in the pan at the base. You will find similar technique used in the analysis of river sediments although a more extended scale is often used for defining classes of river sediments (see page 141). However, in soil studies the pebbles are discarded and the sands aggregated, which minimises the number of sieves required, e.g. the number 3 sieve (0·05 mm mesh). It should be noted that mesh sizes are now more correctly quoted in micrometres (μm) and sometimes as microns (μ). Thus, it is a good exercise to transpose the scales above into these units.

Once the sand has been separated and weighed, the silt, clay and organic fractions can be analysed by various laboratory techniques. The most accurate method uses

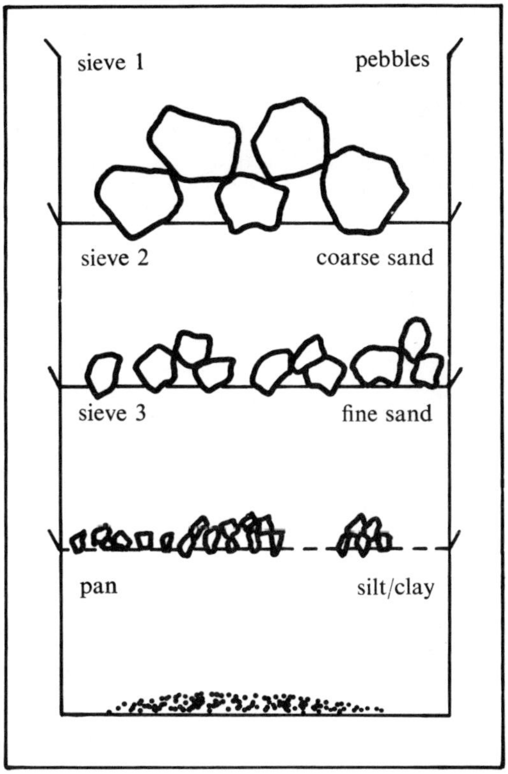

Fig. 4.6 A nest of sieves and the particles retained on them after agitation

an **elutriator**. After weighing the total dry sample caught in the pan, it is then placed into a vessel filled with distilled water. An additional water reservoir is connected to the sampling vessel by a flexible tube so that, on being raised, the velocity of water passing through the sample and overflowing will increase. By controlling the velocity, the fine organic material will be washed out first, followed by the clay and silt respectively. Each is then dried on filter paper and weighed. A simpler technique, sometimes used in this field, is to shake up the whole sample in a glass bottle filled with distilled water until a uniform suspension is obtained. Then, undisturbed settlement is allowed to take place so that the relative proportions can be estimated from the fact that silts fall to the bottom, clay colloids remain suspended and organics float on the surface. In the laboratory a pipette is used for greater accuracy.

Expert soil surveyors are able to estimate the sand, silt and clay fractions with acceptable accuracy by 'working' a moist sample between finger and thumb or 'rubbing' it out over the palm of the hand. The 'feel' and 'appearance' of grittiness and powderiness is taken to indicate the percentages of sand and silt respectively. This technique obviously requires a great deal of practice, but is clearly to be recommended for rapid field work.

Once the percentages of sand, silt and clay have been determined, the information is plotted on to the diagrams illustrated in fig. 4.7. This enables the **textural class** of the soil to be defined so long as several samples have been plotted. It should be noted, however, that these classes do not include the amounts of organic material present; hence, texture is often qualified by adding the terms: 'humose' when 13–25 per cent is organic matter, 'very humose' (peaty) for 25–40 per cent organic matter, and 'organic soils' when there is over 40 per cent organic matter.

Apart from the particle-size distribution, all soils exhibit certain **structures** defined by the degree of aggregate development and by the size and kind of the structural unit produced; usually referred to as *ped*. The following table lists the terms used in assessing the grade of structure:

Table 4.2

1 Structureless	no observable structures or peds, massive and uniform
2 Weak	indistinct peds; easily breaks down when disturbed
3 Moderate	well-formed peds; little break-down when disturbed
4 Strong	very distinct peds; remains aggregated when disturbed

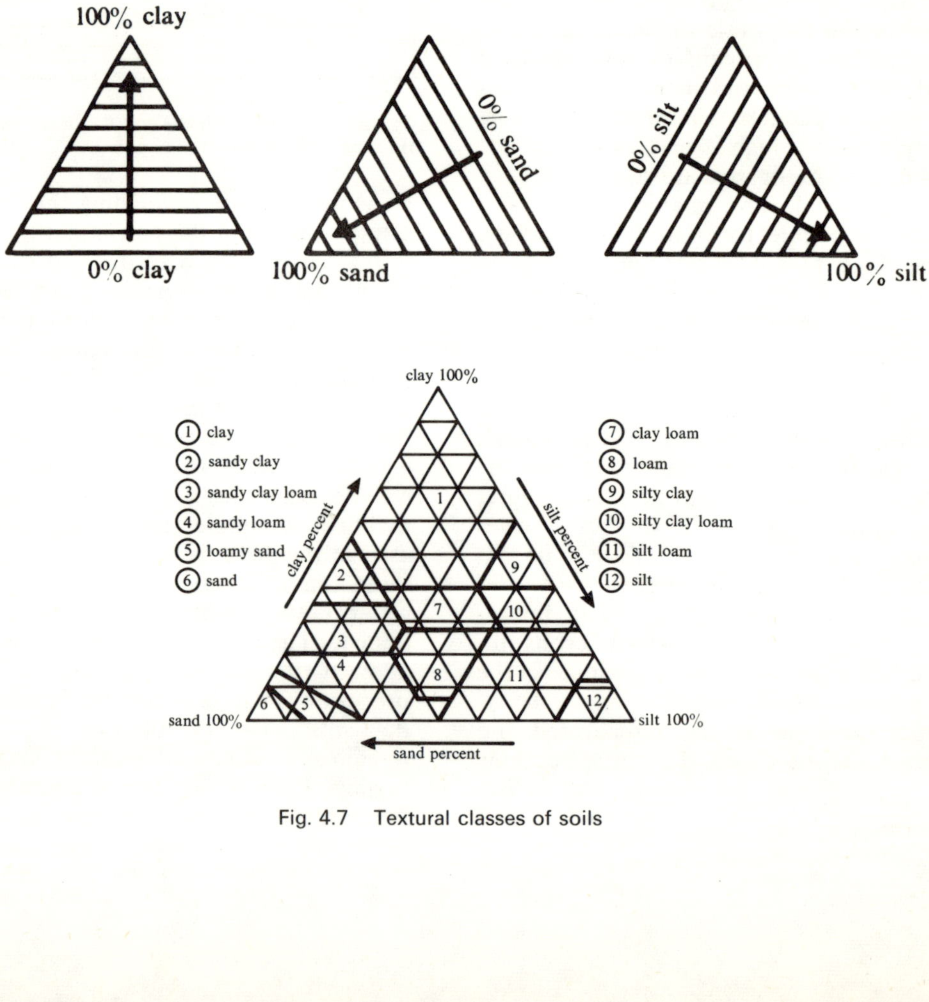

Fig. 4.7 Textural classes of soils

The next table shows how the basic *shapes of peds* are defined

Table 4.3

1 Platy	vertical axis much shorter than horizontal; numerous faces
2 Prismatic	vertical axis more than horizontal; numerous angular faces
3 Blocky	similar axes enclosed by curved surfaces often moulded by adjacent peds; subdivided into angular and subangular
4 Granular	small, roundish peds; few faces or edges, and usually hard
5 Crumb	small granular shapes but soft; like breadcrumbs

Another very useful measure of soil aggregates and fragments is to determine their degree and kind of **cohesion**. You have probably seen farmers and gardeners make this sort of assessment by squeezing a sample in the palm of their hand or between finger and thumb. This is a good indication of the moisture level present and, therefore, the 'condition' of the soil regarding ploughing, sowing and other methods of cultivation. The table below lists the terms used:

Table 4.4

1 Loose	non-coherent and crumbly when moist or dry
2 Friable	when moist, soil crushes under gentle pressure but coheres if squeezed hard
3 Firm	when moist, offers resistance to gentle pressure but crushes
4 Hard	when dry, very difficult to break down
5 Compact	firm and difficult to squeeze
6 Plastic	when moist, can be moulded and rolled without sticking
7 Sticky	plastic, but adheres to the hands

Whilst all of these methods of defining soil structures are subjective, with a little practice and guidance you will become quite skilful. By the simple expedient of grabbing a handful of soil you will be able to comment at length on its characteristics.

The mineralogical and chemical analyses of soils are very important in establishing explanations for soil *formation* in any area. For example, do the minerals present indicate that development took place *in situ,* or are there indications that the soil has been derived from imported debris (see page 154)? The answers to such questions lie in a comparison of the minerals present in the soil and local bedrock. Therefore, geological maps will indicate the nature of the parent materials and topographical map evidence will suggest whether additional materials might have been brought in. In general terms, soils are manifestations of three main processes: the physical and chemical changes resulting from denudation in the presence of moisture and organic matter; the continuous exchanges within the various plant and soil communities established and, lastly, the redistribution of components caused by lateral and vertical water movements within the soil. The latter are particularly important, resulting from downward percolation under gravity and upward leaching following evapotranspiration and capillary action. Lateral movements are dealt with in Chapter 5 and so, since space precludes us doing justice to every aspect above, we content ourselves by mentioning two characteristics of soil moisture which have widespread significance in the plant and soil community. Water is probably the most important factor in energy and mass transfers in the ground.

The retention and storage of water is basically determined by the texture and structure of the soil. Two forces of 'attraction' termed *adhesion* and *cohesion* are involved; the former arises because of the attraction of the solids for water (i.e. their hygroscopic characteristics) and the latter through the attraction of water molecules for each other. Together these forces comprise the **capillary potential** or pF of the soil. As a soil becomes drier the pF increases until the energy required to withdraw further moisture is so great that plant

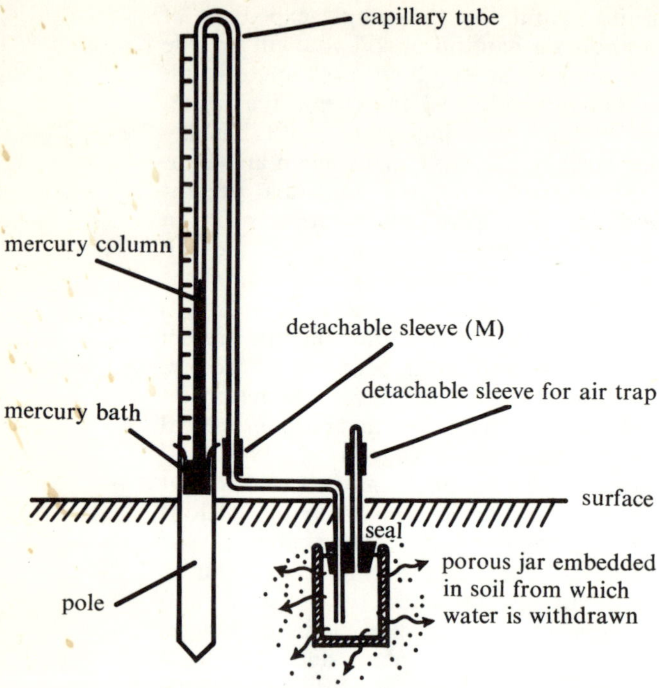

Fig. 4.8 Measuring the pF of a soil

Fig. 4.9 Moisture characteristics in common soils

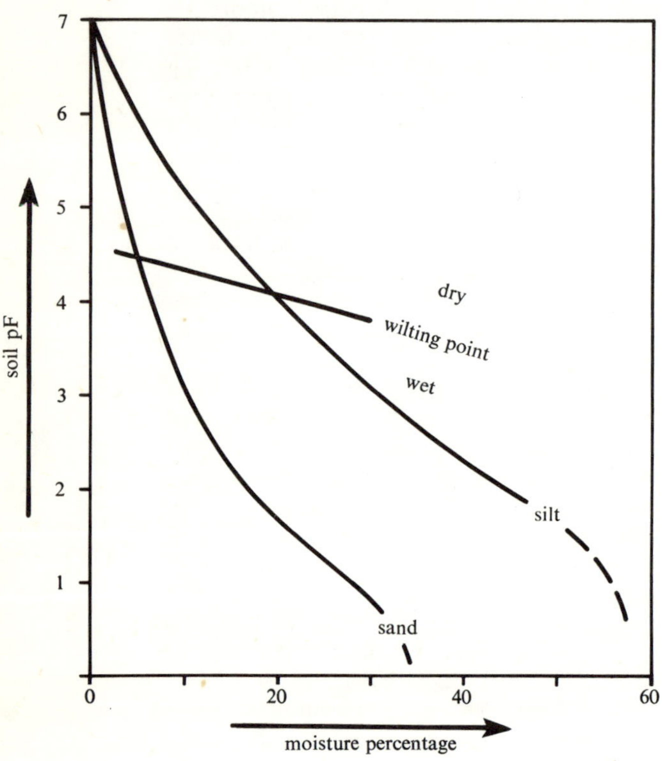

'suction' and evapotranspiration are ineffective. This stage is referred to as the **wilting point**. A useful technique of measuring this critical value in a soil is to dry a sample in an oven at 100°C, weigh it and then reweigh after exposure to saturated air. Clearly, the increase in weight will represent the amount of water taken up by the dry sample. Thus, the **wilting coefficient** can be expressed as the weight of water absorbed to the dry weight of the sample; usually between 10% and 15% for most soils. Obviously, the differences between the wilting coefficient, the actual weight of moisture to dry weight at any time and the amounts of water required by particular plants are crucial to calculating irrigation needs, water deficits and so forth.

Because the pF indicates the amount of energy involved in the retention of soil moisture it can be measured by **tensiometers.** A useful demonstration of such measurements can be made with the apparatus illustrated in fig. 4.8 by balancing the mercury column in a capillary tube against the suction force exerted as water is drawn into the soil through a porous pot buried in the ground. The experimental controls are clearly explained by Ashby (1961, pp. 225–7). Nowadays, gauge tensiometers are used in field measurements and other methods are described later (see page 129). The range of pF values lies between 0 to 7; that is, from mean atmospheric pressure at sea level to 10000 atmospheres. Approximate pF values can be measured by the apparatus shown in fig. 4.8 by taking the logarithm of the height of the mercury column calibrated in centimetres. Fig. 4.9 shows the relationships between pF and the percentage of moisture for two common textural classes of soil. Can you explain the form of the two curves plotted?

In Britain, of course, many soils never reach wilting point because of our comparatively high rainfalls. Indeed, drainage characteristics are often of greater importance to farmers than the likelihood of droughts. The natural drainage of a soil is clearly a function of its texture, structure and the relief of the site concerned. One of the most useful methods of determining the **drainage class** of a soil is by its *colour* and the nature of the *mottling* (various colour patches). Munsell Soil Colour Charts are used to match and standardise particular colours, and table 4.5 below summarises the characteristics of the five main soil drainage classes:

Table 4.5

1 Dark colour; loose with no mottling and usually shallow
2 Brown; little mottling except rusting by old root channels
3 Brown; definite mottling and rusting around most root channels
4 Grey; very strong rusty mottling around all root channels
5 Peat covering grey mineral soil with ochreous mottling

1 excessively drained
2 freely drained
3 imperfectly drained
4 poorly drained
5 very poorly drained

As water moves through the soil, the soluble minerals are taken into solution and removed. Thus, the chemistry of acidic atmospheric water changes as it percolates through different soils, and the removal of soluble minerals from the surface is one of the most important factors governing the colour of various horizons. By measuring the reaction of these soil solutions we can determine whether they are acid, neutral or alkaline. This is done by evaluating the **hydrogen-ion concentration** of the sample, usually called its pH. For those of you who are conversant with chemistry, the pH value is the logarithm of the reciprocal of the hydrogen-ion concentration. Neutral solutions have values of 7 and the scale can range from about 4 for acidic peats to 10 for very alkaline sands; the most common range is from 5 to 9. Clearly, if the solution is strongly acidic, it is capable of solvent action and vice versa. This is an important aspect of weathering by the so-called solution process, especially on limestones.

Carbon dioxide (CO_2), both in the lower air and through the natural processes of plant respiration and decay in the ground, is the main cause of soil acidity. This is why farmers neutralise acid soils after much cultivation by the use of calcium carbonate ($CaCO_3$) 'dressings'. Indeed, the pH of a soil is both directly and indirectly of great nutritional importance, for the availability of many essential minerals is drastically influenced by quite small pH variations. For example, as alkalinity increases, important nutrients like iron, manganese and zinc become less available, and much the same applies to molybdenum as acidity increases. At values below 5 minerals like aluminium, iron and manganese become soluble in sufficient concentrations to be toxic to many plants. Thus the regulation of pH is vital in agriculture. Under natural circumstances, of course, different plants grow best under different pH conditions

Plate 4.1 Measuring soil reaction with a B.D.H. pH indicator kit

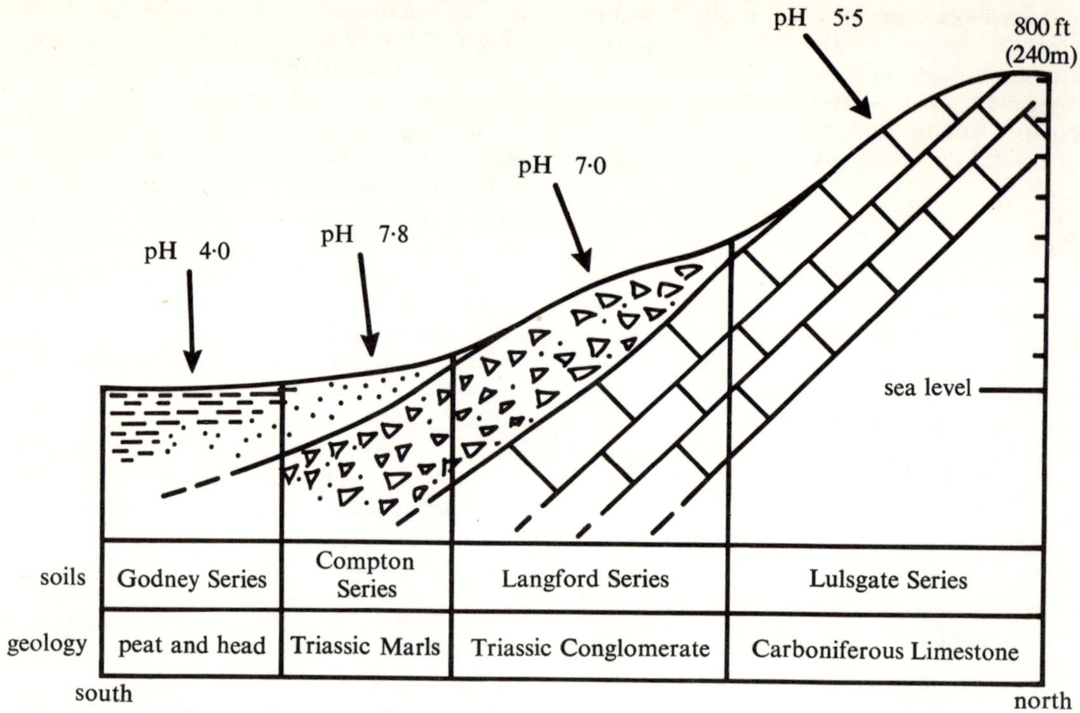

Fig. 4.10 Typical pH values for soils on the southern slopes of Mendip (c.f. Plate i.i)

and thus we have a distinction between *calcicole* (lime-loving) and *calcifuges* (lime-avoiding) plants.

The British Drug Houses Ltd (or B.D.H.) pH colour indicator kit shown in plate 4.1 is one of the simplest ways of measuring soil acidity or alkalinity in the field. You will find that taking pH values of soil samples augered from increasing depths will provide useful data concerning drainage characteristics and soil formation. Also, by plotting the diurnal variations of CO_2 concentrations (see page 68), in the soil against pH readings, you will discover one of the most important relationships in the plant and soil community. Nowadays, it is possible to obtain very sensitive pH-meters using electrical resistance principles; a particularly valuable device for anyone who suffers from colour blindness! Fig. 4.10 shows a typical pH profile taken on a slope on Carboniferous Limestone. Can you explain the variations that have been plotted?

FIELDWORK

Soil characteristics

1 Collect soil samples from the rooted zones of soils in different localities and compare their characteristics on a table like the one below:

	sample number					
	1	2	3	4	5	6
1 textural class						
2 organic percentage						
3 structural grade						
4 ped shapes						
5 aggregate cohesion						
6 wilting coefficient						
7 pF when moist						
8 pF when dry						
9 pH when moist						
10 pH when dry						

2 Can the differences be explained in terms of the locality of the samples?

3 Repeat the tests under varying climatic conditions and before and after different methods of cultivation have occurred. What are the major effects of climates and cultivation?

4A 2 Soil profiles and plans

If you dig a pit to the bedrock and 'clean' a straight vertical face with a shovel, you will be able to study the stratification of the soil at leisure (see plate 4.2). Fig. 4.11 illustrates a typical soil profile and the terms used to describe the different *horizons* that might be present. You will see that there are three horizon *groups* called A, B and C and that, with the exception of the surface litter found mantling most well-vegetated soils, each group may be subdivided. Thus, A_1 represents the uppermost horizon of the soil as such, comprising of mixed mineral and organic matter. The main organic matter has been well decomposed to *humus*, being a complex residue of modified plant and microbial tissues in the same way that compost heaps in the garden undergo biochemical changes. Beneath this occurs the A_2 horizon, or *eluvial zone,* which literally means 'washed out'. Well-drained soils in areas of high rainfall have very distinct eluvial zones because the soluble minerals are leached away by the percolating water. This leaching produces the characteristic light colours usually associated with A_2 horizons as can be seen on plate 4.2. What soil reactions and pH values would you expect to discover in such horizons? Why do you think that in many British soils the A_1 horizons are often poorly developed and thin whereas the A_2 horizons are well developed?

The A_3 and B_1 horizons are basically transitional to the B_2 horizon below. This is termed as the *illuvial* or 'washed in' zone because it normally includes the accumulated minerals which have been brought in from above by percolating water. Thus,

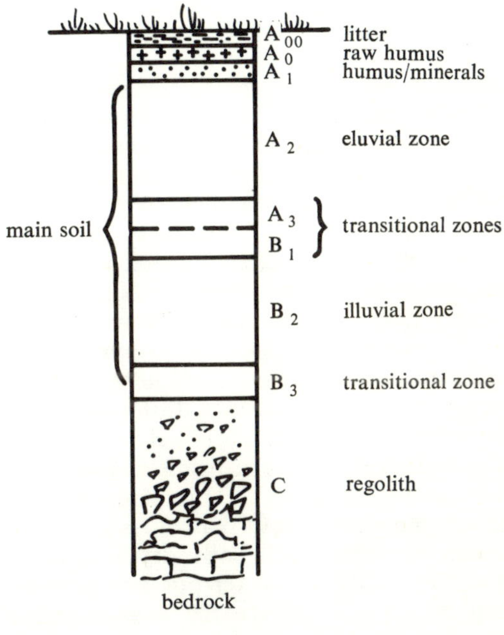

Fig. 4.11 Typical horizons of a podzol

Plate 4.2 A typical soil profile. Can you identify the horizons present and the type of bedrock?

metallic cations like calcium, magnesium, potassium and sodium which have been removed from the A_2 horizon by leaching are often redeposited here. Also, in well drained soils, small amounts of humus, clay colloids and iron and aluminium oxides are similarly translocated from the A_2 to B_2 horizons. The term 'sesquioxides' is often applied to the latter because of their particular chemical formulae, e.g. Al_2O_3 and Fe_2O_3. Clearly, a large number of plant roots can be found penetrating to this horizon to take advantage of the nutrients which accumulate there. In most soils the B_2 horizon is compact and has a strickingly different colour to the leached A_2 horizon. What pH values would you expect in the B_2 horizon?

Beneath the illuvial zone occurs the B_3 horizon, but this is often absent or indistinct. Even when present, however, it is merely transitional to the fragmented parent rock substratum or regolith of the C horizon. The depth of the C horizon can be taken as a useful indicator of the degree of weathering associated with the particular parent rock and local climate, of course.

Thus, you can see that well developed A_2 and B_2 horizons in the profile are manifestations of the general movement of water downwards which, in turn, allows you to comment on the drainage characteristics, of the particular soil and the soil forming processes. The particular process described above is called *podzolization* as it is typical of the *podzol* soils found in Russia. In poorly drained soils and those where upward leaching is common, profiles show different characteristics. Where waterlogging is persistent, what little dissolved oxygen that is available is largely consumed by plants and micro-organisms and anaerobic conditions prevail. This gives rise to thick blue-grey clays called *gleys*. On occasions when such soils dry out temporarily, oxidation of ferrous compounds can take place to give the characteristic mottling already referred to in table 4.5. Often, the upper limit of the ground-water table is marked by a reddish-coloured line of ferruginous materials. Where upward leaching is common through high rates of evapotranspiration, sesquioxides accumulate near the surface of the soil and dry out to form brick-like *hard pans*. Such soils are referred to as *latosols,* and more particularly *laterite,* from the Latin word for brick. Indeed, they are common enough in Mediterranean areas, and other parts of the world with pronounced wet and dry seasons, but are of no consequence in Britain. In areas of low rainfall, of course, there is little downward or upward leaching and soluble minerals like calcium and magnesium accumulate at a depth approximating to that reached by percolating water after the occasional rainstorm. Only those of you lucky enough to visit prairies, steppes and semi-arid areas will see such soils, however, for Britain is dominated by varieties of podzol. In fact, soil moisture movements in response to seasonal climates form the basis for classifying zonal soils on a global scale; it so happens that the general movement in Britain is downward!

By selecting suitable sites on soils with different drainage characteristics and digging profile pits you will be able to measure the features already described. Notes should be made about the location, slope, vegetation or land use, the thicknesses, textures and structures of the different soil horizons and of pH. Hand specimens can be wrapped in plastic bags for determining wilting coefficients, organic contents and so forth in the laboratory. With the aid of geological and topographical maps, and perhaps even a Soil Survey map, you should be able to complete a reasoned analysis of the soils concerned. How do the different soils in your home area influence plant growth, and what are the relationships between rock types and local climate which give rise to these differences?

FIELDWORK

Soil profiles

1 Dig soil pits in reasonably undisturbed soils, identify the horizons and prepare scale drawings of each side by side.

2 For each horizon determine the ten essential characteristics indicated in the previous fieldwork exercise. Can you explain the differences in terms of the locality of the profile and soil moisture movements?

3 Using a trough-shaped metal box with sharp edges, drive this into the face of a soil pit as far as it will go to cover the various horizons of the profile. Carefully dig around the box and cut it out with a sample of the profile still inside. Remove this *monolith* to a laboratory and, by taking actual fragments of soils, stick them to a board with adhesive to make a permanent record of the profile. The surface of the fragments can be coated with a clear varnish. Repeat the experiment annually and compare the monoliths.

Once you have successfully dealt with such questions from field data it is possible to think in terms of simple soil maps and plans. Fortunately, this does not require pits to be dug everywhere. Since you have already sampled representative profiles, you can work outwards from these on traverse lines using an auger to check the areas covered by each soil. Minor differences of texture and thicknesses of horizons are considered as *phases* of a particular soil unit, and only where major changes of structure and profile characteristics occur should you plot the boundary of a soil *series*. Very often, of course, these boundaries are closely related to changes of slope and vegetation; indeed, you can frequently detect the boundaries between different soil series when looking over freshly ploughed fields. Thus, many soil surveyors see vertical air photography as a significant technique for mapping soils and vegetation in more inaccessible areas such as uplands. We recommend that you prepare a slope element map for a small drainage basin with different rock types, and then start digging! Your findings can subsequently be related to vegetation or land use and the sort of climatic data discussed in the previous chapter.

4A 3 Plant transects and quadrats

Earlier in this chapter we concentrated upon the systematic mapping of the larger plants growing in relatively small woodlands. If you have taken our advice and carried out such a survey, you will appreciate that such techniques are out of the question concerning small plants and larger areas. It would take a lifetime to plot the various grasses and herbs established on the average lawn! Thus, the ecologist must adopt sampling procedures, the most common being transects and quadrats.

Since we have already dealt with the principles of choosing particular transect or traverse lines (see page 70), we can concentrate on their uses in surveying vegetation. One simple technique over grassland or moorland is to 'grab' specimens at random points on a walk and record the different species and individual numbers collected. To avoid running the risk of grabbing something unpleasant, however, it is equally as easy to 'throw' a hoop-la ring and record the 'catch'. In order to minimise the subjective element, it is often a good idea to use random number paces between samples. Alternatively, of course, you might have the time to record all species *en route* in a woodland or heathland transect. Whichever method you adopt, having collected the field data you will then be able to plot *frequency histograms* of the occurrences of different species along the transect line. From woodland and heathland transects you will be able to plot the heights and widths of trees and shrubs on *sections*. Clearly, if you can have colleagues and friends walking parallel routes, large areas can be covered and the completed histograms or sections can be superimposed to analyse areal variations. These will make vegetation changes more apparent along particular 'environmental gradients' like slopes. Furthermore, with sufficient point information over a slope element or a drainage basin, it will be possible to plot *isopleth* maps of the *density* of widely-occurring individual plants. In ecological work such maps are called **isonomes**.

Plate 4.3 Quadrat sampling on a slope

Quadrats (see plate 4.3) can also be applied to transects and are the main methods used in the preparation of vegetation maps. The size of quadrat required depends upon the type of vegetation being surveyed and the scale of study. Most quadrats are forms of grid sampling using rectangular wooden frames which can be pegged out (see fig. 2.19). Obviously, the main consideration must be that the dimensions of the *frame* are larger than the vegetation *units* under observation. If the frame is too small, errors will be introduced because of so-called edge-effects in deciding whether or not to include plants overlapping the frame. Similarly, the coarseness of the *grid* within the frame will depend upon the size of the individual plants to be surveyed in any vegetation unit. If the plants were small and randomly distributed, then the sizes of quadrat frames and grids would be immaterial. However, plant sizes vary greatly and random distributions of individuals are rare. Kershaw has shown this to be so even where no groupings or community patterns are immediately apparent.

Having constructed suitable quadrat frames, long pins or needles are dropped from each grid intersection and the number of 'hits' of different species recorded. This is repeated at randomly distributed sites over a uniform slope or in a contiguous line for a belt transect across varying relief. The data from such a survey enables the *frequency* of a particular species to be recorded; that is, the chance of recording at least one 'hit' with a single throw of the quadrat. If the species is found say in 35 quadrats out of a possible 100, then the frequency would be 35 per cent. By recording the *total number* of hits of a species in *one frame* it is possible to determine the percentage frequency of its occurrence at a particular site; for example, 6 hits out of 25 would be a frequency of 24 per cent. As these values vary at different sites, it is possible to evaluate a *mean* figure for a random quadrat survey called the **local frequency**. You will appreciate that this value is likely to differ from the frequency as such because their definitions are not the same. One particular problem in calculating the local frequency is that a single plant with a relatively large spread of leaves may be recorded more than once to the detriment of any species it hides. Thus, it is necessary to distinguish between **local shoot frequency** and **local rooted frequency**: the former records hits to leaves and the latter to individual plants. Obviously, shoot frequency is more closely related to the definition of *cover* given earlier and rooted frequency to that of *density*. Consequently, you must decide *before* making a quadrat survey which measure of frequency you require. It is a good test to have different groups using different measures and then to compare their results. What will higher local shoot frequencies indicate?

It is impossible to predict at the outset how many sites have to be sampled to get representative data of all the species likely to be present. Two methods help us to know when to stop, however. The first is to keep a running check of the mean local frequencies after every five quadrats. This information is then plotted graphically as in fig. 4.12; initially the mean values fluctuate but, after 70 quadrats have become regular. Thus, 70–80 quadrats provide sufficient information in this particular example. The second method is to plot the number of different species recorded as the size of sample increases. Fig. 4.13 shows such information plotted graphically, and it can be seen that after 70 quadrats once again no further species are occurring and there is little point in continuing. Both techniques are particularly suitable when permanent quadrats are envisaged to record changes in the community over long periods of time. Once again, however, we must introduce a cautionary note regarding the second technique which is often referred to as the **minimal area method** as championed by Braun-Blanquet. Very often you will find that your curve 'tails off' gradually

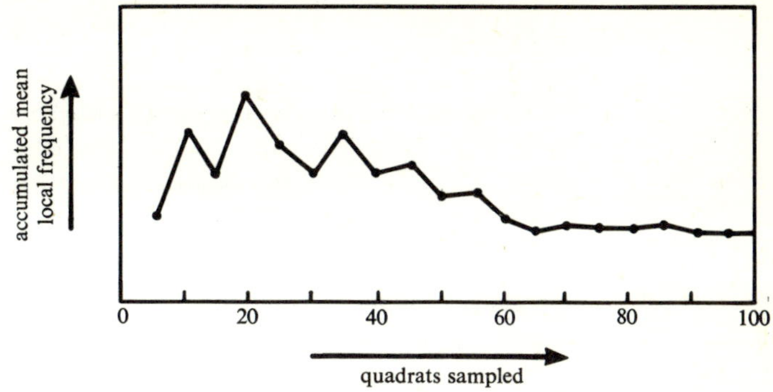

Fig. 4.12 Local frequency method for determining the size of quadrat samples

Fig. 4.13 Species number method for determining the size of quadrat samples

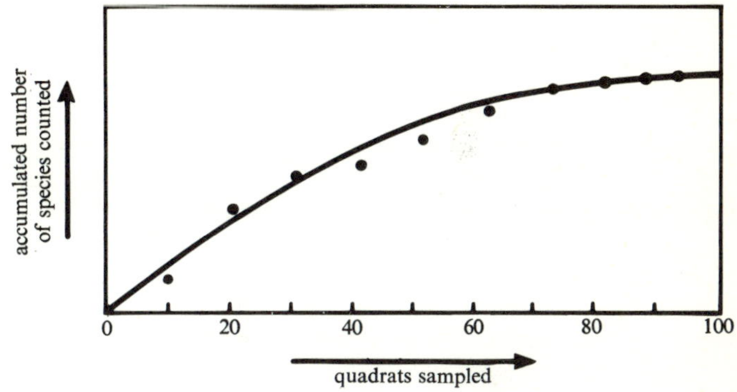

and it is difficult to decide objectively at what point to stop. In theory, of course, you could go on for ever! Thus, such techniques must be treated as guides and not as strictly accurate evidence for the limits of a community.

FIELDWORK

Vegetation mapping

1 Take two contrasting slopes, such as one facing north and another facing south, and complete a quadrat survey of the vegetation on each.
2 Make a species list for each slope with measures of local shoot and root frequencies.
3 What environmental controls explain the differences between the slope data? Assess the influences of climate and soils after

conducting the appropriate surveys on each slope.

4 Prepare large-scale plans of the slopes surveyed locating the quadrat sites. Use these to prepare point data of the percentage frequency of occurrence of individual plants at each site; then, draw isonome maps of the density and cover of each species on tracing paper.

5 Overlay your isonome maps and make an analysis of the vegetation distributions on both slopes. Attempt to prepare community constellations for each slope and explain the differences apparent (see page 106).

4B Energetics and Dynamics

We made the point earlier that no life could exist on earth without solar energy or insolation. Therefore, the transformations of this energy within various ecosystems are vital to support life and enable us to understand the dynamic associations among different life forms in particular communities. **Food chains**, for example, are an obvious means of plotting the linkages between species in terms of energy flows. Can you list the links in the chain which provides you with a cup of tea and a ham sandwich? And, what happens after that?

4B 1 Energy transformation and flows

The laws of thermodynamics already mentioned show that the energy entered and leaving a closed system like a plant or an animal must balance, given suitable time scales. We lose energy, for instance, by heat losses and the work we do. That energy is replaced mainly by the food we eat. Similarly, the solar energy absorbed by plants is lost by the chemical energy of growth and reproduction and the heat energy of respiration. Energy stored in the tissues of either plants or animals is eventually released on their death so that the net primary production of plants, the total consumption by other organisms and the decomposition of all tissue are in dynamic equilibrium. In fact, most food chains have a fundamental linear relationship from plants to herbivores to carnivores. Complex variations of this simple chain give rise to **food webs**.

Various techniques have been used to trace food web patterns in different communities, the most effective being the introduction of particular radioactive isotopes which can be tracked through the systems concerned. Another method is based upon the fact that, for animals at least, there is a progressive decrease in numbers along any food chain. This gives rise to the **pyramid of numbers** and the **pyramid of biomass** concepts. All species with similar feeding habits are grouped together in their particular ecological *niches*; with the former, the numbers of each niche are counted, whilst with the latter, the dry-weight of each group is measured. The main objections to these methods are that the number pyramid takes no account of individual size variations and the biomass pyramid ignores the fact that different life cycles affect the weight of material at any time. Thus, many ecologists now prefer to measure the *energy* utilised by each feeding group per square metre per annum. This technique is referred to as building a **pyramid of energy**. Sometimes, the energy is measured indirectly through observations of growth rates, but with suitable communities, samples are burnt under controlled laboratory conditions and the heat energy released is evaluated. This somewhat drastic technique is referred to as **bomb-calorimetry**. Other methods use the calorific value of food rations consumed, since it is clearly out of the question to burn many animals however useful the results. What are the mean energy levels of different feeding groups such as dairy cows in milk, laboratory mice, hamsters and your class at school?

Such approaches have great significance to physical geography because the feeding groups or communities are located at different levels spatially. Fig. 4.14 shows a

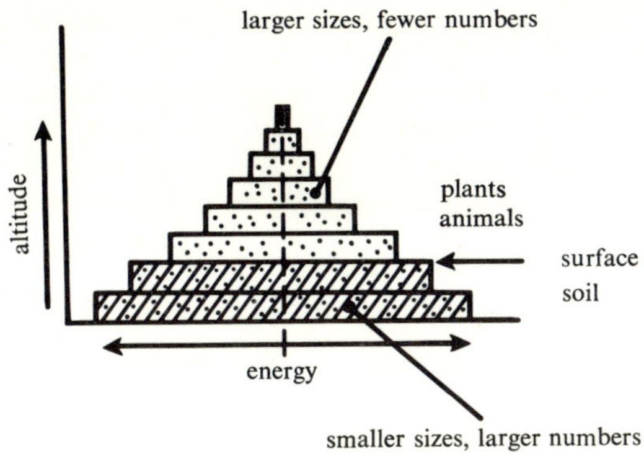

Fig. 4.14 Features of an energy pyramid for the biosphere

typical energy pyramid; the soil harbours countless micro-organisms, small invertebrates and many burrowing animals at the base of the pyramid whilst the birds of prey are the 'overlords' at the apex. You might reflect upon the obvious symbolism associated with the choice of the eagle or albatross as military emblems throughout history and the ancient myth of the phoenix rising from the funeral pyre. On the other hand, you should consider the present-day disadvantages of being at the end of the food chain because the effects of pesticides become relatively more severe as one passes along the chain.

We mention these techniques, not because you will have the resources to carry out exact investigations but because they underline the significance of soil communities when studying dynamic associations in ecology. Many laboratory experiments to determine energy pyramids indicate that as much as 90 per cent of the flow of energy through an ecosystem can be attributed to micro-organisms in the soil. An interesting field experiment to test this somewhat startling conclusion was conducted by Edwards and Heath. Cut leaves were sealed in nylon bags with various mesh sizes ranging from 7·000 mm to 0·003 mm and buried in cultivated plots. The former were large enough to give access to the vast majority of soil organisms whilst the latter could only be entered by micro-organisms. Periodically the bags were unearthed and the percentages of leaf areas that had 'disappeared' carefully measured. After nine months the leaves in the 7 mm bags had been consumed but those in the 0·003 mm bags were virtually untouched. The conclusion was that the micro-organisms needed larger animals to break down their food mechanically, and possibly chemically, before they could assimilate it. In other words, they required the community as much as we do regarding food supplies. You should try this experiment in different soils.

The breakdown of soil by earthworms was first recognised by Charles Darwin who estimated that there could be as many as 125 000 per hectare. Recent counts, however, have recorded ten times that figure in humose soils. You can get some idea of the amounts of soil they turn over by collecting and weighing the worm casts which appear on a lawn over a certain period. The distribution of casts can also illustrate a lot about the communities that exist as well as the nature of the soils concerned. Farmers will tell you that the presence or absence of worms are indicators of 'good' and 'bad' soils. To sample whole communities, of course, requires great patience and skill. One of the most successful techniques to evacuate the organisms from a soil sample is to use a Tullgren funnel. An ordinary 40 watt electric light bulb in a shade is trained upon the soil so that the organisms escape through gauze and drop into a dish or beaker filled with clear alcohol. Using a pipette, samples are extracted and placed on a glass microscope slide which has been gridded to hold given amounts of alcohol. These so-called Rafter cells can then be examined under a microscope and quadrat techniques applied to count the organisms present. Thus, by using quadrat methods at the minutest of scales it is possible to make useful comparisons between the densities and frequencies of organisms and other edaphic and biotic aspects. For example, the micro-organisms at different soil horizons, under different pH and pF conditions and other features of the soil atmosphere and climate. Numerous techniques are explained by Dowdeswell in *Practical Animal Ecology*.

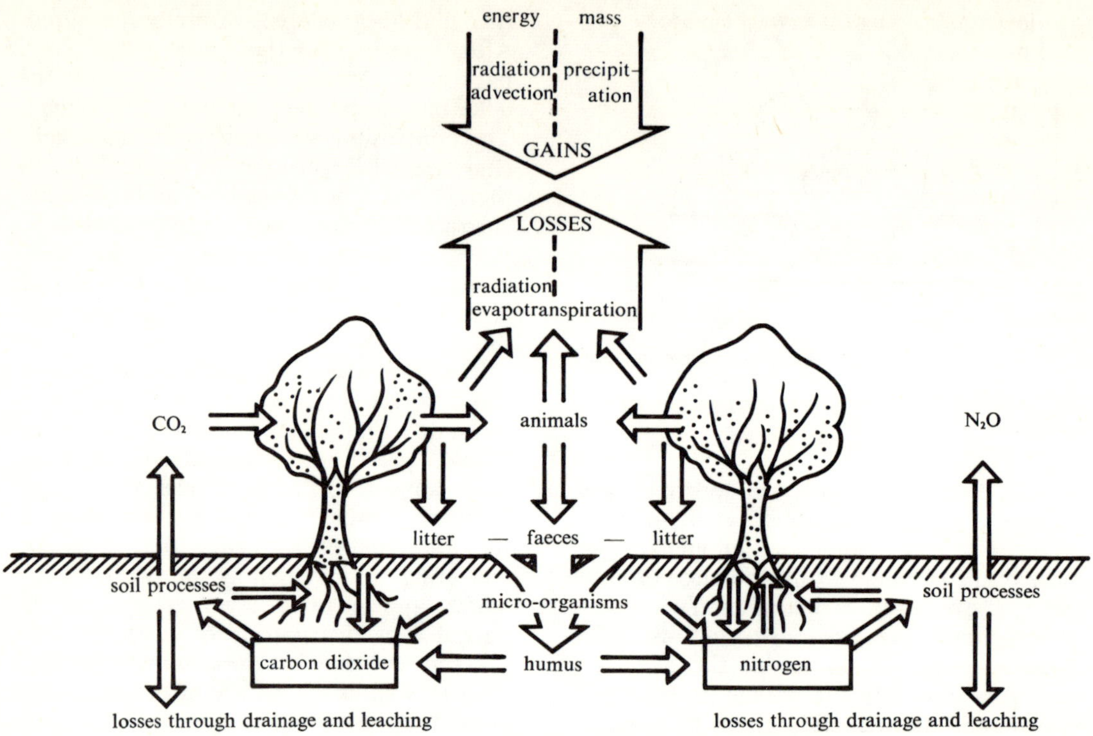

Fig. 4.15 The carbon and nitrogen cycles

Even in soils with low organic contents, the potential energy available reaches staggering proportions. At the Rothamsted Experimental Station it was calculated that over 10^6 joules per hectare per annum were dissipated from an untreated low-producing soil whilst the corresponding figure for one treated with farmyard manures was about 150^6 joules per hectare per annum. Much of this energy is transformed by the continuous oxidation of carbon in the so-called *carbon cycle*. The other great biochemical process in the plant and soil community is the *nitrogen cycle*. Fig. 4.15 shows the basic relationships in both cycles and, although their consideration is beyond our objectives in this book, we hope that you appreciate their significance in the dynamic relationships which characterise the plant and soil community. Clearly, if we seriously impair any link in either cycle through bad farming or indiscriminate waste disposal, we endanger the community and the benefits it affords us.

4B 2 Linkages and associations

In nearly all ecological studies we are attempting to *relate* and *associate* different field measurements of the sort already mentioned. Having decided that the data collected is valid, attempts can be made to *correlate* any number of measurements to discover what might be generally *expected* from what has been *observed*. The simplest techniques can be applied to **bivariate** correlations where two sets of data are portrayed graphically. Fig. 4.16 illustrates how this can be done using scatter diagrams. You should test this technique regarding say pH and the frequency of the occurrence of particular plants or even between two separate species. The scope for work of this sort is endless and by using the more complex methods of **multivariate analysis**, inter-relationships within communities can be discovered.

A statistical rather than graphical method

of determining possible relationships between two sets of field data is to use **contingency tables**. Here the numbers of quadrats in which two plants occur *singly* or *together* are counted out of the total sample. For instance, if in a total of 100 quadrats species Y occurred alone 20 times, species Z alone 12 times, both were found together at 40 sites and neither appeared 28 times, then a two-by-two contingency table or matrix can express this as follows:

Table 4.6

		SPECIES Y		
		+	−	totals
SPECIES Z	+	40	12	40+12
	−	20	28	20+28
	totals	40+20	12+28	= 100

Since it can be shown by a Chi-square test what the expected value for the joint occurrence of species Y and Z should be, it is possible to compare this with the 40 occasions on which the two were actually *observed* together. If the expected value is higher than 40 in this case, then the two species are negatively associated, whereas if the reverse holds then positive association exists. With the data from Table 4.6, the expected value is 12·9 which, with one degree of freedom, indicates that the probability of the observed 40 occurring by chance is less than 0·1 per cent. This shows a strong positive association between species Y and Z. You can check this calculation by using the formula for a Chi-square test given in Toyne and Newby or other statistics texts. When you carry out such examinations of your own, however, you must be sure that the size of the quadrat frame used is suitable for the particular plants that you are attempting to associate.

By developing the Chi-square technique it is possible to examine the inter-relationships of many species in complex communities, and then to search for the *causes* of any associations apparent. Two arrangements of the data can be adopted: the first is termed as **multi-dimensional** and the

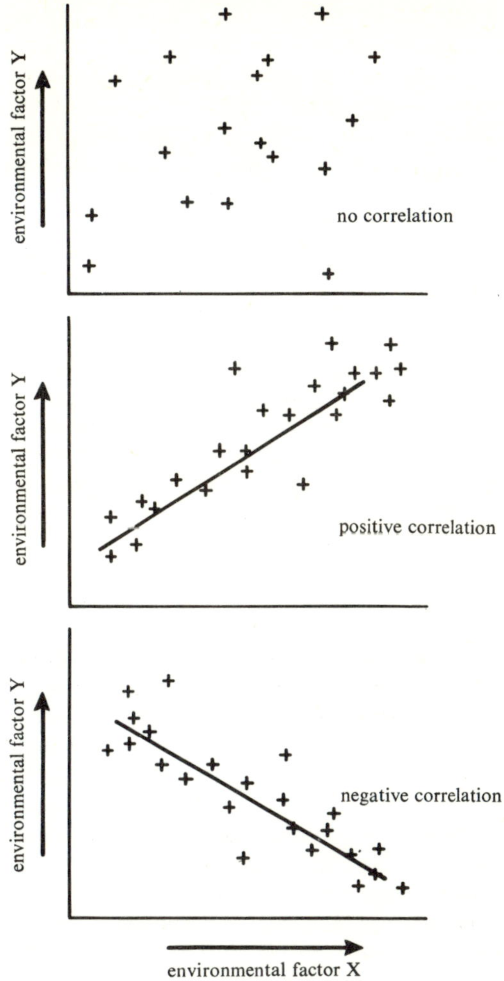

Fig. 4.16 Correlating two environmental factors on scatter graphs

second as **hierarchical**. With the former, the presence or absence of species is recorded as above and the positive (+) or negative (−) associations between every pair calculated by Chi-square tests. The results are then recorded on a complete matrix as indicated on fig. 4.17a. This information enables a constellation to be drawn as in fig. 4.17b to illustrate the linkages between the species wherever positive associations have been recorded. Those which have high associations are grouped close together in the constellation and those with negative associations are set apart. In the example shown, the visual effect of the constellation enables the community to be divided into three distinct groups or sub-communities: group 1 on the left shows strong inter-relationships amongst B, E, G and H; group 2 in the centre shows associations between D, J and I and group 3 to the right shows that A, C and F form a small sub-community only associated with the others through D. Which are the dominant species in the whole sample? Can you express the three groups in terms of Venn diagrams?

The geographical import of the above approach is considerable. For example, if the length of the links are drawn to *scale* according to the magnitude of the expected or Chi-square values (the reciprocal is usually taken), the constellation has *spatial* significance. This can be compared with an actual distribution map showing the density or frequency of the plants concerned.

The same approach also allows a *hierarchy* of associations in the whole community to be compared with the frequency of occurrence of each plant. This will often enable you to see possible causal relationships within the community regarding other environmental aspects like soils, topography and local climates. Thus, you can see that statistical approaches can be correlated with cartographical ones. We have suggested sufficient for you to attempt ecological surveys of your own in a small locality, and we hope that you see scope for applying similar techniques to other branches of physical and human geography. Other methods can be found in the references and further reading sections.

4B 3 Ecosystems

There is so much discussion of man's relationships with ecosystems nowadays that we must mention the role of the physical geographer in such studies. After all, this is the area where the links between physical and human geography are the strongest and, as geographers rather than environmental scientists, we feel that our subject's particular contributions have far-reaching possibilities. Ever since the Reverend Thomas Malthus propounded his pessimistic views in *Essays on the Principle of Population* as long ago as 1798, many scientists have sought to contribute positive solutions to how man might live in greater harmony with his natural environment. You will be only too aware from the recent explosion of literature and debates on such topics that controversy about Malthusian principles still rages.

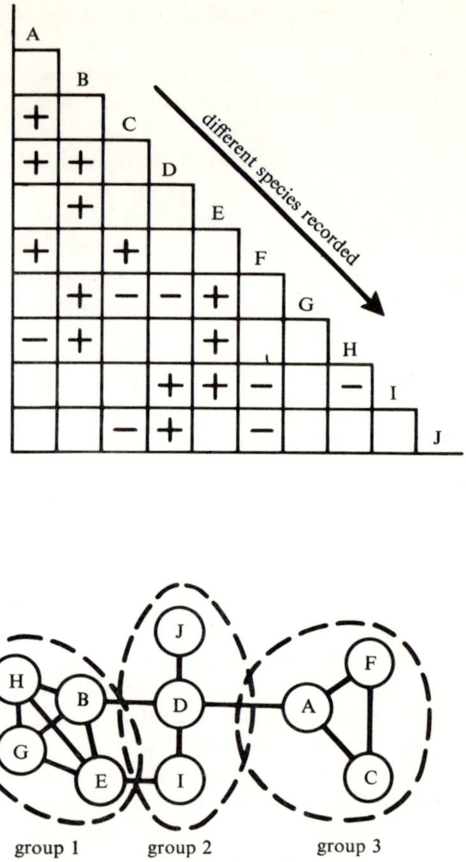

Fig. 4.17 Chi-square and constellation methods of determining plant associations

4C Ecology

4C 1 Trophic levels and efficiency

As world demands for energy increase exponentially, it becomes more important to explore ways of using the energy potentials of various natural ecosystems more efficiently. Only small percentages of solar energy are used by the ecosystems upon which we depend for our food, and so to waste this potential upon widely-spaced crops and many domestic farm animals which dissipate heat energy is clearly inefficient. One of the techniques used to measure the relationships between solar energy receipts and plant productivity is to weigh the amounts of litter produced per unit area per annum in natural forests and woodlands. Bray and Gorham have shown that this production varies with latitude, being to the order of 12 metric tonnes per hectare per annum in equatorial regions, about 4 t/h.p.a. at our latitudes and barely 1 t/h.p.a. in tundra areas. Thus, the influence of climate is considerable, and it is clear that the greatest potential for food production lies within the tropics where regeneration is rapid because the energy content of any *trophic level* in the energy pyramid there is much greater than elsewhere. Obviously, growing crops in heated greenhouses is an attempt to boost production above the natural energy levels for our particular climate. You should examine the biomass of a plant community grown in a greenhouse and various local climates out-of-doors, paying special attention to the dry weight of litter produced. Also, you should reflect upon the ecological advantages of intensive animal rearing like broiler chickens, even if you disagree with the ethics of such farming, or alternatively the wastefulness of grazing European cattle on tropical grasslands.

Since the regeneration of plant communities represents an indirect measure of trophic levels and efficiency, another technique of analysing local differences in ecosystems is to record the rates at which plants and animals recolonise cleared plots. As it takes a long time to get satisfactory results, however, you can often carry out useful studies related to your discoveries in historical land use changes in a district. The results of the following survey undertaken on Codden Hill near Barnstaple, North Devon, give you some idea of the scope for such links between physical and human geography.

Fig. 4.18 shows the basic land use changes on this isolated hill between 1843 and 1967. Using quadrat techniques, each land use

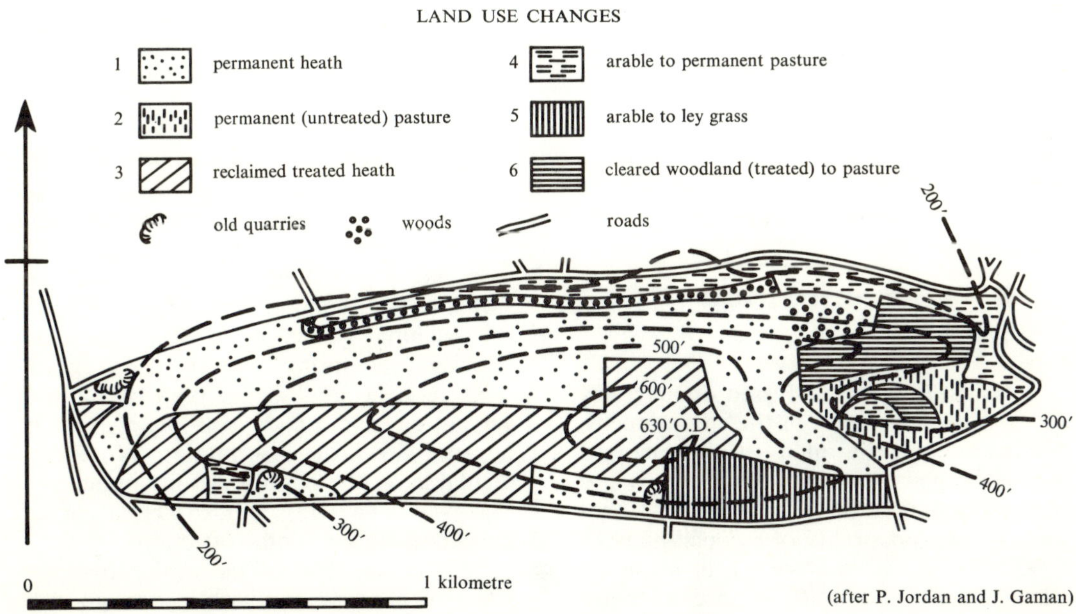

Fig. 4.18 Major land use changes 1843–1967 on Codden Hill, North Devon

zone was studied for the occurrence of nine 'indicator' species. Table 4.7 shows the percentage frequency of these plants in each zone sampled:

Table 4.7

Indicator Species	Percentage Frequency in					
	1	2	3	4	5	6
Graminae	88	100	100	100	100	100
Rubus	7	80	0	8	0	55
Trifolium	2	52	89	80	95	73
Erica	24	0	0	0	0	0
Vaccinium	47	0	0	0	0	0
Calluna	19	4	0	0	0	9
Bryophyta	77	67	78	64	5	41
Ulex	55	0	13	0	0	59
Pteridum	67	28	0	9	0	68
Soil pH	5·5	5·0	6·0	5·6	5·7	8·0

Although this survey has certain limitations, it shows quite clearly the *higher species numbers* to be found on unfarmed or neglected land. Thus, whilst certain species are unaffected or even favoured by particular types of land use, it is possible to discern the degree to which reversion to heathland is occurring. You will have seen distinct changes in vegetation along man-made boundaries, and it is a good test to plot the encroachment of natural vegetation across fence lines into reclaimed fields. Farmers can usually give you very valuable information as to how rapidly such processes occur in different fields. Can such evidence be related to local climates and soils?

4C 2 Environmental balance

A great deal is now discussed about interfering with the *balance* of the environment. We have found that many students regard man as being on one side of the scales and the environment on the other and that there is little possibility of reconciling both sides. The ecosystems approach, however, shows that this is not so since all components are linked within the biosphere *as a whole*. Thus, the concept of balance is different in scientific usage than in everyday language where one is normally referring to two things only. This distinction is important because it shows that our use of the term is perhaps better defined as being a particular **state of equilibrium.** Most ecosystems are in a state of *dynamic* equilibrium as discussed at the beginning of this chapter, which means that they have a constantly changing system condition. If man modifies one part of a link in the system he can expect the balance to be restored by a response in another part. It so happens, of course, that these compensating responses are not always to our liking; a situation which appears particularly true in many aspects of medical geography. Therefore, by understanding the complexity of ecosystems (if this is actually possible) it is hoped that *prediction or projection* of likely responses elsewhere will be facilitated.

There is now a growing body of evidence to show that, by growing diverse crops and *not* completely clearing the weeds, productivity and yields actually increase. If one then considers the costs that would be incurred in clean-weeding, the cash benefits are even greater. Polycultural systems have more to commend them than monocultural ones in ecological and probably economic terms because they can be organised to *resemble* or even *replace* the natural plant and soil communities with few ill-effects and many benefits. This is what we mean when we talk of environmental balance. If you study the events whereby many great monocultural systems have come to grief, you will appreciate the points we are trying to make. You will also see why plant and soil studies at community levels have more far-reaching claims upon your work in physical geography than is usual when ranging over a global survey of natural vegetation and major soil regions. It is because so little information is readily available on the plant and soil community that we have stressed the techniques of measurement and data collection.

This is one field of physical geography where you are virtually obliged to go outside and collect the information required for yourself.

5 Catchment Hydrology

Until recently hydrology has been but a small aspect of hydraulic engineering; ever since man first wanted to control and divert rivers for irrigation or power production, hydraulic engineers have been studying the mechanics of water flow and its relationships with man-made structures. However, water has become an increasingly valuable resource and *hydrology*, the study of the occurrence, circulation and environmental relationships of water, has grown as a result. Some data presented by Valentine show just how large the amounts of water used by modern society have become:

to produce 1 tonne of steel	225000 litres
to produce 1 loaf of bread	1700 litres
to produce 1 hen's egg	700 litres
to produce 1 restaurant lunch	8 litres

A useful class or school project would be the quantification of one's individual or corporate water use on these lines.

The average annual *rainfall* over Britain is 904 mm. However, such an average figure is virtually meaningless in view of the great geographical variation of rainfall.

CLASSWORK

What is the approximate area of Britain? Multiply this figure (in square kilometres) by the average annual depth of rainfall. Be careful about the units in which your answer is expressed. Is this water supply enough? If half the rainfall evaporates is it still enough? Refer to the wilting coefficients discussed in the previous chapter.

Now look at the distribution of annual average rainfall in Britain on an atlas map. From where will the water supplier take his largest reserves? Where is the biggest area of need? How can man store water?

Similarly, great is the variation exhibited by the *losses* from rainfall before it becomes *streamflow*. Such losses are mainly by evaporation and transpiration, from open water (or bare soil) and plants respectively; their combined effect is called *evapotranspiration*. These spatial variations are both the reason for the geographer's interest in hydrology and the drive behind the hydraulic engineer's inventiveness in distributing water from areas of plenty (high rainfall and/or low losses) to areas of need. In the case of Britain the solution to the supply and demand problem is to feed water from the highlands of Wales, the Lake District and Scotland to the populous and relatively dry areas of central and eastern England.

The hydrologist is not only interested in spatial variability but also in changes with time. Most temperate areas exhibit seasons of greater and lesser rainfall or losses. Some witness, for short periods, the extreme conditions of *flood* and *drought*. Thus, the measurements made in hydrology pay particular attention to time and often need to be continuous to detect the natural variability inherent in the system. Regular measurements allow the hydrologist to discover the *processes* which effect the behaviour of rainfall, losses and the flow of rivers and, once enough data has been collected, to *predict* the future behaviour of the hydrological system under study. Thus, most nations have official organisations to collect such data. In Britain large-scale hydrological measurement or **hydrometry** was only started in 1963 with the Water Resources Act, although, of course, the Meterological office has organised rainfall recordings for many decades. Since the Act, hydrometry has been the prerogative of the River Authorities, organised around the major river basins of England and Wales, and the Purification Boards in Scotland. These have recently been grouped into larger scale regional water authorities to match similar changes in Local Government.

5A Hydrological models

The basic framework, or model, for the study of hydrology is the **hydrological cycle** (fig. 5.1). The source of rainfall over continents is the moisture evaporated from land, oceans and seas. After precipitation and partial loss by evaporation it returns to the oceans by a variety of flow routes; over hillslopes, in streams, or through porous and permeable rocks. The latter route is followed by that part of the total rainfall which *infiltrates*. If it passes through the soil and enters the rocks below it becomes *ground water*. Evaporation from open water or bare soil and transpiration via the cells of plants is a short circuit to the whole cycle. The scale of system represented by the cycle is global. Because the same mass of water is conserved (there are no true 'losses') the system is a closed one in this respect but there are, of course, energy overlaps with other global systems.

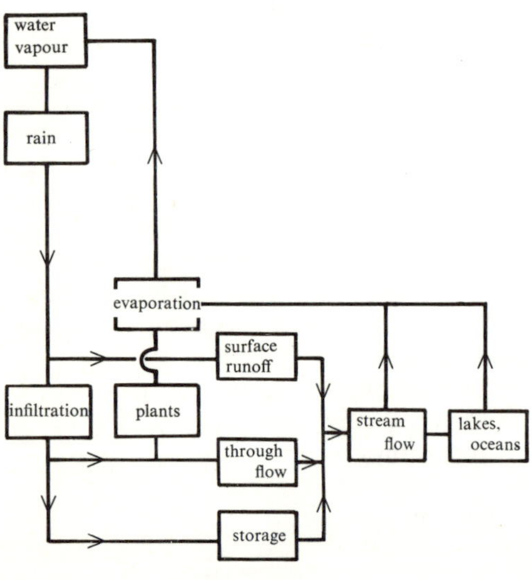

Fig. 5.1 Flow diagram of the hydrological cycle

5A 1 The water balance

Although the hydrological cycle is the framework for hydrological measurement and its theoretical guide, the scale of study is not yet the global one which it implies. Rather it takes as its unit the **drainage basin** or river **catchment**; the former term is the internationally accepted version of the latter which, however, is preferred in Britain. It treats the drainage basin as a system quantifying input (rainfall) and output, (streamflow and losses) over a variety of time scales. The basic time scale is that of the **water year** and during such a period a **water balance** is established according to the equation:

$$Q = P - E \pm \Delta S$$

Q	P	E	ΔS
(flow in stream)	(precipitation)	(losses by evapo-trans-piration)	(changes in storage)

For example,

$$187\,\text{mm} = 645\,\text{mm} - 458\,\text{mm} \pm 0$$

(as established near Hull, Yorkshire, by Pegg and Ward (1971))

The term **storage** is the key to the use of a water year stretching from October 1 to September 30. Obviously, when rainfall infiltrates the soil and rocks of the drainage basin it may take some time to reach the stream as ground water. Thus, while being measured as part of the input rainfall it takes some months to register in the output streamflow measurements. On a large scale the variability of this stored volume would upset the water balance equation. However, by beginning the water year in October, the period of maximum input (winter) occupies the first half of the year, allowing spring, summer and autumn of the next year for all output to occur (fig. 5.2) and leaving ΔS equal to zero. Of course a drainage basin chosen for the quantification of a water balance equation must be clearly defined and not 'leak': no underground

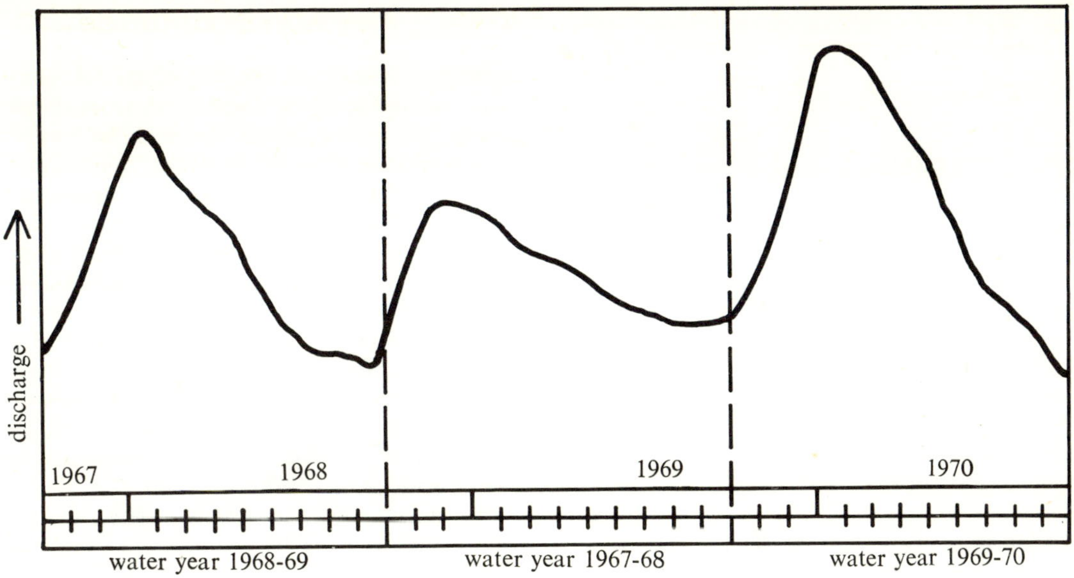

Fig. 5.2 Calendar years and water years, showing how the latter divide the annual hydrographs of river flow at base-flow when annual storage is at a minimum

flows must leave the basin without measurement. Basins on rocks such as chalk often leak in large quantities.

Rainfall not subject to loss or storage becomes streamflow and is called *runoff*. For any water year the amount of runoff may be expressed either as that depth of rainfall which actually reaches rivers, or as a percentage of rainfall. Fig. 5.3 shows that, for major river basins in Britain, the amount of runoff as a percentage of rainfall may vary from 10 to 90 per cent. The location of the basins on the map and figures for their runoff will suggest some of the factors behind the differences. The processes by which rainfall runs off to streams will be dealt with later. For the moment we can make a geographical conclusion from fig. 5.3 that it is rivers with steep slopes, thin soils and a dense stream network which produce most runoff—in other words those in western Britain. These *static* variables of drainage basins which affect the water balance are dealt with more fully under *drainage basin morphometry*. They do not alter over the period of a storm or between storms.

Fig. 5.3 Mean annual runoff as a percentage of mean annual rainfall. What static factors may explain the variations?

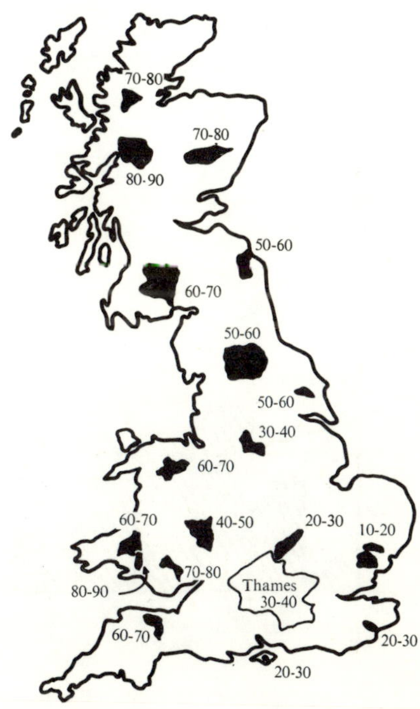

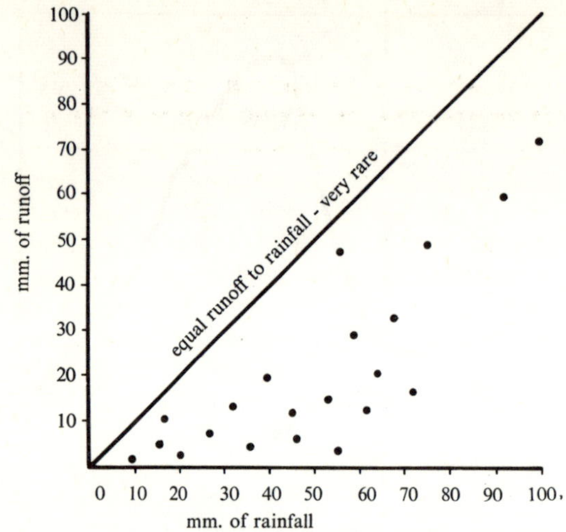

Fig. 5.4 The varying relationship of rainfall to run-off for individual storms on a single catchment. What are the dynamic catchment variables involved?

5A 2 The storm hydrograph

The percentage of runoff does not only vary spatially across Britain. It varies with each storm (see fig. 5.4) on a particular basin. Thus, while 40 per cent runoff may be the annual figure for a river in the Midlands, during the year there may have been storms which produced 80 per cent runoff, or 20 per cent runoff. Study of such variations is achieved by taking individual **hydrographs** from storms over the basin and performing on a time scale often as short as one day, the basic calculations of the water balance approach. The volume of rainfall can be obtained from records for the storm in question. The volume of streamflow is not the total below the graph drawn by the flow recorder. This includes a natural background flow which represents the slow

Fig. 5.5 The hydrograph of storm runoff. Calculations of storm runoff volumes are made easier by summing the average discharges during each hour. For areal rainfall calculations see page 119

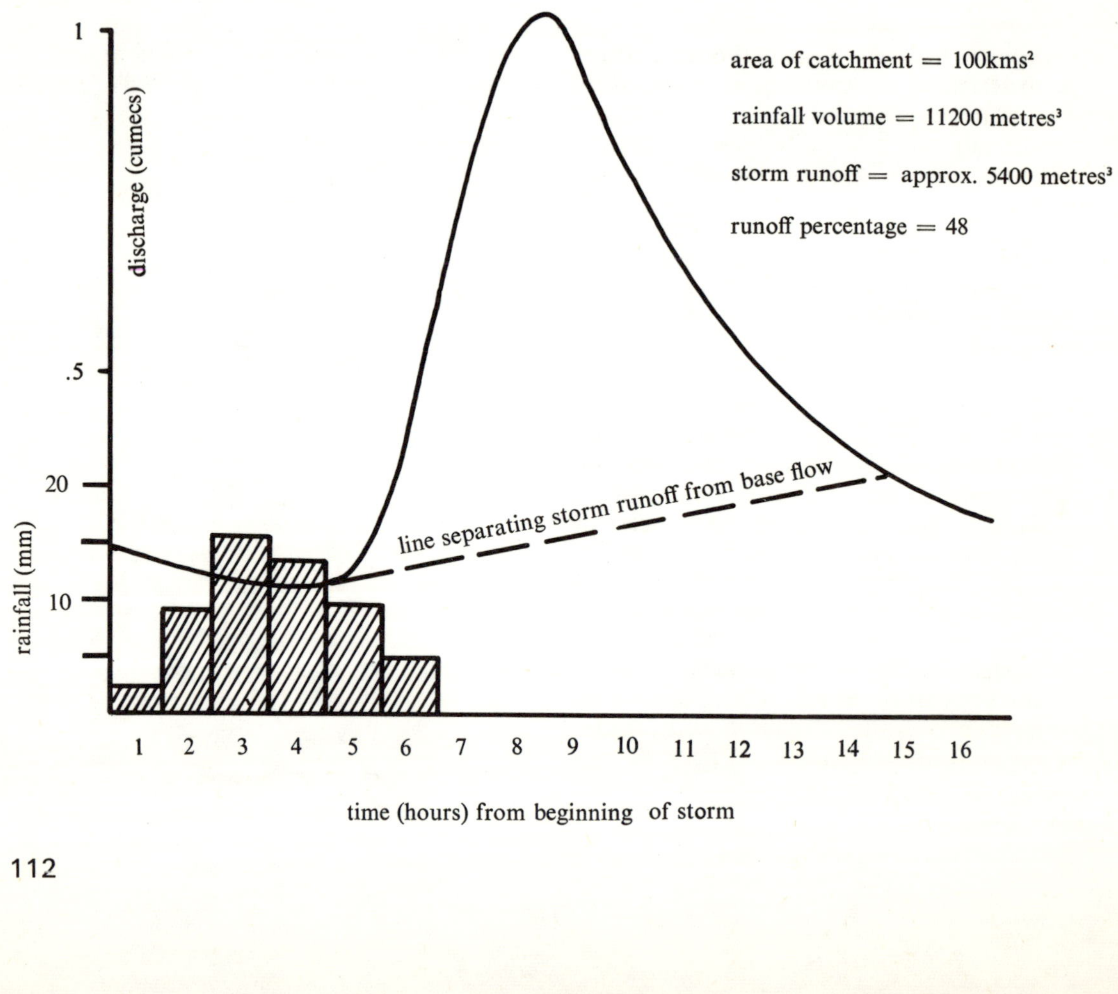

release from storage of previous rainfall.

The *runoff hydrograph* is therefore separated according to a variety of arbitrary methods from the background base flow. The volumes of rainfall and runoff are then compared (fig. 5.5) and the runoff percentage obtained. The factors governing the amount of runoff on this short time scale include those static variables which affect the water balance and, in addition, certain *dynamic* ones, so-called because they change from storm to storm. Dynamic factors to be considered are:

a) *Total rainfall* Other things being equal, the storm with the larger total input will cause a larger streamflow response. It is the other dynamic variables below which result in this relationship becoming non-linear.

b) *The intensity of rainfall* If 5 mm of rainfall falls in one hour it produces more runoff, other factors being constant, than if a similar amount fell in five hours. This is basically because infiltration of rainfall into the soil is quite slow and, if further rain falls quickly, it runs off.

c) *The antecedent rainfall* One of the most impermeable surfaces on which rain can fall is one already saturated with water. If, therefore, a storm occurs after some weeks of wet weather (high antecedent rainfall), more runoff is likely to result than after a drought. In the latter case we may think in terms of the basin's storage capacity (in the soil and rock) being refilled before runoff will occur.

d) *Changes in loss rates* The losses through evaporation and transpiration to which rainfall is subject are not constant throughout the year. Clearly, atmospheric temperature, windspeed and other factors discussed in chapter 2 (which affect evaporation from bare surfaces) vary. So too does the plant cover on the catchment. Botanists may be able to inform you of the area of leaves available as transpiring surfaces on a deciduous tree in summer. These are, of course, absent in winter.

FIELDWORK

How does the amount and intensity of rainfall affect the behaviour of streams in your area? As well as streams, investigate the behaviour of gardens and parks; what happens when heavy rain falls on already-sodden ground? The behaviour of gutters and drains in streets is also helpful. Which is worse, a thunderstorm in summer or slow, steady drizzle in winter? Which leads to most flooding? How quickly do puddles disappear?

5B How runoff occurs

The above list of dynamic variables which affect runoff give us some insight into the processes which cause runoff to streams, a topic which has recently been investigated with renewed vigour by hydrologists. Only brief mention is possible here but the topic crops up again in the section about slope development, since water flow is a major item to consider in the geomorphology of slopes. The present investigations of runoff on slopes has been prompted by the discovery that *surface runoff* from slopes occurs infrequently and from only a part of the total drainage basin. Surface runoff was an integral feature of the work of Horton on infiltration (see Dury, ed. 1970). Once the capacity of the soil to absorb rainfall (*infiltration capacity*) was reduced to zero by prolonged rainfall, drops of water would begin to collect on the surface of slopes and thereafter descend to the nearest stream, eroding surface debris. This model may well apply to the clay badlands from which Horton drew examples but on more permeable slopes, particularly those with vegetation and consequently a layer of leaf litter, the soil has been proved to transmit water laterally down the slope by *subsurface* routes. Thus, it is only in very severe storms or in surface concavities—(where the soil is saturated much of the time)—that infiltration capacity is exceeded and surface runoff occurs. The runoff produced via these subsurface routes is described as *throughflow* (Kirkby, 1969). The relationship with the surface runoff model is shown in fig. 5.6.

Unfortunately there is little alternative to walking the slopes and channel banks of a small drainage basin during a storm if an observational study of runoff is required.

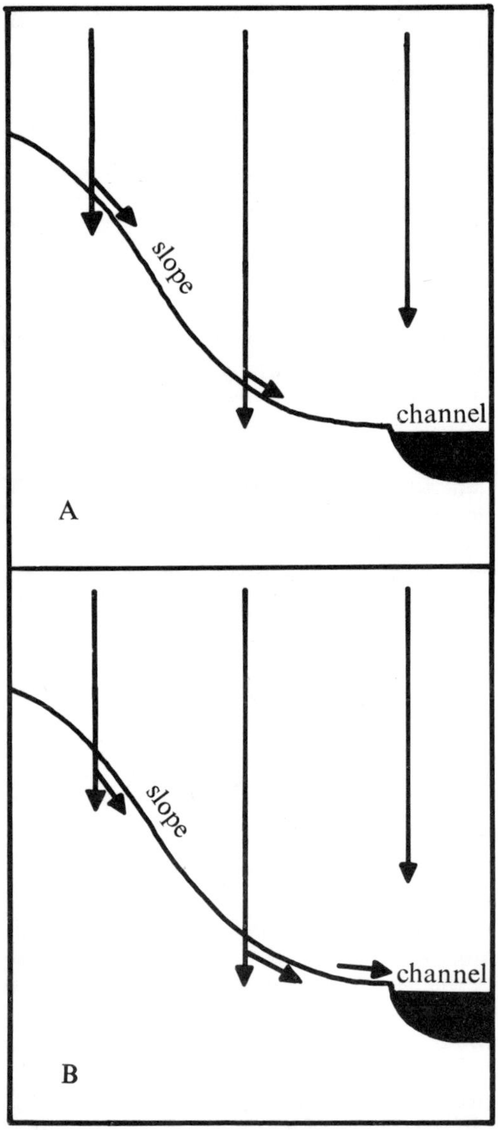

Fig. 5.6 The routes by which runoff occurs, according to the surface runoff model (A) and the throughflow model (B). Notice that some surface runoff is presumed in satured areas near the channel in B

It is fairly certain that few glimpses of surface runoff will be obtained although small ditches and gullies which are normally dry may be running with water. If possible, soil samples and measurements of the stream level should be taken and analysed using the techniques described below. Try to detect rising water level against a marker in the stream and observe where runoff has its origins. Throughflow may be detected in the channel banks, oozing from the soil, although its main contribution may not occur until after the peak level of streamflow. It is worth looking for natural *pipes* which have been described in a variety of soils (plate 5.1). These carry considerable volumes of rainfall which has infiltrated a permeable surface only to meet an impermeable one lower down the soil profile. We may call this *saturated throughflow* and it responds quickly to rainfall. If a rainfall record is available for the storm in question, calculate how much of the rain has fallen on the stream surface itself. Since no delay in runoff is caused by this channel precipitation it may be important in contributing to an early flood peak.

The various routes by which runoff reaches a stream are shown in fig. 5.7 according to the time they take to become represented in the streamflow hydrograph. One of the remarkable features of saturated throughflow in pipes is that, despite travel through the soil, it can be relatively rapid.

To follow up the concept of delays in runoff varying with the route taken to the stream, a laboratory basin model may prove helpful. The drainage basin is considered as consisting of a series of reservoirs; these may be constructed from glass, polythene or tin. Each has a different capacity and represents storage of rainfall on the surface, in the soil and in the rocks below. Each can overflow by two routes, one at a fairly low level into the reservoir below and one near the top into the stream (see fig. 5.8). It will be noticed that a reservoir is included to represent *channel storage*. This occurs in nature because of the irregularities and roughness of the stream channel profile. Water added, to simulate rainfall, at the top of the model

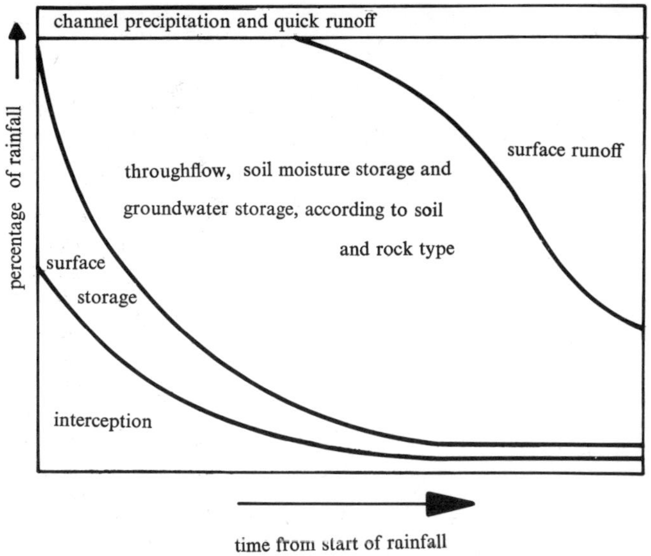

Fig. 5.7 The components of the rainfall-runoff process as they vary in proportion during a storm. The diagram is purely conceptual and is not drawn to a rigid scale

Plate 5.1 Runoff reaching a channel bank by subsurface route. The small natural pipe near the penknife is discharging all the water seen at this point—none is moving overland through the grass

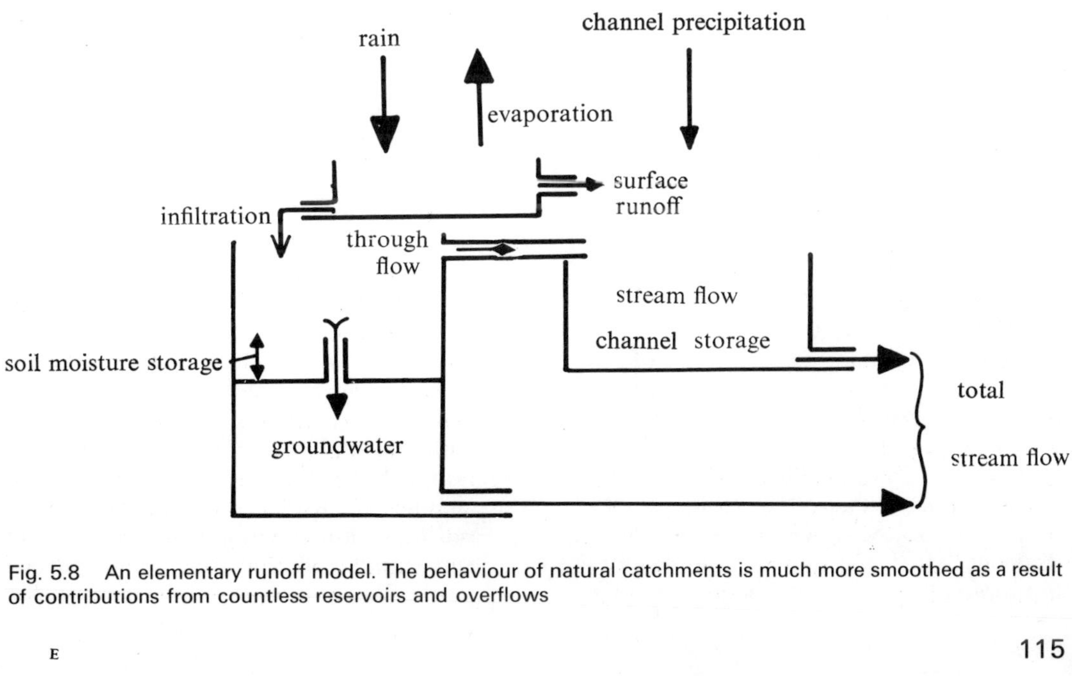

Fig. 5.8 An elementary runoff model. The behaviour of natural catchments is much more smoothed as a result of contributions from countless reservoirs and overflows

results in a response by the 'catchment' at delayed intervals. Of course the natural processes are much smoother.

Some examples of laboratory experiments on soil type and moisture movement are given in chapter 4. Other runoff experiments can be performed more domestically. Various textures of bath sponges and flannels may be used on a sloping surface such as a draining board to simulate soil or vegetation and its effect on water supplied from above. Scale model catchments have been constructed in the laboratory by researchers investigating the effects of permeability and rainfall intensity.

5C Measurement and Data Collection

5C 1 Drainage basin morphometry

Morphometry includes the quantitative expression of the dimensions of all types of landforms. Comparisons between groups of features can be made using morphometry and interpretations of their genesis or age often result. We have already seen that the runoff characteristics of drainage basins are related to certain static variables; these are measurable from maps of these basins because they do not change rapidly. If such relationships can be established mathematically they may be used to predict the runoff from catchments in which the expensive and time-consuming business of hydrometry cannot be pursued—in other words the majority of catchments! Thus, drainage basin morphometry has an applied value to hydrologists.

There are numerous variables which characterise drainage basins. Those of most value are those easily derived from maps which are readily available and those which should intuitively have a large bearing on runoff. Thus, while the parameters of the hypsometric curve (see Chorley, 1969) may describe the slope properties of the drainage basin, the applied hydrologist is far more likely to use a simple gradient measurement. In the end, of course, a morphometric variable is only as good as its value to predict runoff and so experiments must proceed using a great variety of such variables in an effort to select the best. Considering the processes of runoff however, the following appear, logically, to be most important:

a) *Area* The importance of a well-defined watershed in delimiting the drainage basin has been stressed. This watershed defines the area from which the total output from the basin will be drawn and it is not surprising that the runoff characteristics, particularly maximum flows, of a basin will be related to area. Since area is an extremely variable property of basins, the scale at which it is measured is not critical; differences of a fraction of a square kilometre make little difference to the output of two adjacent basins unless they are very small. Another way of expressing the general magnitude of the drainage basin is the **order** of the biggest stream. Fig. 6.1 shows how the smallest streams are called first order (1) and when they join they form a second order (2) and so on. Where a (1) joins a (2) they do not form a (3). A (3) is formed by two second orders.

b) *Geology and soils* If the area of a basin determines the broad volumetric properties of rainfall input and streamflow output the amount of storage or delay in runoff is likely to vary with the type of surface upon which the rain falls. Unfortunately there are no widely-available maps of geology or soils which are immediately capable of hydrological interpretation but recently a map of soil infiltration capacities has been produced and the values given have been found to be successful in predicting flood flows.

c) *Gradient of slopes* Obviously the amount of water which is delayed by storage en route to the stream will vary not only with soil and rock type but with slope, both of the hillsides and the stream channels. Consider rain falling upon a red-brick house with a slate roof: rain runs rapidly off the steeply-pitched impermeable roof, less rapidly off the less steeply pitched impermeable guttering and not at all off the vertical but highly porous brick wall. Obviously both the type of surface and its slope are important but, in view of the scarce data on the hydrological properties of soils and rocks, slope becomes highly important. Because of the strong relationship between valley-side slopes and channel slopes it is possible to measure the gradient of the mainstream alone to quantify slopes in the basin as a whole. An even coarser slope measure, involving the height of the watershed and the altitude of the streamflow measuring station (both obtainable from the *Surface Water Yearbook*), is shown in fig. 5.9b to bear quite a strong relationship to maximum annual flows.

d) *The stream network* Though rain may fall upon an impermeable, steep slope its velocity down the slope, by throughflow or surface runoff, will not be as great as in the stream channel. Thus, the distance involved in reaching a stream will be an important variable in runoff studies. Looking at this another way, the degree to which streams thread their way into all parts of a basin will be an important variable to quantify. This is usually done by measuring *drainage density*, or the length of streams per unit

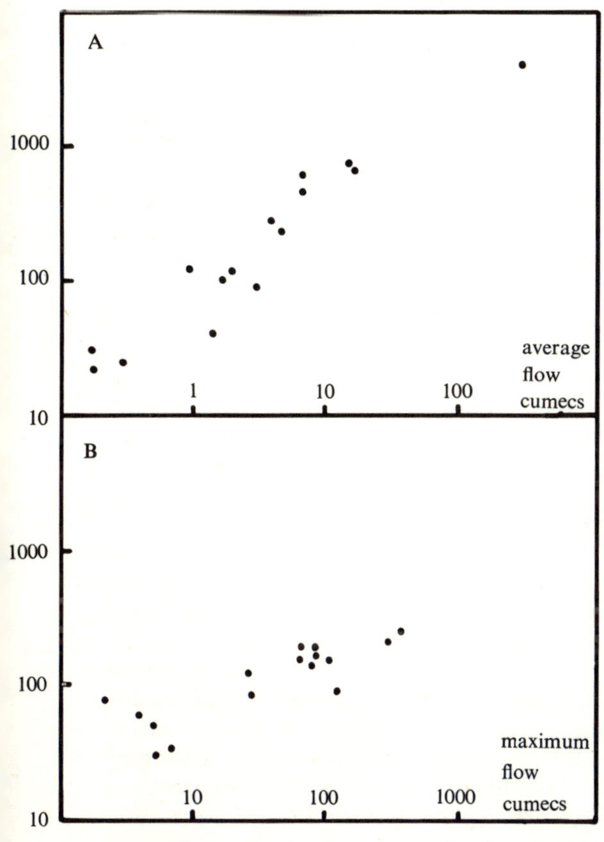

Fig. 5.9 Relationships between (a) average flow and basin area and (b) maximum flows and simple gradient derived by random sampling of stations listed in the *Surface Water Yearbook*

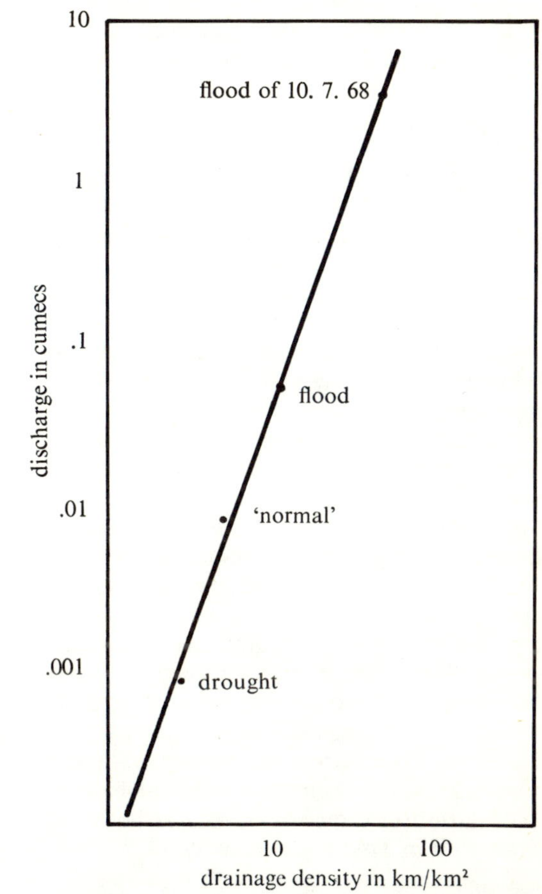

Fig. 5.10 Field measurements of drainage density and stream discharge, Swildons Stream, Somerset. The graph reveals how the stream network extends during flood in proportion to streamflow. Is there an analogy with roads and traffic over the years?

area of the basin (total stream length, divided by area). An easier measurement is that of **stream frequency** (the total number of stream segments, or junctions, divided by area). Whichever is chosen it is important to measure stream network properties from a map which reliably depicts the distribution and length of streams as closely as possible to conditions in the field. A quick investigation of the network displayed on the Ordnance Survey 1:25000 map for any upland basin and that shown for the same basin on the 1:63360 map will show the underestimations of drainage density and stream frequency liable from using the latter. The measurement of drainage density from field mapping, if possible in both flood and drought, should be undertaken as a project and compared with published maps at various scales. Comparisons may also be made with aerial photographs from which some more recent maps are made. Fig. 5.10 shows the relationships between streamflow and drainage density discovered by *field* mapping during very different weather conditions.

Clearly there are large areas of overlap between the elements of drainage basin morphometry. Large basins tend to have more gentle slopes; steep basins often have high stream frequency. It is important to sort out these inter-relationships statistically, firstly to inform ourselves of certain laws governing the dimensions of drainage basins and, secondly, because the regression relationships used to predict runoff from morphometry require that the independent variables (the morphometric ones) be unrelated. We can measure the degree to which morphometric measurements are linked by correlating the results obtained for each variable in a series of basins. Calculation of the correlation coefficient is dealt with in Toyne and Newby.

CLASSWORK

1 Using data from the *Surface Water Yearbook* (H.M.S.O.) it is possible to plot average yearly flows against area (fig. 5.9a). What is the approximate correlation coefficient by ranking? Such relationships have been used by engineers to predict flows from other basins. Check your findings by predicting the average flow of another basin in the Yearbook without looking up anything except its area.

2 If each member of the class or group selects a small drainage basin (covering, say, one sheet or less of the 1:25000 map) a lot of morphometry data can be quickly obtained. What is the area of each basin? Measure its stream frequency or drainage density and obtain some rough gradient figure for the main stream. Using everybody's data, what is the correlation between area and total stream length or between gradient and stream frequency? Choose different physiographic regions if possible.

There is clearly a case in our analyses of morphometry and streamflow for dropping those variables which are highly correlated with basic measurements (such as main-stream length, highly related to area) from further analysis. The remaining variables are then used in a multiple regression analysis with the desired statistic of streamflow. Rodda (1969 p. 412) gives several examples of equations used to predict flood discharges.

One of the characteristics of drainage basins coming under scrutiny from hydrologists in the recent past has been the amount of *urban development* in the basin. We have already mentioned gutters and drains in terms which reveal our consideration that they are definitely part of the hydrological system. Urban hydrology has concentrated particularly on assessing the effect of large impervious surfaces such as roads, airfields and car parks on the rapidity of runoff. Virtually no infiltration can occur in such areas. If drains and sewers are well designed, rainfall is quickly conveyed through the system after it falls. If the drains and sewers empty into natural streams they make the latter rise more rapidly and to a higher peak than in non-urban areas. Obviously one town can do its neighbour downstream some harm in this way, particularly if the natural part of the basin also responds quickly. If not, it is possible to separate the 'urban flood' from the natural flood in the streamflow hydrograph.

At the opposite end of the scale, well-established forests with a deep, leaf litter delay runoff to some extent and make a much smoother hydrograph. Clearly, land-use is an important hydrological variable and gives us another example of man's great responsibility for management of natural systems.

5C 2 Rainfall inputs

It is perhaps surprising that we are still using varieties of a tin-can to measure rainfall! Design and siting of rain gauges vary from country to country, yet how often have we compared global figures for rainfall in regional geography! The chief reason for concern about this situation is that, although weather forecasting and recording can tolerate rainfall inaccuracies, the hydrologist who is quantifying the water balance or examining the volume of storm runoff requires extreme accuracy in the volume of rainfall input to the drainage basin. There are two main disadvantages to the hydrologist in the present state of rainfall recording:

a) The projection of the raingauge above the ground surface introduces a general underestimation of rainfall because turbulent air currents around such a projection divert the stream of falling droplets. Rodda (1969, p. 130) has shown that a ground-level gauge, protected from extra raindrops splashing into it by a grid-like frame, collects a far more accurate amount of rainfall. Ward (1970, p. 395) has shown how simple experiments can prove the effects of turbulence on rainfall.

b) Most hydrological studies require the calculation of the precipitation over the total *area* of the drainage basin. This has to be done from data collected at a few *point* locations. Fig. 5.11 shows the two

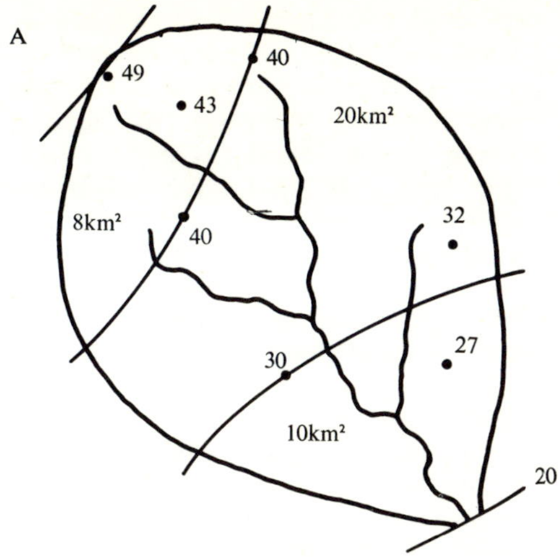

i point figures guide plotting of isohyets

ii areas between isohyets calculated

iii weighting of mid-isohyet figures by areas

iv rain $= (\frac{8}{38} \times 45) + (\frac{20}{38} \times 35) + (\frac{10}{38} \times 25)$ mm

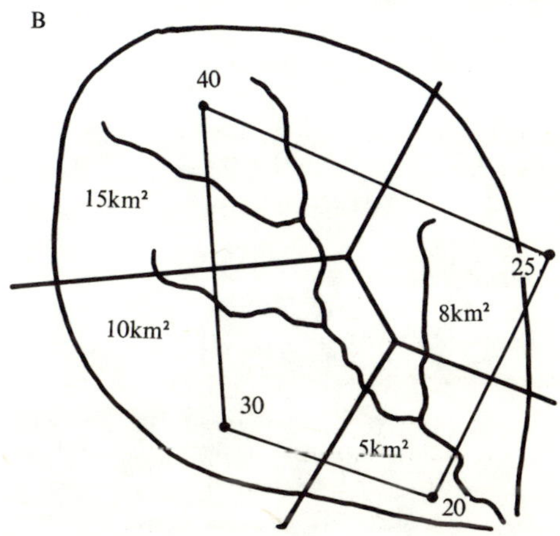

i stations linked by lines

ii lines bisected by perpendiculars

iii perpendiculars joined and areas within each station polygon calculated

iv rain $= (\frac{15}{38} \times 40) + (\frac{10}{38} \times 30) + (\frac{8}{38} \times 25) + (\frac{5}{38} \times 20)$ mm

Fig. 5.11 Calculation of areal rainfall from point values by (A) Isohyets, (B) Theissen Polygons

main methods of calculating a real figure for rainfall, Theissen Polygons and isohyets. A further method involves the calculation for each gauge of the proportion of the **annual average** for that gauge represented by the rainfall for the storm under consideration. For example, a gauge may record 10 mm during a storm and have an annual average of 1000 mm. Obviously the storm percentage is 1·0. The average of these percentages is then used to multiply the basin average annual rainfall (obtainable from the *Surface Water Yearbook*; gauge annual averages are available from the Meteorological Office) to give the basin input for the storm.

Two types of raingauges are used in hydrology—the **daily gauge,** of which there are over 6500 in Britain and the scarcer **autographic** or **recording raingauge** (plate 5.2). The autographic is so-called because it graphs each increment of rainfall on a paper chart. The charts last for various periods, the most common being one day. Thus, at 09·00 hours there is no need to measure what has collected, as with the daily gauge (plate 5.3). The basic aim is to allow calculation, from the charts, of hourly rainfall amounts throughout the day. One reason for this is to allow calculation of **rainfall intensities** which are expressed in millimetres per hour. Another is to make possible the collection of data on the precise time interval in hours between a rain storm and the resultant increase in river flow. This interval may be much less than one day and so daily rainfall totals are of little use.

The autographic gauge has its chart changed at 09.00 hours G.M.T. (and B.S.T. in the summer) each morning to standardise with the daily recordings. A standard daily

Plate 5.2 shows the siphon and chart mechanism of an autographic raingauge, the collecting funnel having been hinged open. Normally the base of the recorder is buried. 5.3 shows the daily raingauge with its funnel removed and the day's catch being measured manually

gauge is usually sited next to the autographic gauge. Thus it tends to record slightly less rainfall. The turbulence is caused by the larger height and girth of the autographic gauge (necessary to house the clock and chart). The total on the chart, when compared with the daily total also reveals whether the complex pen motions have been performed correctly. While daily raingauges may be purchased, or imitated by a school or college (Gregory and Walling, 1971, give useful advice on this) the autographic type is expensive and records of hourly rainfall for a local area are probably best borrowed from the River Authority, local council, or from the Meteorological Office archives. The Meteorological Office's publication *British Rainfall* is an extremely useful source of data on rainfall. It has been published for many years; the latest volume is that for 1964. A supplement for the years 1961–5 is available. While daily gauges are suitable for water balance calculations an autographic record is essential for short-period studies. The way in which an autographic rainfall chart is analysed to hourly rainfall is shown in fig. 5.12. Once derived, the hourly totals may be plotted as a bar graph or a cumulative curve.

FIELDWORK

Since most of the class will find it convenient to measure rainfall near their homes, use simple funnel gauges (of the type described by Ward or Gregory and Walling) to assess the rainfall variation in just one or two storms over the neighbourhood. Try to ensure standard height and exposure to wind. Plot isohyets through the mapped data. Do these isohyets give reasonable estimations of rainfall in areas where rain was measured but the result was not mapped? Does rainfall vary with altitude and site exposure?

Before moving on to the measurement of streamflow it is as well to know a little of the behaviour of rainfall. The broad pattern of annual average rainfall over the British Isles will be well known to those who have studied physical geography at lower levels. The reasons for the distribution in terms of westerly airstreams, depressions and orography will therefore be understood. It is seldom, however, that the hydrologist would require such information. His interest is in the storm, that period of heavy rainfall likely to produce a runoff 'event'. Many of you will realise that the distribution of storms (especially thunderstorms) is the opposite to that of annual average rainfall, being concentrated in southeast areas of Britain where convection is higher in summer. The rainstorm tends to be limited in time and space. Generally speaking, the more intense a storm, the shorter time it lasts (perhaps the reason for the popular impression that a burst of heavy rain is a

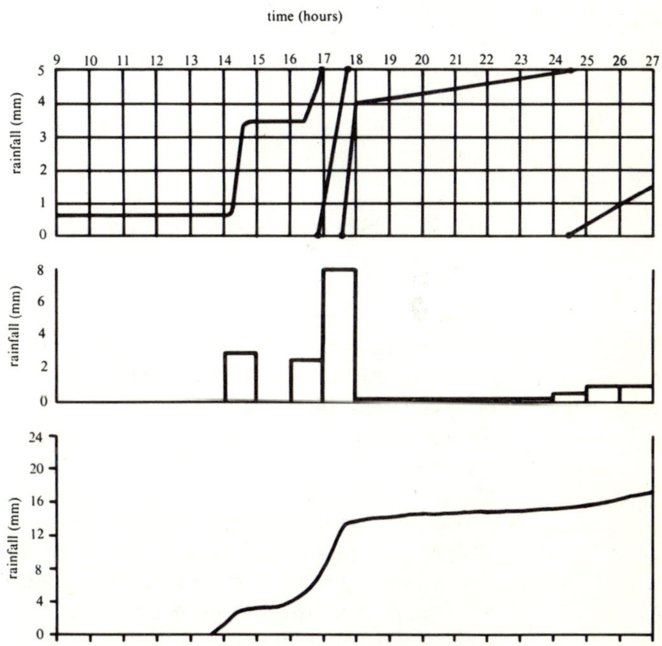

Fig. 5.12 A. Extract from an autographic rainfall chart, showing only 18 of the 24 hours. We have used dots to mark the points at which the siphon operates, returning the pen to the base of the chart
B. Bar graph of rainfall compiled from (A)
C. Cumulative curve compiled from (A)

'clearing-up shower'). This **intensity/duration** relationship is quantifiable after the fashion

$$I = \frac{a}{t + b}$$

where I is rainfall intensity, t is its duration and a and b are fixed locally by experiment.

Similarly the **depth/area** relationship shows that storm centres of heavy rain are fairly local and reduction factors can often be applied to the rainfall from one storm accruing to an area of, say, 100 square kilometres as opposed to one of only 10 square kilometres. Heavy rainfall rarely falls over large areas.

One final measurement problem in dealing with rainfall is that of frozen rain or *snow*. The **water equivalent** (or density) of snow may vary from as little as 500 kg m^{-3} when freshly fallen to 600 kg m^{-3} in old snow. Core samples should be taken in snow and melted if correct interpretations of total precipitation are to be made. In areas of prolonged winter snow the water equivalent of the total cover is essential data in the prediction of the snowmelt floods which occur during the spring thaw.

5C 3 Streamflow outputs

The output of runoff from the drainage basin system is best measured on the main channel at the lowest point of the drainage area. It has been emphasised that all runoff, however delayed by storage, from the basin delimited for study should pass the measuring point. Obviously, small upland catchments on impermeable rocks are frequently chosen for research because of the ease of delimiting their watersheds, the lack of natural leakage underground and the general absence of human interference such as mill sluices and water-meadow systems.

One way to measure or gauge the flow of a river is to visually inspect and note the level of the water surface each day with reference to a handy datum such as a riverside tree, or a bridge upright. We refer to the river's level as its **stage** and a brightly-painted *stage pole*, marked out vertically in centimetres and metres and driven hard into the bank will provide a more accurate measure of how the level changes. Though stage recordings can be matched up in a qualitative way with rainfall, quantitative work requires the conversion of stage readings to *flow* or *discharge* figures, expressed in units of volume per unit time. While water suppliers tend to measure flow in millions of gallons or litres per hour or day, hydrologists express flows in cubic feet per second ('cusecs') or using SI units, cubic metres per second (cumecs). In the United States it is often found that, to allow direct comparison with rainfall depth, flow is expressed in the equivalent depth of rain over the catchment represented in one second's volume of water passing the gauging station.

To measure riverflow requires a volume measurement and a time measurement. The volume of water cannot be measured directly in the normal way by filling a container of known capacity—imagine the ridiculous picture of attempting to pass the flow of the Thames into a series of litre buckets! Instead the cross-sectional area of the stream is measured and the length dimension needed to calculate volume is combined with the time measurement in the recording of stream velocity. Discharge may thus be viewed as that 'block' of water, of x square metres cross-sectional area and y metres long, which is passing the measuring position at y metres per second.

To measure cross-sectional area requires only elementary surveying. The result may be simply obtained by multiplying the width of the water surface (not the banks) by the average of three or more depth readings across the stream. The number of depth readings will depend upon the irregularity of the stream bed and the width of the stream.

Velocity measurement requires, at the very least, a stop-watch and a float. The float's progress down a measured section of stream should be timed. It is important that the float should move at the same velocity as the water and something which floats slightly submerged, like an orange, is best. The inaccuracies of the method are mainly

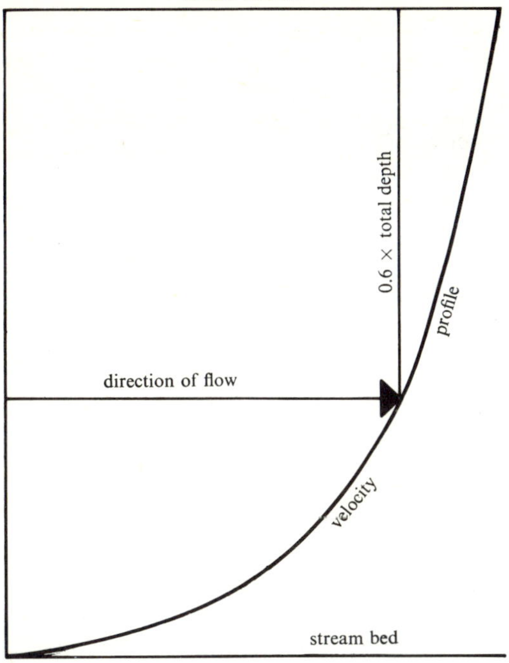

Fig. 5.13 The velocity profile of a typical stream. Average velocity occurs at sixth tenths of the depth. Why does this not correspond to the median velocity?

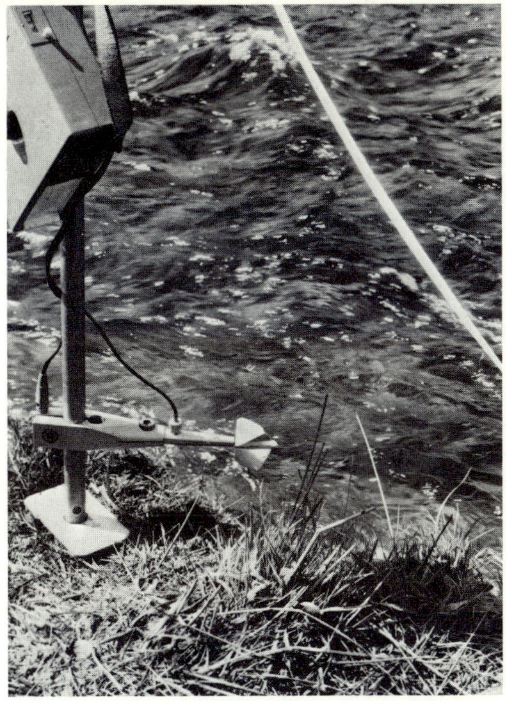

Fig. 5.4 A small current meter. The revolutions of the propeller are recorded by the control box (top left) held by the operator who wades in the stream with the rod. A tape measure is seen stretched across the stream to divide it into the measuring sections. The propeller is fixed at 0·6 of the depth of the stream

due to the variability of velocity down the stream profile (fig. 5.13). The average velocity is required and a surface float does not sample this. The depth at which average velocity is thought to occur is approximately 0·6 of the stream depth (measured from the surface down). The surface layers move more rapidly than this and it is suggested that surface float readings be multiplied by 0·8 to remove this exaggeration. Since velocity also varies across the river (see Dury, 1963, p. 23) a series of recordings should be made, usually along long profiles selected between the sites of depth measurements used to compute cross-sectional area.

To obtain an accurate picture of stream velocity a **current meter** (plate 5.4) is used. It consists of a propeller which revolves at a speed varying the velocity of the water, in which it is immersed at 0·6 depth. The propeller is mounted on the upstream end of a fish-shaped weight connected to a hand-held rod. The reader who enjoys construction and simple electronics may be able to construct one. The person holding the rod stands as far as possible downstream of the propeller so as not to cause turbulence during measurements. A wire connects the meter to a dial on which the number of revolutions are recorded. Another operator uses the dial and stop-watch to calculate velocity at the point in question. On very large rivers the current meter is slung on a cableway which is stretched across the river and from which the meter can be lowered to any chosen depth.

It is also possible to measure velocity with dyes, common salt or by using the ripples produced around obstructions in the stream. However, these are less direct than velocity metering and may, in the case of dyes, pollute water supply.

It is probable that you have always talked of highland streams as being fast-moving, even 'youthful'. Try to prove that the relationships in your mind between small streams and high velocity, or large rivers and 'sluggishness' are mistaken. As in the case of rainfall measurement, the class or group has a great potential for collecting data from a wide area. To all who have enjoyed 'boat races' with floating objects in streams the collection of a lot of simple velocity measurements will prove no problem, especially on holidays. Try to adopt a standard float method and beware of artificial influences on riverflow such as lock-gates. Is there a difference between 'small' and 'large' streams? Is there a greater difference between the velocity at one point on the river in flood and in drought? As Dury (1963) points out, when the river is up to the top of its banks there is hardly any variation of velocity between the small upland tributaries and the large main stream. *Take care* when observing or measuring large streams, especially at high flows.

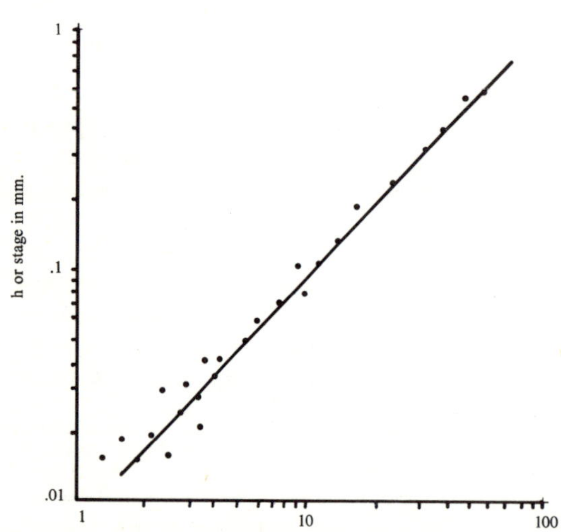

Fig. 5.14 Streamflow rating curve for the River Axe at Cheddar. Notice the scatter of points around the regression line which averages them

The aim of gauging a stream's flow is not to do so on every visit but to establish a **rating curve** composed, over a period, from several individual meterings. This curve, usually a straight line on graph paper with logarithmic axes (fig. 5.14) describing the relationship between stage (which can be read off the pole in the stream) and discharge (measured by the methods described above on occasions at which a particular flow is in operation). The relationship is of the form:

$$Q = K\,(h_1 - h_o)^n$$

where Q is the discharge, h the recorded stage, h_o the level of zero flow in the stream, K a coefficient and n an exponent. Such *power functions* plot as straight lines on log – log graph paper.

Obviously, the most difficult part of the rating curve to establish reliably is that dealing with floods since the conditions are likely to be poor for fieldwork or the river may even be unapproachable. To obtain a good rating curve the section of the river at which measurements are made should be as stable as possible—with a firm bed and high regular banks. This ensures that cross-sectional area does not change with time of year or during the passage of a flood. There may be problems of summer weed-growth in the stream which occupies a large amount of the channel cross-section. In order to obtain the stable conditions required most river authorities choose sections at which bedrock is exposed in the stream bed and banks, or where man-made regularisation exists. In some cases they concentrate the bed and banks to ensure stable conditions. Where a stable section is not possible it should be frequently surveyed. Cross-sectional data is also useful for fluvial geomorphology (see chapter 6).

Once a rating curve has been established (20 or 30 measurements are often enough), stage readings can be converted to discharge at a glance or computed from the equation of the regression line which best fits the measured points.

It is possible to design a structure which, when built across a stream allows calculation of the river flow across it from theory. This is because its regular geometry gives a

Plate 5.5 shows a rectangular weir with a sharp metal crest bordered by concrete wing walls. A stage pole is shown in the foreground. Plate 5.6 shows a temporary 'V' Notch weir constructed from steel, 'Dexion' and polythene

standard cross-section and an identifiable relationship with the velocity. Such structures are *weirs* and *flumes*, built of steel or concrete. The major types of weirs are *broad-crested*, usually made of concrete and spanning the whole river, and *sharp-crested* made of steel plate, with a central rectangular (plate 5.5) or 'V'-shaped notch (plate 5.6) through which most of the flow passes. The following equations are those governing the relationship of flow to the depth of water passing over or through these sharp-crested weirs:

$$Q = 1837\,(L - 0{\cdot}2h)h^{1{\cdot}5} \quad \text{(rectangular)}$$
$$Q = 1336\,h^{2{\cdot}48} \quad \text{('V' notch} - 90^{\circ}\text{)}$$

(h is the stage measured in metres above the base of the notch at a position just upstream of the weir to avoid the 'drawdown' of water as it approaches the weir, L for the rectangular weir is its width in metres, Q is discharge in litres/second—a more convenient unit for small streams.)

For very small streams a 'V'-notch weir can be constructed from any piece of sheet metal, polythene (see plate 5.6), or even wood (see Gregory and Walling, 1971, p. 285). Long-term work on small streams should not be undertaken without permission from the local River Authority, who may also be able to give advice on methods. In urban areas it may be possible to estimate flow from roofs, gutters and so on.

So far, it has been assumed that we have been forced to visit the stream and measure stage (converting it to discharge where possible) once a day. However, unlike the raingauge which, when emptied at 09.00 hours G.M.T. gives us the total volume of rain over the previous 24 hours, the stage pole or weir will not inform us of the possibly great variations of flow between our visits. Either visits have to be made more frequently or the equivalent recording method to that of an autographic raingauge be designed. Some streams, such as those fed by springs in Chalk areas, do not fluctuate quickly in level and daily readings adequately characterise the steady rise and fall of discharge. Some prior knowledge of the natural variability of the stream is therefore necessary before adopting a **sampling interval** at which to record flow. It has long been the aim of those engaged in hydrometry to predict the necessary sampling interval from catchment characteristics; basins with small areas, steep slopes and impermeable soils needing the most frequent recordings because of their 'flashy' flow variations (fig. 5.15). The use to which the measurements are put will also affect their interval: calculation of mean annual discharge may be made more accurately with daily data than calculation of mean monthly discharge.

The answer in most cases is to record the level of the river continuously on a roll chart. The chart rotates on a clockwork drum while the pen moves up and down according to the level of the river, which it follows, at a scale of 1:10 or so by means of a series of gear wheels and a float. The float is housed in a **stilling well** below the recorder, thus protecting it from damage in the stream. The water level in the well is equal to that in the stream because they are connected by

Fig. 5.15 Daily measurements (shown by circles) of river levels on (A) a flashy stream—where they badly misrepresent the actual fluctuations—and (B) on a stream of slower response, where they may be quite accurate

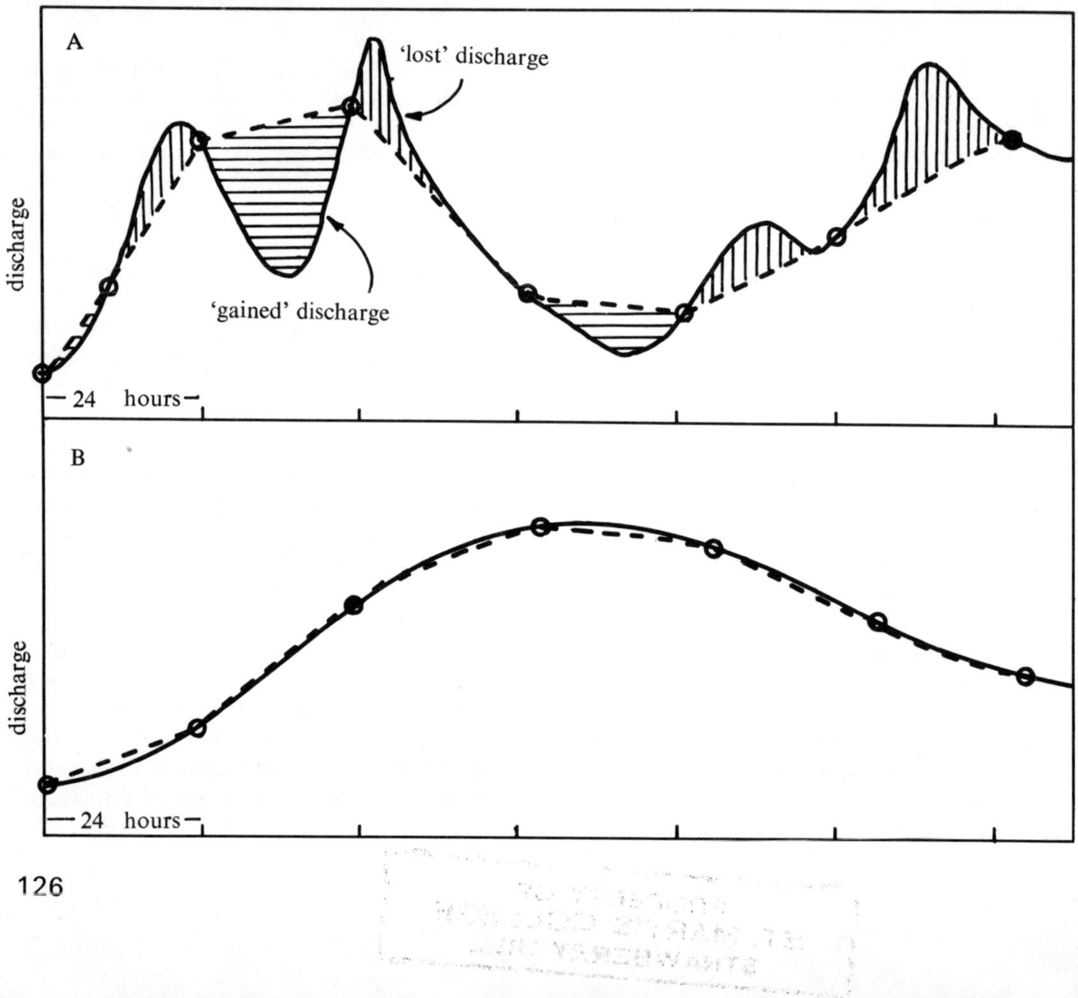

an inlet pipe. The charts are usually changed once a week. They may be marked in stage units or flow units. If you can obtain charts marked in both, construct the rating curve linking the two scales. What is its approximate equation? Most river authorities nowadays also record the level of the river at 15-minute intervals on a punched paper tape, the pattern of holes in the tape representing a computer code, allowing quick calculation of flow statistics from the tapes by the water resources and research organisations.

Having mentioned above that flow measurements are often impossible or very perilous in high floods, it is worth ending the section on measurement by describing a method by which estimations of flood flows can be made after the flood has subsided. The maximum stage can be recorded by the 'crest stage gauge' described by Gregory and Walling. Most streams, when they flood, leave debris scattered on the banks. These *rack marks* approximately follow the surface of the river at the highest part of the flood. The downstream slope of the line of rack marks is a useful indicator of the energy of the stream at the time and hence the velocity. The cross-sectional area can also be estimated from the rack marks and so discharge can be computed. The equations for discharge in such circumstances are various, the most popular being the **Manning equation.** As well as cross-sectional area (A) and slope of the rack marks (S) the equation uses the hydraulic radius (R) and a figure called Manning's 'n'. The hydraulic radius is the cross-sectional area of the stream, divided by the 'wetted perimeter' (the length of bed and banks over which flow occurred, measured across the channel). Manning's 'n' describes the roughness of the channel and therefore with hydraulic radius, characterises the reduction (by friction on bed and banks) of the velocity potential represented by the slope of the water surface. Manning's 'n' increases from 0·020 for earth-lined canals to 0·040–0·050 for mountain streams with rocky beds. The full Manning's equation is:

$$Q \text{ (discharge)} = A.\ \frac{R^{\frac{2}{3}}.\ S^{\frac{1}{2}}}{n}$$

It may, of course, be used at lower flows too, when the slope is calculated from a survey of the actual water surface in the channel. Slope is expressed as a fraction in the Manning equation.

CLASSWORK

Calculations involving the Manning equation provide a useful field exercise. If discharge is already known, from a nearby gauge, it may be possible to compute Manning's 'n' for your local stream. How rough are the bed and banks? Is the value the same as you would have chosen by observing the channel?

5C 4 Losses and storage

So far losses and storage have been considered as the most difficult items to quantify by measurement in the water balance. Annual losses from evaporation and transpiration are most frequently calculated by subtracting runoff from rainfall while storage changes are assumed to be zero because of the choice of the water year for measurements. The major problem in the use of the drainage basin over a year to calculate evapotranspiration by subtraction of streamflow from rainfall is that any errors in rainfall or streamflow measurement are transferred to the calculated figure for losses. If rainfall totals are typically underestimates because of air turbulence around the standard gauge the calculated figure for losses will be similarly underestimated. Equally, if the assumption about the changes in storage being zero is invalid because of a very wet summer, some of the rainfall which has yet to reach the stream will be calculated as part of the basin losses when this is untrue. It will flow into the stream during the early part of the next water year and lead to an underestimate of losses then!

The catchment which is instrumented only to measure rainfall and streamflow is usually referred to as an example of the *black box approach*, 'black' because we know only about inputs and outputs. Where the methods described below are used at various sites in the basin to measure losses or storage the approach is that of the *white box*. The measurements are a check on (and are checked by) those calculated or assumed by streamflow and rainfall measurements alone.

In contrast to the fairly standardised methods of rainfall and streamflow measurement there is a large variety of equipment for determining evapotranspirative losses and moisture storage; some controversy surrounds the choice of the 'best method'.

Evaporation is the escape of fast-moving water molecules from the surface of a body of liquid in relation to radiation energy exchanges (see chapter 2). As already mentioned, if the body of liquid were always a lake, pond or pan, the measurement of evaporation would be relatively simple. It is, however, greatly complicated by the very large variety in the type of water body and water surface. As well as occurring in large still masses like lakes, water occurs as moving streams, as films on soil particles and in the cell fluids of plants. Perhaps the greatest spur to the measurement of losses directly in hydrology has come from the assertion that forests reduce streamflow by increasing losses by evaporation. The term *transpiration* is given to the evaporation of water from the leaves of plants through openings called *stomata*. This process can well be demonstrated by covering a growing plant with a polythene bag, or by weighing a freshly-cut leafy twig on a sensitive balance and observing the loss of weight as it wilts. Wilting occurs when transpirational loss of water is not balanced by fresh intake from the roots. The large total surface area of leaf surfaces in a forest provides an evaporative surface far greater than that of the ground surface below. There is also the volume of rainfall *intercepted* by the leaf canopy to account for. It is while trapped by the leaves of trees and plants that much of the direct evaporation of rainfall occurs. Interception is a familiar phenomenon to those who have soaked their friends by shaking the leaves of trees or shrubs after rainfall!

FIELDWORK

1 What is the effect of the plants in the garden, park or school grounds on the local hydrology? Use simple funnel gauges to assess the volume of rainfall intercepted (by placing some on open ground and some beneath trees, shrubs or crops). Does the amount of interception vary seasonally?

2 Can you calculate how much water is tied up in the actual cellular matter of plants during spring and summer? What is the difference in weight between an actively-growing leaf and a dead, fallen leaf of the same species and size?

Since both evaporation and transpiration occur simultaneously and are physically identical the combined term is usually applied. Attempts to measure evapotranspiration have come from botanists and meteorologists as well as hydrologists which is why the topic frequently crops up in this text. Most botanists and hydrologists adopt an empirical approach, using instrumentation to assess volumetrically or gravimetrically the amount of water lost from various experimental tanks or plots of vegetated ground. However, it has been said that 'the meteorologist has the advantage in the use of the basic philosophy and principles of his parent discipline—physics . . .'. Thus, the meteorological approach is more theoretical, employing measurements of solar radiation, windspeed and other factors causative to the basic process of the evaporation of water. Dr H. L. Penman is the author of one of the best-known formulae for calculating evaporation from such measurements. Though the result applies to evaporation from open water, he performed experiments which produced suitable reduction factors for the conversion of these figures to those for *potential evapotranspiration*. For example, the average conversion factor for grassland in southern England is 0·75. The word 'potential' is applied because the reservoir of water which plants transpire is in the soil and as this declines so does the rate of transpiration. Thus the rate predicted by theory will be the potential rate, the actual rate being much less in cases where water is not continuously available. Clearly, as was seen in the last chapter, the nature of the soil and the root systems of plants are as important to evapotranspiration as the leaf system.

Turning to direct measurement there are three main types of instrument, the **evaporation pan** from which evaporation is measureable by the lowering of the water level, the **atmometer,** consisting of water-filled glass tubes with an open end from which water evaporates through a porous membrane (signifying the soil/plant surface) and **lysimeters.** Lysimeters are columns of soil and vegetation (usually grass or crops) which have been removed as carefully as possible, surrounded by a waterproof material and reinserted in their former positions. The base of the surrounding material is perforated to allow percolation of soil water into a collecting vessel lowered into position before the reinsertion of the column of soil. Rainfall measurements are taken nearby and on each visit the difference between the amount of percolation and rainfall is calculated. This figure represents evapotranspiration providing there has been no loss of stored soil moisture. This can be verified by weighing the column of soil (if it is contained in a tin it can be removed and hung on a spring balance) or by resting it, in the hole, on a water-filled innter tube connected to a manometer which records changes in the pressure of the column on the inner tube.

Conjecture as to the precise influence of forests on streamflow and hence storage in reservoirs for water supply in forested uplands was largely the reason for setting up a research unit now known as the Institute of Hydrology. The Institute is conducting white box, water balance studies in adjoining forested and grassland catchments at Plynlimon, Wales and also directly measuring evapotranspiration from coniferous forest in East Anglia (see the Institute's Annual Reports). The Institute of Hydrology has also developed a fully-automatic weather station which records, in computer-compatible form, the variables needed for calculating evapotranspiration by the Penman formulae.

Perhaps the most direct way to measure the moisture in a sample of *soil* is to remove it in a waterproof bag to the laboratory where it is weighed before and after drying in an oven. However, once sampled, the site is destroyed for ever and the aim with most measurement in this chapter is to choose representative sites for repeated measurements. Instruments which can be inserted into the soil include the **tensiometer** (illustrated in the previous chapter), which measures that tension set up between a porous pot and the surrounding soil (see page 94) and the **neutron probe** which emits fast-moving neutrons. The neutrons are slowed by contact with the hydrogen ion in the water molecules and the probe gives an estimation of soil moisture on the basis of measuring the 'slowed' neutrons. The Institute of Hydrology has done much of the development work on neutron probes. However sophisticated instrumentation becomes to measure soil moisture, it is still frequently calibrated by reference to the gravimetric method of laboratory drying. Simple estimates of soil moisture variability can be gained by feeling the soil under various conditions or at depth. If a stake is driven into the soil and removed the hole left may well contain water from the surrounding soil after a few hours.

Fluctuation in **groundwater levels** is relatively easy to measure by lowering into wells and boreholes floats or water-sensitive probes on wires whose length on contact with the water can easily be measured. The basis of the electrical methods of well recording is the completion of a simple circuit by contact with water. Such a device is easily made. The rise in the water level in response to rainfall will be much delayed in some rocks, while in others it may be surprisingly rapid. Another way of obtaining the same information is to gauge small streams issuing from obvious springs. *Karst hydrology*, in areas of crystalline limestone, has demonstrated to us the importance of joints, faults and bedding planes in rock. Such fissures greatly influence the permeability of the rock and while in *porous* rocks there may exist a general *water table* whose surface follows the smooth lines of the surface topography, in *fissured* rocks such a surface is often absent, water flowing at a variety of levels depending on the availability of fissures. (The difference between porosity and permeability is that the former depends on the presence of interstitial voids, or pores into which water may flow between the grains of a rock, such as Chalk; the latter refers to the movement and storage of water in the joints and faults of a rock, such as quartzite. Clearly, some porous rocks are permeable too, and vice versa, for example sandstones.)

Karst hydrology has also demonstrated ways of **tracing** water from surface streams underground, using dyes and lycopodium spores, and from the soil surface to springs using chemicals containing trace elements. Because of the great increase in atmospheric tritium following the beginning of nuclear tests in 1952 it is possible to date some of the spring water emerging from very retentive aquifers in Britain as having entered the ground as rainfall before that date.

5D Using the data

Though quite short records of rainfall and streamflow are adequate for investigations into the processes by which runoff occurs, **statistical analysis** in hydrology requires adequate sample from which to work, which usually means a long record. For example, one year's records of rainfall and streamflow may be enough for a researcher to make useful new conclusions about whether surface runoff or throughflow predominates in a small catchment. The same year's records will hardly suffice to provide an adequate basis on which to predict floods. We often draw a line between **physical hydrology** and **parametric hydrology** and it is the latter to which we now turn, using as it does the statistical parameters of streamflow records, mainly for predictive purposes of value to the engineering profession. Hydrology touches upon civil engineering in several ways, the most obvious being the construction of bridges, dams and stormwater sewer networks in cities. All such works require large capital investments and it is likely that the builders will ask about the liability to failure or damage from flood or drought. Some of the ways in which parametric hydrology can provide the answers to these questions from long hydrological records are presented below.

5D 1 Flow duration curves

Supposing that it takes a river flow of a certain volume to maintain the operation of a mill-wheel or turbine, what is the likely proportion of any period of time over which flows equal to, or above the critical figure will operate? Obviously this question is frequently asked in the water supply industry too, and in other commercial enterprises where the variability of streamflow, particularly around a certain critical figure, is an essential item of information. Statistics will provide an answer in the **coefficient of variation** which expresses the *standard deviation* of the data as a percentage of the *mean*. However, the graphical construction of the **flow duration curve** allows us more flexibility, since it is not always in fluctuations around the mean value with which we are concerned.

Fig. 5.16 shows the flow duration curves for two streams from their records collected over 15 years. To draw the curves the chart records for the periods in question have been inspected and the average daily flows (either the average of the 24 hourly flows, or a simple estimate) tabulated within certain **class intervals,** for example 0–0·5, 0·5–1·0, 1·0–1·5 cumecs, and so on. Next we need to know the proportion of the total number of observations (i.e. days) upon which various classes of flow occurred. This is the duration of that particular flow. For example, if a flow of between 1·5 and 2·0 cumecs was recorded on 183 days in a leap year it would represent a duration of 50 per cent. When the duration of each flow class has been calculated, curves such as those in fig. 5.16 are drawn, cumulatively and beginning at the high values of flow which occur

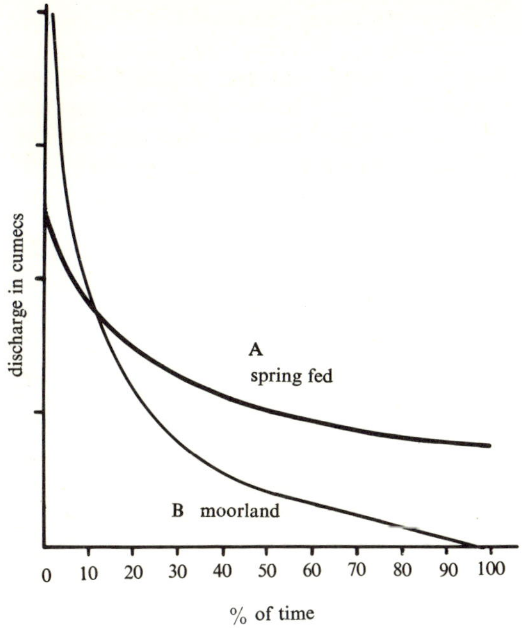

Fig. 5.16 Flow duration curves for two contrasting catchments

very rarely. Thus, the completed curves have an abscissa from which the percentage of any time period over which a critical flow is *equalled or exceeded* can be read. The importance of a long record is obvious—a few months' records would hardly include the full range of variability and so curves based upon such a small sample could not be used to predict the future variability of flow.

Fig. 5.16 shows the contrast between two typical river *régimes*. That fed by springs fluctuates over a much smaller range than that dominated by runoff from moorland. Spring flow is generally released gradually from a natural groundwater storage reservoir in the rock interstices. Rapid moorland runoff reacts to rainfall almost directly, resulting in almost alternate flood and drought. We call the latter streams 'flashy'. Many urban streams are flashy. The two flow-duration curves serve to remind us of the meaning of storage in hydrology. In dealing with water balances and hydrograph studies it was seen that there was both a rapid and a delayed response to rainfall. In the hydrograph studies these two components were separated to obtain the volume of runoff. Thus, while flow duration curves 'a' and 'b' are for separate streams they also represent the two types of response within catchments which are neither wholly moorland nor wholly fed from springs.

5D 2 Flood recurrence intervals

Another use of long hydrological records is the calculation of the statistics of extreme conditions. Though the same methodology could be applied to droughts it is most often used for flood prediction. It is important to distinguish between prediction and **forecasting.** The latter relates the occurrence of an event to a *real time*: 'skies will clear in the early afternoon', and so on. Prediction merely allows us to relate the magnitude of the event to the frequency with which it will happen. In statistical terms we calculate the **probability** of the event happening, just as we have already calculated above the probability of a certain flow being equalled or exceeded, based on the number of times this has occurred so far.

When hydrologists speak of flood recurrence intervals they say that the biggest flood in a certain record of flows is, say, the 20-year flood. It has a recurrence interval of 20 years. The meaning and methodology of this are quite simple. The annual maximum flows for the period of record (this can also be done with daily rainfall figures) are **ranked,** with number one the largest annual peak, number two the runner-up, and so on. The recurrence interval is calculated as follows:

$$\text{flow X's recurrence interval} = \frac{\text{(number of peaks in the list)} + 1}{\text{rank in the list of flow X}}$$

The main interest in flood prediction for engineering is normally the very big flood, perhaps of a 100- or 250-year recurrence interval. Obviously we cannot usually calculate this from our records, which are seldom over 10 years in length because hydrology is relatively new as an organised science. We are therefore forced to **extrapolate,** using the recurrence intervals we calculate from our list to derive figures for the flow of enormous floods never so-far recorded. The statistical theory by which this is done is the **Theory of Extreme Values.** It has been used by Gumbel (see Dury, 1969, pp. 322–5) to prepare graphs on which the relationship between peak flows and their magnitudes plots as a line. This line can be drawn in by eye (fig. 5.17), or calculated by regression analysis. Either way it can be extrapolated beyond the points we plot from our tabulated, ranked flood peaks. However, it must be stated that such extrapolation can only be correct and such an analysis could not be used as any more than a rough guide by engineers. Estimates by extrapolation ought to carry a figure for the **confidence limits** (Toyne and Newby, p. 65) within which the flow estimates may vary. Dury (p. 324) has said that only those flow estimates for recurrence intervals up to half the length of record used (e.g. 25-year flood for 50 years of records) may be used with any accuracy. This is not discouraging for the geomorphologist who, as will be shown in the next chapter, is usually concerned with floods with a recurrence interval of just over a year (see page 160).

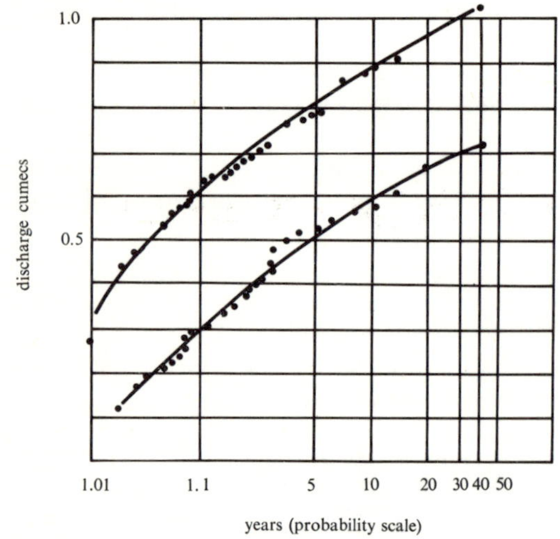

Fig. 5.17 Flood recurrence intervals for two small streams. The lines curve and were plotted by eye through the points. An alternative would be to calculate the straight line equations which would fit the points

Obviously one of the major aims of long-term hydrology is to make long records available to the statisticians for such analyses.

5D 3 Use of hydrographs

In an earlier part of this chapter the calculation of the amounts of runoff from various periods of storm rainfall was demonstrated by separating the flood runoff from the background, or base flow, and relating it to the volume of rainfall input to the basin during the storm. Though this volume has been shown to vary with the storm, according to dynamic catchment variables like antecedent precipitation, engineers sometimes need to assume that the volume of runoff is constant in one catchment. Similarly the time between the input of rainfall and output of streamflow (the **lag time**) is deemed constant, as is the period between the beginning and end of storm runoff (the **time base**). These terms are clarified in fig. 5.18.

There have long been arguments amonst hydrologists about the degree to which these parameters of the flood hydrograph can be assumed to be constant and there-

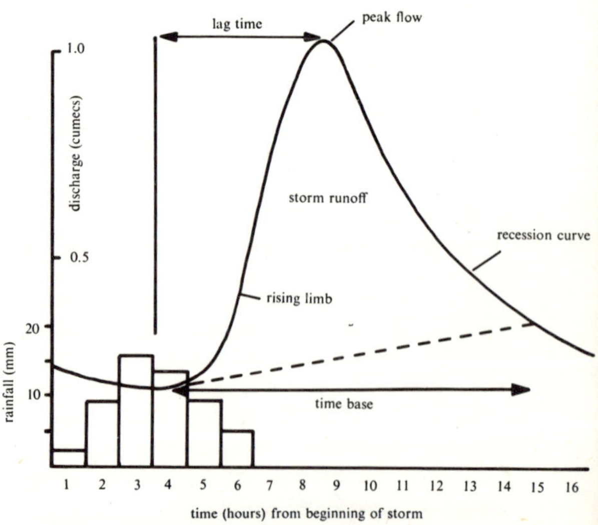

Fig. 5.18 Parameters of the streamflow hydrograph which are used by engineers for prediction purposes

fore the reliability with which they can be predicted. Suffice it to say here that the prediction of hydrograph parameters has been used for engineering purposes for some time with quite a degree of success. Most frequently engineers use hydrograph analysis in conjunction with the calculation of recurrence intervals for flood peaks. Thus, as well as the peak flow which may occur at their dam or bridge site, they can calculate the delay between the rainfall and the flood and how many hours the engineering works will have to stand up in flood conditions. Most frequently this information is needed for catchments which have no record of flows. Thus, another important use for basin characteristics like area and slope, derived from maps, is in relating them to the shape of the hydrograph defined by its lag time and its time base. Rodda (1969, p. 417) gives examples of such predictions.

CLASSWORK

1 How often do big floods occur in your area? If your memory is too short consult the local newspaper. High flood levels are often marked by stonemasons on bridges, riverside cottages or even inside buildings. Look for these. Most parents will remember the floods which resulted from the thaw after the severe winter of 1947. Readers themselves in the Bristol area and in southeast England may remember the thunderstorm floods of July and September 1968.

2 How much does a bridge cost to build? Do people who live near rivers pay extra house insurance because of the risk of floods? How would you fix the premium for such a house if you were the insurance company?

5D 4 Trends in rainfall and streamflow

While we are all familiar with the drastic changes of climate and physiography during the very long periods of geological time we tend to assume an unlikely constancy in conditions at the present day in spite of hearing our parents say that 'summers are not as warm as they used to be', and so on! One use of long records in both climatology and hydrology is to try to prove this constancy, or if it is disproved, to establish and be able to predict the *trends* shown by the data.

Such analyses usually begin by plotting the data we have on a graph whose abscissa represents the years over which it has been collected. A pattern may be immediately obvious. Usually, however, it will only appear vague unless some statistical work is done. Since the data points will often contain fluctuations within the broader patterns we have to try to remove these. One way is shown in fig. 5.19, using **running means.** To diminish the merely seasonal fluctuations in the stream issuing from the Severn Tunnel Great Spring, twelve-monthly running means have been calculated. The first twelve months of data points are averaged and this average plotted against the end of the final month. The next twelve months' data to be averaged are those ending in the month following the one just plotted. The plot of successive twelve-monthly mean values is much less fluctuating and allows us to see clearly that the data presents a slight trend to increased flow. We call this process one of **smoothing.** It reduces the influence of the irregularities which become 'noise'. Of course the trend can also be presented as a regression line through all the raw data points. The equation for such a line can be tested for the significance of its slope and hence the significance of the trend can be quantified and prediction of values far in the future be made with the equation.

Not all trends in natural phenomena are simply linear. Many aspects of physical systems exhibit cyclic behaviour. There are said to be several examples of weather cycles, although none has been proved from instrumental records. The analytical methods used to decipher cyclic variations are complicated. The more advanced statistical texts deal with them under such

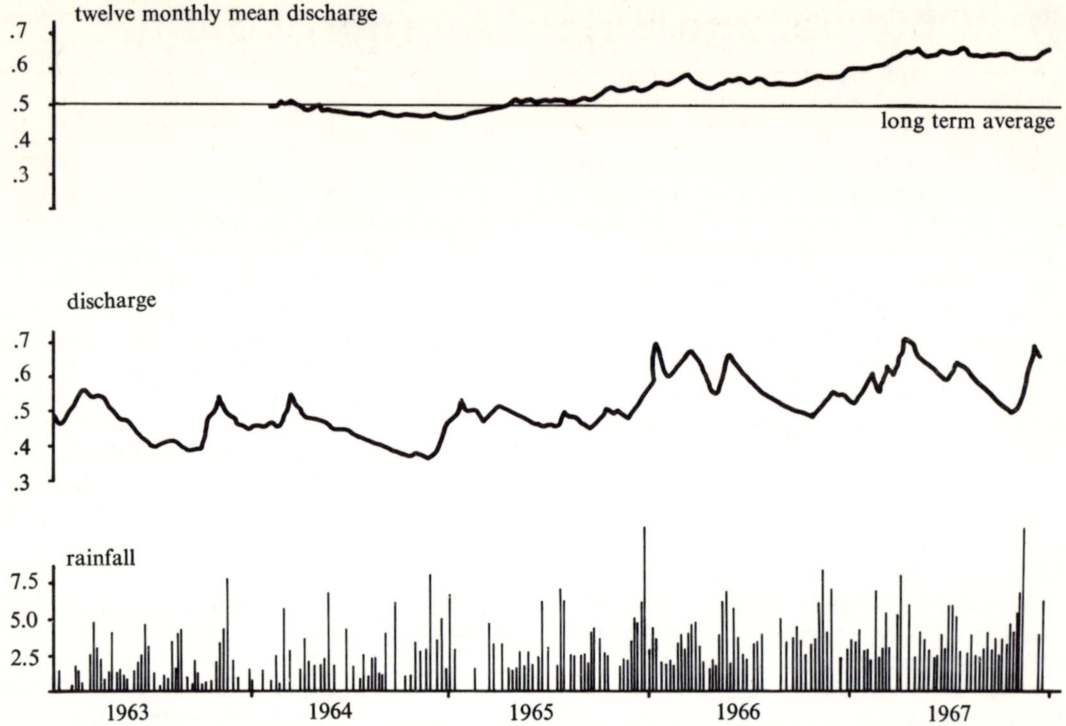

Fig. 5.19 Running means as a method of smoothing discharge measurements and detecting trends. Notice how unconverted discharge itself is a smoothing of the rainfall pattern

headings as **harmonic analysis, spectral analysis** and **Fourier Series analysis,** and some of their uses were mentioned in chapter 3. The Fourier theorem postulates that no matter how complicated the fluctuations in the data, they can be accounted for by superimposing a number of much simpler mathematically derived curves. Basic to such analyses are sine curves and cosine curves. Spatial cycles can also be analysed by the same techniques, meanders and sea waves being two examples.

More simple investigations of weather cycles can be made by comparing each year's mean rainfall or temperature with the long-term average and recording a 'plus' or 'minus' sign according to its departure above or below that average. One then examines the pattern of 'pluses' or 'minuses' to detect regularities or lack of them.

6 Channel and Slope Geomorphology

The recent increase in the attention paid by geomorphologists to the landforms which compose river basins has to some extent been a corollary of the growth of hydrology. The 'normal' fluvial development of landforms, in spite of being the local case to most researching geomorphologists in Europe and America, had been largely subordinated to the spectacular cases of glacial and arid erosion during the descriptive phase of geomorphology. Some say that the almost universal acceptance of the oversimplified fluvial models of W. M. Davis was the cause of this lack of attention. However, after a post-1945 phase of interpreting upland remnants in peneplains, British geomorphologists have moved on to the instrumentation of the slopes and channels which transmit water and materials from the drainage basin, in an effort to quantify erosion. Far from being a duller pursuit than the study of deserts and glaciers the 'normal' aspects of geomorphology have proved exciting in scope and application. Dury's collection of essays (1970) provides an insight into the changing approaches to fluvial geomorphology. One of the most important aspects has been the change in scale of operation, from 'temperate areas' to a drainage basin or valley-side plot.

6A Channel forms and Dynamics

In the new phase of fluvial geomorphological studies we have adopted a new framework for study. Instead of using terms like 'youth', 'maturity', 'old age' and erosion 'cycles' we pay more attention to drainage basin *dynamics*. The reasons behind the general abandonment of the Davisian model have been given by Chorley (1965) and Small (1969). By dynamics we mean **processes** of landform development rather than the **forms** they produce; these are now the theme of investigation. To the engineer, dynamics is the mathematical and physical study of the behaviour of bodies under the action of forces which produce changes of motion in them. In the case of fluvial geomorphology the 'bodies' are clearly the particles (from large boulders to small silt grains and molecules in solution) which cloak hillslopes and enter river channels; it is necessary to deal with the 'forces' in a little more detail. They result from the flow of energy and mass, as in all natural systems.

The *potential energy* of water falling on a drainage basin is that imparted by gravity operating through height. In flowing downslope and downchannel towards a base level, usually the sea, this energy becomes *kinetic energy* and performs *work*. In some cases the local scale study reveals other base levels—lakes or pools. We can discover in many of the typical forms in drainage basins the action of the laws governing energy and work discovered by mathematicians and physicists. Our measurements will fall into two main groups, those of forces and those of bodies, in order to calculate the work performed. For example, to compute the amount of suspended sediment leaving a river basin it is necessary to sample the sediment and to measure the river flow.

It is important to realise that slopes compose most of the fluvial landscape. Both hydrologists and geomorphologists are beginning to investigate the less obvious movements of water and materials on slopes where water is no less important than in channels, merely less abundant relative to solids. Conceptually it is helpful never to forget the channel below when considering processes on a slope and always be mindful of the contribution of slopes

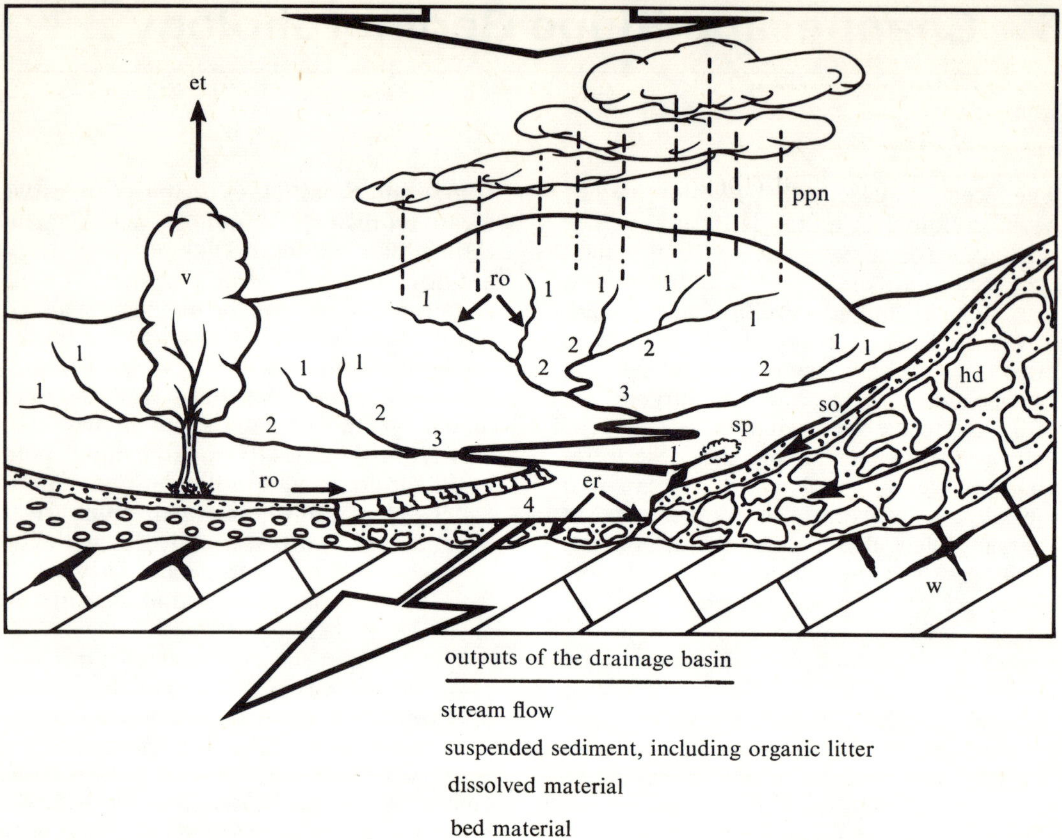

1, 2, 3, 4 = orders of stream channel. The smallest streams are first order. When two join the resulting stream is second order and remains so until joined by another second order stream when it becomes third and so on.

ppn = precipitation et = evapotranspiration ro = runoff v = vegetative cover sp = spring

so = soil (water & material movement) hd = historical deposits (water & material movement)

w = weathering of bedrock er = erosion by channel processes

Fig. 6.1 The drainage basin as an integrated system of channels and slopes

when investigating channels. We are dealing with a drainage basin system (fig. 6.1).

Since the predominant force dealt with below is the flow of water it is worth digressing slightly to stress that there are two basic modes of water flow which are relevant to erosion. *Laminar* flow occurs by layers of water molecules shearing one over the other. It is uncommon in natural streams because of their depth and velocity but it will be encountered in groundwater flow. Where velocity exceeds a critical value (dependent on viscosity and channel roughness as well as depth) the flow becomes *turbulent*. The two types of flow are shown diagramatically in fig. 6.2. Since velocity and depth are critical factors it is clearly possible for a channel to show increasing

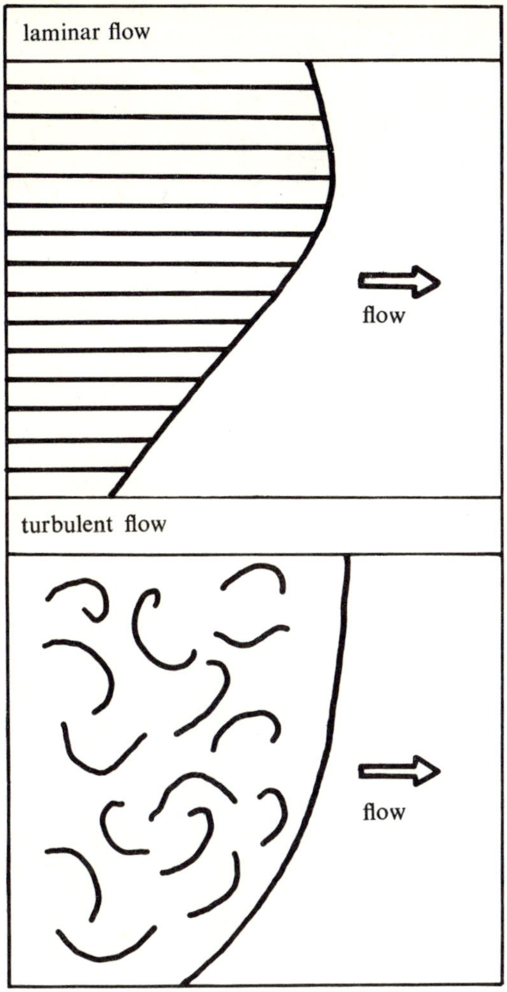

Fig. 6.2 Laminar and turbulent flow, showing the differences in velocity attenuation with depth

degrees of turbulence in flood. Turbulent flow is all important in the erosion and transportation of particles by streams. The greatest factor in determining the kinetic energy of the flow is its velocity. As has been seen in the Manning equation used for calculating flows, velocity is in turn a function of stream gradient, water volume, water viscosity and the roughness of banks and bed.

6A 1 Typical profiles and plans

Davis made famous many of the terms which he used in his work. One was the **graded profile** which, he deduced, was the shape adopted by the longitudinal profile of the 'mature' river, a shape which demonstrated the balance between erosion and deposition—a balance only obtained by 'mature' rivers. Even before Davis, Gilbert had stressed the importance of the balance between processes which is a feature of all parts of a river's length and throughout time. Grade is not a late acquisition. Gilbert's deductive essay, illustrated by field work, may be read in Dury's collection. Real progress began when the United States Geological Survey began to actually measure profiles in the field and from maps. As well as measuring gradient they worked on stream velocity, width, depth and the roughness of bed and banks—all the factors with a bearing upon stream energy. They concluded that a downstream decrease in channel slope is necessary to preserve the balance of all these factors, such a balance being a requisite of the laws of dynamics. Again, the conclusion was that such a profile does not signify the onset of any particular state in stream development. The changes of flow and material load which accompany the stream's passage from watershed to base level may be accommodated by alterations in any one of the factors mentioned. Hence small-scale irregularities in the profile may be the result of the river changing width or depth instead of gradient in response to, for example, increased flow or sediment load. Yet, overall, there is a steady decline in gradient, accompanied by decreased roughness, increased width and of course increased flow. However, as we have already proved, velocity remains fairly constant. It is concluded that the apparent steady decline in gradient downstream is merely the **most probable** or **mean state** situation. This is an interesting conclusion because it means that apparently regular forms can be produced by the operation of the *laws of chance* on random occurrences.

There are two fixed points, the watershed and base level, but between them the profile is the one governed by probability. Many of nature's patterns are similarly based.

CLASSWORK

Using dice it is possible to generate another *most probable* pattern in the drainage basin. Beginning at a series of equally-spaced points down one edge of a piece of graph paper, move your pen forward, to the left or to the right according to whether you throw one of a chosen pair of numbers. For example, a 1 or a 6 may mean forward, 2 or 5 left and 3 or 4 right. Your *random walk* should be marked on the graph paper as you leave one edge of the paper (the 'watershed') and head for the other. How soon before your 'streams' join up? Played on a large scale the resultant pattern quickly resembles that of the various rills and streams in the fluvial landscape. Obviously at no stage is one allowed to move backwards for this would break the one *physical constraint* of the system—that of downhill flow.

When you have completed the random walk drainage network see if its properties correspond to those of natural river networks. Draw graphs of the number or length of streams of a certain *order* (see fig. 6.1) against the order number. Choose a logarithm scale for the number or length of streams.

Mathematically the most probable profile may be described as an **exponential** or **semi-logarithmic** curve: semi-logarithmic because one variable, (altitude), changes by its logarithm as the other, (distance), changes arithmetically. Thus, it plots as a straight line on semi-log graph paper (fig. 6.3) and an equation may be fitted. Such equations have been used to extend the graded profiles of upper reaches in British rivers to their former base levels. The mathematical curve of the graded profile is seen to flatten off at about this former base level. The exercise is clearly one of *extrapolation*, the use of a trend in data one has collected to predict

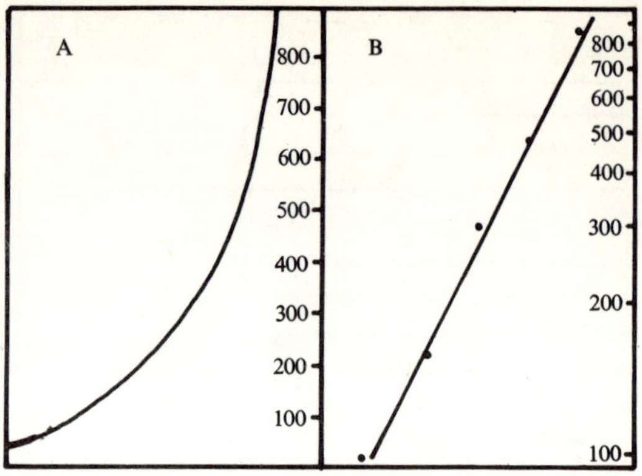

Fig. 6.3 The 'graded' stream profile (A), replotted with a logarithmic altitude scale (B)

the values of data required beyond its limits. What example of extrapolation was given in flood prediction?

The fact that one river profile may exhibit several reaches, each of which appears graded to a former base level above the present one, suggests that such *polycyclic* profiles can be separated statistically in the following way. Since the long profile may be approximated as an exponential curve the frequency curve of the logarithm of elevation should be normal. A probability scale straightens the normal curve and so a plot of elevation on a log scale against the percentage of the profile below each elevation on a probability scale (fig. 6.4) breaks up the stream profile into straight line segments. However, great care should be taken in interpreting such polycyclic profiles as indicative of several erosion 'cycles', each with a separate base-level. They are frequently the result of climatic changes or of local structural and geological controls on the erosion of the channel bed.

There is obvious scope in this topic for field and mapwork projects at school or college level. The class can combine effectively to record far more measurements of stream channels than professionals are allowed, both by taking rods and tapes to a nearby stream, and by diligent work from Ordnance Survey maps. Those at a scale of 1:25000 are best, but it should be remembered that several problems present themselves on maps, not the least of which is the inevitable shortening of the stream's length,

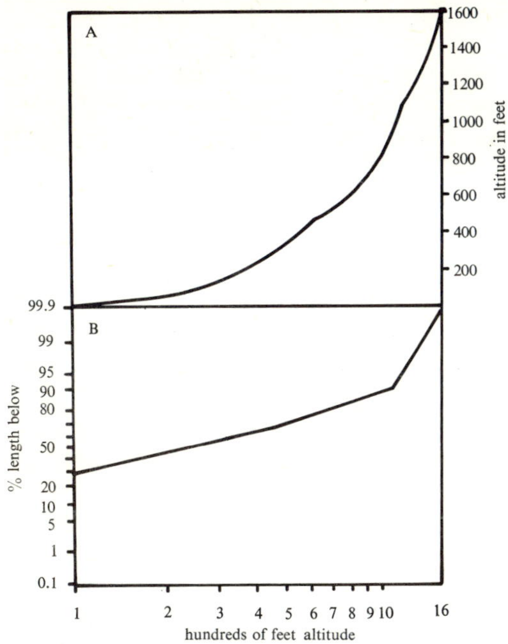

Fig. 6.4 (A) The profile of a polycyclic upland stream showing three partly graded profiles. (B) A logarithmic/normal probability plot using data from (A), showing division into three clear segments

which arises from drawing up a long-profile along a straight edge when the stream being followed may meander quite tightly. A section line taken down the valley floor, without following the stream, may produce different results: try it! Measuring channel dimensions from maps is less profitable because, at the smaller scale of operation, there is an inevitable plottable error; we require an accuracy best obtained by 'live' recordings, i.e. at a scale of 1:1. Fieldwork is also more likely to record significant changes in time. As has been shown by Leopold, Wolman and Miller (1964), the relationship between channel width, depth, velocity, load, roughness and slope often changes more at one location on the river during the passage of a flood than between different stations up and down the same stream at any one time. Figs. 7 to 22 on page 245 of their book show the conclusions drawn from temporal as well as spatial consideration of the river channel. Patterns in time have become more important to the student of physical geography since he started studying processes and recording measurements. It is important not to confuse the effects of space and time. W. M. Davis regarded the 'youthful', 'mature' and 'old age' reaches of a river as spatial demonstrations of a temporal sequence. By surveying our local stream we may well discover that some of the features he mentioned as typical of 'youth', 'maturity' and 'old age' may be demonstrated to us within short distances or in short periods of time. Our small stream may become 'a Thames' during floods!

Another typical and regular feature of river channels is the tendency to **meander.** River courses always exhibit various degrees of sinuosity—indices of the degree of sinuosity usually compare the distance along the channel between two carefully chosen points of inflection to the straight-line distance. The ratio between these distances may be compared from map measurements. Where sinuosity is regular enough to be termed meandering the true regularity may be assessed by taking the field measurements shown in fig. 6.5. The

Fig. 6.5 Dimensional measurements of meanders

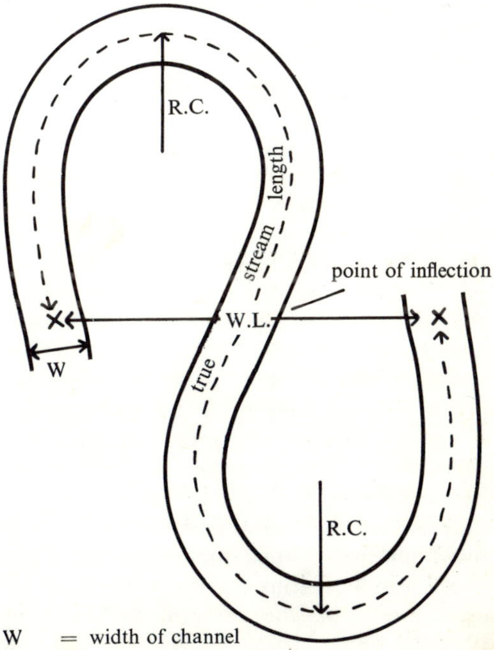

W = width of channel

W.L. = wave length

R.C. = radius of curvature

sinuosity of meandering reaches seems to keep values of between 1.3 and 4.0. There also seems to be a fairly regular relationship between meander wavelength and channel width, a modal value of around 10 being found. The relationship between meander dimensions and stream discharge has been used to calculate the effects on river flows of past climates (see chapter 8). The mathematical expression of the typical meander curve is that of a sine-generated curve, one in which the rate of change of curvature is minimised, as seen in chapter 3 (fig. 3.19). Herein lies the key to the regularity of shape. Langbein and Leopold's essay in Dury's book (page 250) shows how it was attempted to fit four mathematical curves to the stream meander: the parabola, the sine curve (which many natural wave forms take), the circular curve and the sine-generated curve. Only the latter has the property of keeping the rate of change of curvature at a minimum (minimum variance). Thus a highly efficient flow round the bends is possible. This means that the work necessary by a stream (in bank erosion) to effect curvature is also minimised. Like other open systems the river behaves in such a way as to minimise the variations in its properties by distributing them evenly throughout.

Even in relatively straight reaches of rivers, especially during low flows, the main thread of water tends to swing regularly from one side of the channel to the other. This can be demonstrated in the field by means of floats, or by observing the more turbulent area of flow. Also apparent from simple tape-measurements is the regularity shown by *riffles and pools*, alternating shallows and deeps in the stream's long profile. If the distance between riffles is measured it may be found to have similar relationships to other dimensions of the channel as those shown by meanders. In this case the explanation may be due to erosion and deposition during high flows since meandering reaches show similar features. Observations of the appearance of the stream, possibly by photography during a variety of flow conditions, would help to clarify the relationship these features have to flow and sediment movement. What makes a straight reach of channel with riffles and pools becoming meandering is not known but the nature of the bank material is clearly important.

FIELDWORK

Take a meandering section of stream and spend a day measuring width, wavelength, sinuosity, velocities and so on. As well as by using the numerical data which result, a far more lucid picture of stream channel processes results from the concentrated observation involved.

Try also to see where meanders occur. Are they always in reaches with low gradient? Is gradient the most important variable or is it the calibre of the bank material? The quick formation of meanders in road-bank gullies or in laboratory tanks should dispel any association between meanders and the 'old age' of rivers!

6A 2 Work in channels

It is stressed above that at least two of the features of the behaviour of rivers are the result of the relationship between the flow of water and the work which it performs. We now examine some of the processes which bring about this behaviour. Most of the energy imparted to the stream by gravity is transformed in overcoming the friction of the bed, banks and internal turbulence, some 95–97 per cent being lost in this way. However, in times when flow volume and velocity increase there is generally enough energy available to raise particles from the bed into *suspension* or at least roll them along the bed as bed load. Gravity is also acting on the particles, pulling them back to the bed, at a rate expressed by their settling velocity in still water. The relationship between *erosion transportation* and *sedimentation* (deposition) of particles to kinetic energy (represented by its largest term, velocity) is shown in fig. 6.6. Once *eroded* from the channel bed a lower velocity is capable of *transporting* the particle. The reason for the greater resistance to erosion of small particles is that they have great cohesion and on the channel bed they impede the flow of water less than larger particles. Cohesion is obvious in mud pies and the smoothness of a muddy stream bed may be seen on estuaries at low tide.

Having grasped the concept of the stream network as comprising lines of energy expenditure by erosional and depositional processes, we turn to the measurement of the work thus performed by this system. Before describing the techniques we can use, two points should be emphasised:

a) Solid material on the bed and banks of a channel has been supplied by the bordering slopes, as well as from channel erosion. The process by which the material has been prepared is mainly that of *weathering*.

b) Often no further comminution in size occurs once the sediment has entered the stream. However, erosion of the channel bed and collision between the particles themselves results in comminution by the processes of *abrasion* and *attrition* respectively. Gilbert's essay (in Dury, ed. 1970, pp. 95–108) is a masterful piece of theorising on the relationship of these processes.

6A 3 Size and shape of sediment

We first took note of the particles on the bed and banks of the stream channel in estimating the resistance to flow which they might effect (roughness). This was for use in the Manning equation. Size and shape analyses are performed too in order to assess the rate and nature of comminution.

Sediment size has always been measured along a rather indefinite ordinal scale: boulder, pebble, sand, silt, etc., as already mentioned in chapter 4, concerning soil texture. This has been expanded by a variety of workers, trying to ensure standardization, to an interval scale based upon the intermediate diameter of the particles. The intermediate diameter is that measured across the particle's B axis, the A axis being the longest and the C axis the shortest. Two interval scales are presented in table 6.1 to represent size classes. The ϕ scale is equivalent to minus the logarithm (to base 2) of the diameter in millimetres. It is more convenient to use because it presents whole numbers in contrast to the fractions on the millimetre scale before transformation and has the same effect as the logarithmic binary scales mentioned in the Introduction, that of increasing and standardizing the interval between classes.

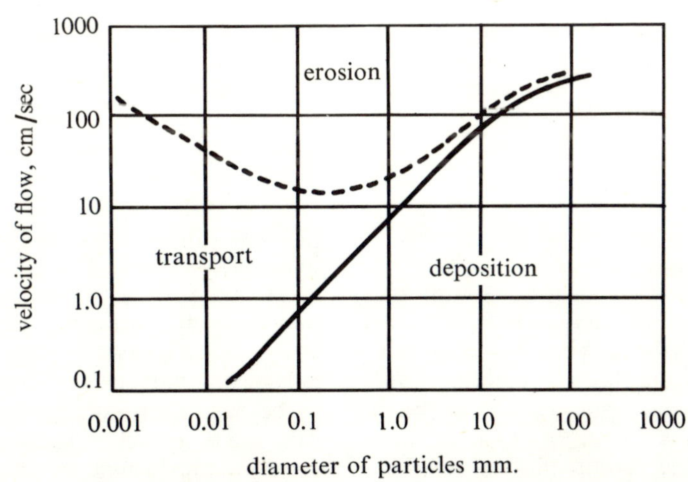

Fig. 6.6 The theoretical relationship between particle size and stream velocity (after Hjulstrom)

Table 6.1 Size ranges of sediment in stream channels

Type	*mm*	*phi (ϕ) units*
boulder	more than 256	more than −8·0
cobble	256 to 64	−8·0 to −6·0
pebble	64 to 4	−6·0 to −2·0
granule	4 to 2	−2·0 to −1·0
very coarse sand	2 to 1	−1·0 to 0
coarse sand	1 to 0·5	0 to 1·0
medium sand	0·5 to 0·25	1·0 to 2·0
fine sand	0·25 to 0·125	2·0 to 3·0
very fine sand	0·125 to 0·0625	3·0 to 4·0
coarse silt	0·0625 to 0·0312	4·0 to 5·0
medium silt	0·0312 to 0·0156	5·0 to 6·0
fine silt	0·0156 to 0·0078	6·0 to 7·0
very fine silt	0·0078 to 0·0039	7·0 to 8·0
coarse clay	0·0039 to 0·00195	8·0 to 9·0
medium clay	0·00195 to 0·00098	9·0 to 10·0

Between the clay particle and the ions of a true solution there exists a size range of *collodial* material and some erosion products are known to be transported by rivers in this form.

In fluvial geomorphology there have been few studies of the particle size of suspended sediment because of the obvious sampling problem. Most attention has been given to bed load, relating its size distribution to channel geometry or flow velocity. Since bed load is invariably coarse in the streams selected for study it may be sampled by hand.

FIELDWORK

The most important particles on a channel bed are those of the surface layer, forming an interface with the flow. In a convenient size stream a sample of 100 pebbles, cobbles and boulders should be taken randomly as follows. The person appointed to do the sampling dons his boots or waders and walks about in the stream picking up the stone which falls beneath the toe of his boot each time he takes a step. He should try not to exercise any conscious choice for one stone or another. Each of the hundred is passed to the bank where his colleagues measure the three diameters of the stone, note its rock type and describe roughly its shape. This may be done in several streams or at different locations on one stream. Velocity measurements should also be made and the channel gradient surveyed.

What are the relationships between the measurements? What is the source of the bed material? Has it fallen from the banks or been eroded from the bedrock? What is the velocity and how does this compare with that calculated from roughness and gradient?

There are, however, difficulties with the fieldwork method. Though it is not known what the precise relationship is between easily measured dimensions of particles and streamflow, it is clear that the *number* of particles in various size ranges is not likely to be so important as the *total weight* in each class. Hence it is the weight of sediment in each range which is usually employed for more detailed investigations of fluvial, glacial or coastal sediments. It

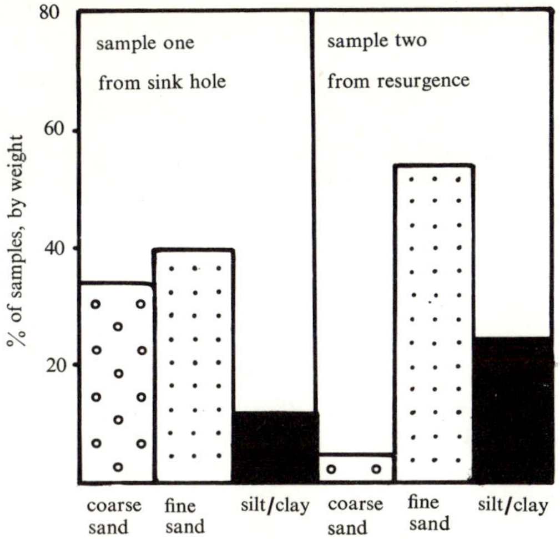

Fig. 6.7 An example of sieve analysis in fluvial geomorphology, showing the changes in sediment size during its passage through a cave system

clearly requires more work and often the removal of a large mass of sediment to the laboratory. It may, therefore, be more convenient to spend less fieldwork time and merely take a *grab* sample from the stream bed in a bucket or bag. The total sample is weighed on return and passed through a series of *sieves*. By agitation (with a dry sample) or washing with a jet of water the sample passes down through the nest of sieves (see chapter 4), each successively finer mesh trapping material fine enough to pass the sieve above but too coarse to pass through to the one below. To ensure thorough shaking a mechanical device is often used. Certainly shaking for a lengthy spell is needed to ensure a reliable result. Identical shaking periods are used for each sample. The number of sieves depends on the requirements of the investigation. Fig. 6.7 shows the changing ratio of three sieve divisions during the passage of sediment from the sink-hole end of a cave to the resurgence of the water. Usually more sizes are used. The coarse material at the entrance was concluded to have abraded the cave floor to produce the finer sample at the resurgence. The percentage in each class is obtained by weighing that material collected on the mesh of each sieve.

In more detailed investigations employing sieves of several mesh sizes, the results may be subjected to more sophisticated analyses. The simple histogram presentation is replaced by plotting the size of material along a phi scale against the cumulative percentage, by weight, of the total sample in each class. Such a graph makes it more easy to calculate the statistics of central tendency and dispersion, for example the mean size of the sample which is equal to the sum of the size values for 16% and 84% divided by 2. Further examples are given by King (1966, pages 279–85) and in our section on historical deposits in chapter 8. The use of a probability scale on the ordinate of the graph means that a normally distributed sample will plot as a straight line. This is helpful since large departures from normality often signify that the sediments are worthy of further investigation.

We may summarise our uses of the size ranges of fluvial (mainly bed load) sediments as:

a) The study of relationships between particle size and streamflow or other channel characteristics.
b) The detection of the downstream changes brought about by abrasion or attrition. Leopold, Wolman and Miller (page 193) show the steady reduction in bed material size downstream in the Mississippi. Care should be taken in such studies that the changes observed are not merely the result of *preferential transport* of a certain shape or size of material. Thus if angular sand grains are preferentially transported the careless researcher could well conclude that sand gets more angular as it moves down stream; of course quite the reverse is true.

Though reduction in size indicates comminution by impact during fluvial transport, one cannot conclude this on size evidence alone. Thus the measurement of **particle shape** has become important to the study of the mechanical aspects of erosion. Generally, particles resulting from weathering are *angular*. The major exception to this rule is that grains from sandstones which have already been rounded once by transport, before deposition and lithification, are well rounded from the start. Since the most unstable section of the particle is its corner this becomes progressively removed and the particle becomes *rounded.* Again we have the example of an imprecise ordinal scale requiring quantification to allow more objective use. A variety of scales has been devised for measuring shapes on the continuum between *angular* and *rounded.* They often use the geometry of the perfect circle or sphere as a basis for comparison (see King, 1966, pages 291–6). The example shown in fig. 6.8 is not one described by King. Fig. 6.9 shows its uses. Values of the index for a few typical shapes can be given on a chart and this is a quick way to assess the angularity of a large number of particles, rather than measuring each one according to the method shown. Obviously, particles of sand size and smaller require mounting on a microscope slide for analysis. Photography may then be used so that actual drawings and measurement can take place later on a flat surface (see plate 6.1). It is especially convenient to perform microscope inspection with a polarising microscope so that mineralogical analyses can be performed simultaneously. However, the techniques are difficult and used mainly in geological research. Problems may also be encountered in measuring the angularity of large boulders such as those shown in plate 6.2. The origin of the sediment is, however, important and the larger particles should be analysed into major rock types and compared with the rock outcrops traversed in the bed of the stream. Do different rock types produce different sizes and shapes of sediment? How does this affect the stream? The mineralogy of fluvial sediment has been useful to prospectors since the days of the gold rushes and is now a refined geochemical technique.

Much of the study of abrasion and attrition and their effect on the size and shapes of sediment, has been conducted in labora-

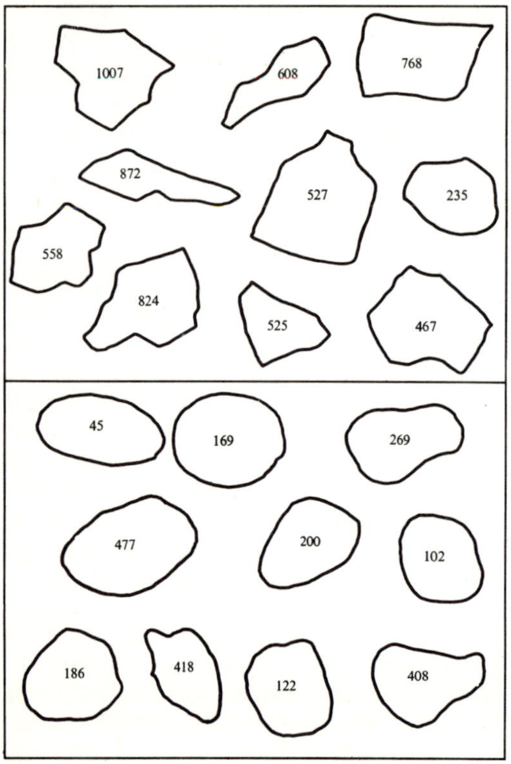

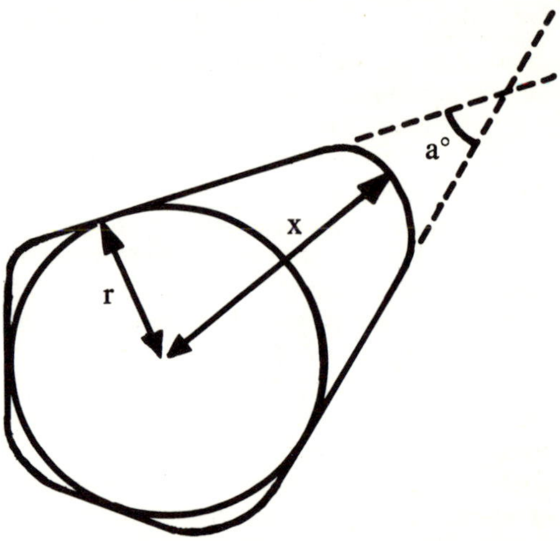

Fig. 6.8 Lees' angularity index. The angularity of each corner of the grain equals $(180° - a°) . \frac{x}{r}$ *r* is the radius of the maximum possible circle which fits the figure. The angularity of the grain is the sum of all its corners

Fig. 6.9 The angularity of the sand particles sieved in Fig. 6.6. Are the angularity figures for the two samples statistically different?

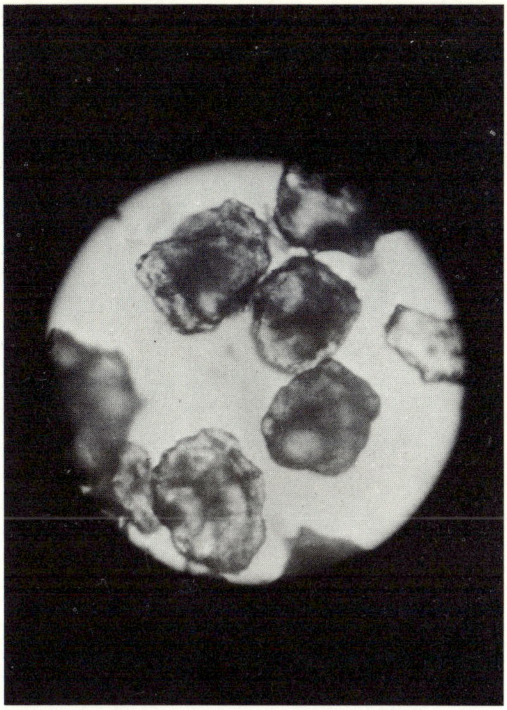

Plate 6.1 shows a microscope slide of fine sand (x200). Using tracing paper calculate the angularity of the grains. Plate 6.2 shows sediment of a much larger size on the bed of the River Tees. Notice how rounded the boulders are, only a few kilometres downstream of their origin. However, can we be sure that historical processes are not the cause?

tory experiments. These have the advantage of being free from processes like frost or glacial action and preferential transport, which may have affected pebbles collected from streams in the field. Such important factors as velocity can be controlled. At the simplest level a closed vessel containing water and sediment of known size and shape can be agitated until changes in one of the variables is suspected. Mechanical revolution of such an apparatus can be achieved with a motor such as that which drives the turntable of an old record player. Do not expect the changes to be rapid or profound unless soft material is used. Such methods have a practical application in the testing of aggregates used for road construction. The shape of sedimentary particles is also mentioned in chapters 4 and 8.

6A 4 Sediment load and flow

By measuring size and shape of sediment, mainly bed load, we have been concerned with changes which occur with site. The work done by the stream has been assessed by reference to the changes it has wrought between the sediment at one location and that at another downstream. The only way, however, to measure this work in terms of the total movement of material out of the drainage basin is to sample the sediment suspended in the flowing water or moving along the bed.

Suspended sediment is measured by sampling the concentration of sediment in the water. The problem is similar to that of discharge measurement—the whole stream

Fig. 6.10 The US.DH.48 suspended sediment sampler

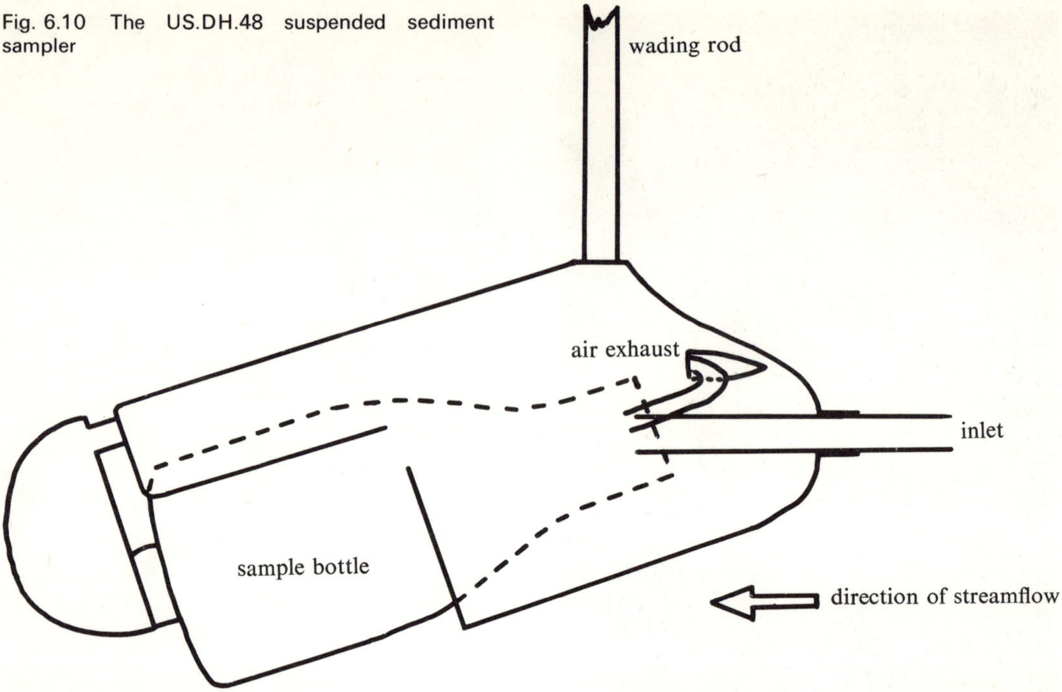

cannot very well be collected in a bucket and the sediment allowed to settle out! Instead, as with current meter measurement of stream velocity, a few representative locations across the stream are chosen and a specifically-designed bottle sampler lowered into the stream (fig. 6.10). Because the distribution of suspended sediment is variable with depth (like velocity but less regular) the sampler is gently lowered and raised in the water until the bottle is filled. Each vertical zone of the stream thus contributes equally. This is **integrated sampling** since the water and sediment in the sample will be an integration of the variety of concentrations which exist with depth. The sampler is specially designed not to deflect or concentrate the flow of water into the nozzle and it fills evenly because air is allowed out through the exhaust. Once filled the bottle can be removed and a clean empty one put in the metal 'fish' which forms the body of the sampler. The sampler is raised and lowered on a rod held, in small streams, by someone wading in the stream (plate 6.3). The sampler is held well upstream of the person operating it so that he does not stir up extra sediment.

A simple version of the sampler can be constructed from any streamlined bottle (a milk-bottle will do) fitted with a rubber bung. Through the bung are bored two holes. Into one is pushed a length of copper tube to be the water inlet. This is a straight tube. Into the other is pushed another piece of copper tube, this time bent back. This forms the air exhaust from which bubbles will be seen to emerge as the instrument is lowered into the water, tubes upstream, with a retort stand and clamp borrowed from the chemistry laboratory. Details are given by Gregory and Walling (1971).

Materials and apparatus used by the chemist may also be useful for determining the concentration of sediment in the sample. Very accurate balances are required as sediment concentrations are usually low. After shaking well, the sample should be filtered through a previously dried, weighed glass-fibre filter paper. A vacuum pump may be used to speed up filtration. After it is complete the paper is carefully removed from the funnel (a Buchner type is best) and again dried in an oven before reweighing. On both occasions the paper should be left

Plate 6.3 A US.DH.48 sediment sampler being used in a small stream. The operator stands well clear of the instrument (as in current metering). Both current metering and sediment measurements may be done on large rivers using bigger instruments which are lowered from bridges or cableways.

for a while after emerging from the oven to take up moisture from the atmosphere—or unstable weights will be recorded. The change in weight of the paper is the mass of sediment collected in the bottle of water. The volume of water is known. Concentrations are usually expressed in milligrammes per litre. At low flows it may require very precise weighing to detect the very small amounts of suspended sediment.

If flow measurements are taken at the same time the relationship between flow and suspended sediment can be observed. A typical example is shown by one of the lines in fig. 6.11. Rivers become *turbid*—discoloured by sediment—in flood and the increase in load is rapid. Obviously, if the flow is measured continuously and recorded on a chart, a year's flow data may be used to calculate the total amount of suspended sediment transported out of the drainage basin in that year. This has been done in the United States to assess the effects of various agricultural practices in soil erosion. In Britain the effect of urbanisation on sediment load is being studied. From the geomorphologist's point of view this is a very good way to calculate the rate of erosion in the basin—such rates are too slow to spot with the naked eye! We will return to this point, and the importance of floods.

FIELDWORK

Using a single milk-bottle sampler try to establish the relationship between rainfall, streamflow increases and the amount of sediment in the stream. If the cloudy colour of the stream is found to be related to the concentration of sediment, one may well be able to 'guess' the concentration or build a simple device to measure turbidity better.

Is more sediment carried in suspension while the stream level is rising or falling? Does it vary seasonally? What is the result of calculating the volume of sediment removed during your period of observations? Express it as the depth of soil which it would represent if laid all over the catchment which the stream drains. Is the catchment ploughed land, pasture or urban?

One element of sediment load whose concentration has not been measured so far is *bed load.* Since this consists of quite coarse particles which move on or near the bed of the stream, sampling with a bottle is out of the question. There are, however, formulae which link the suspended sediment concentration obtained from the above procedures to the total load, suspended and bed. There are two simple field methods of directly measuring bed load which are recommended. One is to install a box or *trap* in the bed of the stream, into which bed load will tumble over a known period of time. The trap can be removed for weighing, sieving and geological analysis (see Gregory and Walling, 1971). Another method is to proof the bed of the pool which forms upstream of a weir (for discharge measurement) with a sheet of polythene. As bed load encounters the more slowly-moving water in the pool it is deposited and the polythene can be removed, enclosing the load, after a period. Such a stilling-pool can also be produced by damming a small stream with cobbles. Both methods, however, disturb the streamflow and may lead to deposition of suspended load too. Since spectacular movement of boulders often occurs on the beds of streams during floods it may well be that your bed load sampling will be done after such an event. If it has been a big flood the bed load may well be on the normal banks of the stream, across a road or in someone's garden! For simple calculation of bed load concentrations a figure of ten per cent of the suspended load concentration has been assumed, this being the average value discovered in the United States where frequent measurements are made. However, measurements in Devon, one of the few experimental areas for bed load, give figures of only one or two per cent (University of Exeter).

6A 5 Dissolved load

Though the chemical load of rivers is not generally visible, unless it appears as a deposit in your kettle at home, it is far from being unimportant in the general fluvial lowering of the land surface. Some rock types attract the attention of geomorphologists measuring erosion because the dominantly chemical nature of their erosion allows the rate of erosion to be easily measured. Areas of karst limestones and chalk have received much attention. However, most mineral substances are susceptible to chemical solution and, as well as bicarbonates from limestones, there are chlorides and sulphates, in addition to the metallic ions of sodium and calcium. In over half of the major physiographic regions of the United States and Canada the dissolved load of rivers exceeds the solid load over the passage of a year. The dynamics of solutional erosion differ from those of mechanical action. Solution is favoured in most cases by the presence of acidity, mostly as provided by vegetation, and contact between the reactants over a large surface area. As with soils, this acidity can be measured with pH indicator papers, or with a portable meter. Try measuring the pH of water oozing out of peat. Thus, groundwater, frequently enriched with hydrogen ions for passage through an acidic soil or peat and moving slowly through pores or joints in the rock, effects most solutional erosion. From studies in limestone areas it appears that the site for much of this solutional erosion may be just beneath the soil. The classic *limestone pavements* of Malham and elsewhere in the Pennines are thought to have developed under a soil cover which has since been eroded away. Obviously, the effects of such erosion are really dramatic only in the case of karst limestone areas where the characteristic closed depressions, or *dolines*, are often formed by localised solution around joints in the rock. Dolines, of course, frequently lead underground to the most spectacular of solutional features, *caves*.

The example of karst erosion studies is useful because of the relatively simple field and laboratory methods used to obtain the concentration (in milligrammes per litre) of calcium carbonate. Since dissolved limestone is equally distributed throughout the vertical stream profile, one dip of a bottle will adequately sample the stream at that site. In this case a polythene bottle of about 250 ml capacity is best. Care should be taken in transporting the bottle to the laboratory where *titrations* with a chemical known as EDTA are performed. Simple reagents are used but care is necessary to establish the colour change which gives the result. Samples from the sink-hole and limestone spring will reveal the changes due to underground solution of limestone in between. If possible, drips in caves should be collected and analysed. One can often obtain the permission of show-cave owners to do this without special equipment. pH measurements also reveal the gradual neutralizing effect which the solution of limestone has on natural acidity.

Chemistry textbooks present the precise laboratory methods. Titration can also be used for analysing water to detect other minerals but calcium carbonate measurements in karst areas are the most easily linked to discharge measurements (see fig. 6.11) and to local landforms.

It is worth remembering that some parts of the chemical load of rivers are contributed by man—industrial pollutants in big rivers and agricultural chemicals in small streams. It is possible that useful studies of pollution may accompany those of solutional erosion.

FIELD AND LABORATORY WORK

There is another technique by which the dissolved load of streams may be quickly calculated. A sample of stream water is evaporated to dryness using the procedure employed in chemistry. The concentration of all dissolved substances is equal to the weight of residue divided by the volume of stream water in litres (answer in milligrammes per litre to allow comparison with other methods). How does dissolved load vary with flow? What rocks, other than limestone, yield the most dissolved load?

For those in urban areas the pH and biotic content of water will yield data on the degree to which it is polluted.

6B Slope forms and Dynamics

It was stressed earlier that most of the landscape is composed of sloping land. However, until the recent development of instruments capable of measuring slope processes, the less perceptible movements of water and materials which occur on slopes made most geomorphologists concentrate on stream channels. Indeed, much of the early theorizing about slope development was conducted without precise measurements of the form of slopes. Man considers that his eye is a good estimator of the shape and angle

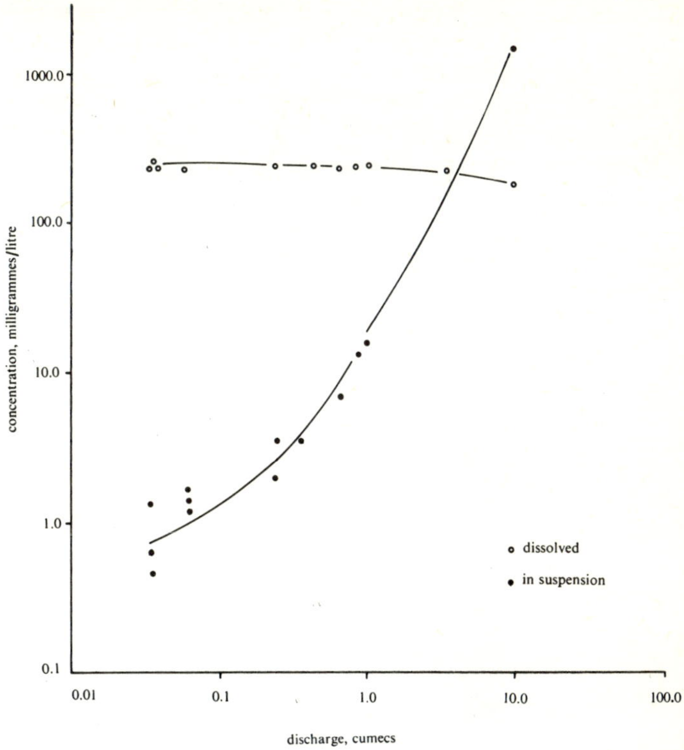

Fig. 6.11 The relationship of erosion products to streamflow, often called solute and sediment rating curves; taken from the River Axe at Cheddar

of a slope. Those of you who have toiled up hillsides which appeared gentle from the road below will realise that visual perception is far from an accurate aid to measurement of such features. Chapter 3 has already mentioned climatic work on slopes. Waters (1958) encouraged geomorphology students to map slopes in the field by means of a technique called **morphological mapping**. In this one is forced to actually tramp over the land and draw in lines representing sharp or gentle *breaks of slope* (changes in angle) on a large-scale map. The difficulty experienced in successfully completing such a map (fig. 6.12) without putting in lines which slope two ways simultaneously (or cross) reveals to the field student the complexity of slopes. If completed correctly it will give a good visual impression of the integration of slopes towards the channel network—a functional necessity of the drainage basin.

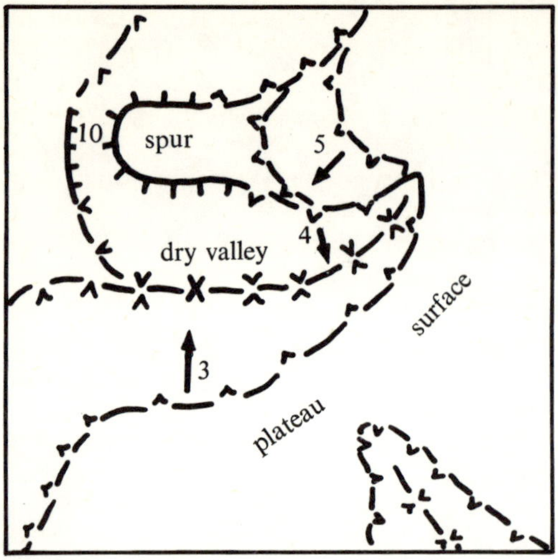

Fig. 6.12 Morphological map of an area of downland. For symbols see R. S. Waters' paper

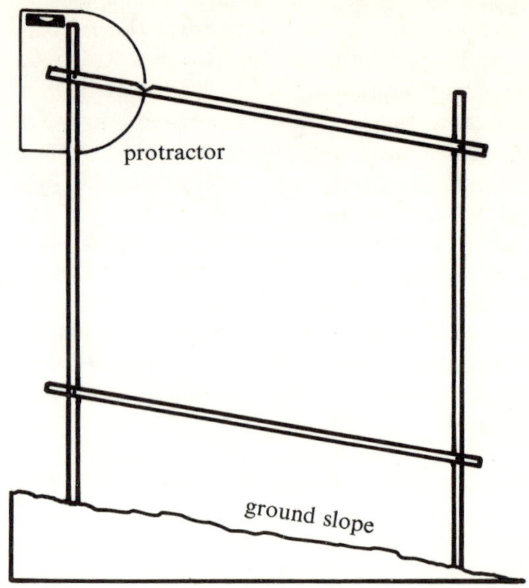

Fig. 6.13 The slope pantometer (after Pitty)

Another useful field exercise is to add angles to the map, measured with an Abney Level or simple clinometer, sighted on a fellow student or eye-height object, preferably upslope. As indicated in chapter 3, it is important to measure the maximum slope of the valley side, not the apparent slope. In other words, one measures the angle along a line at right angles to the contours. Such isolated measurements are not as useful, however, as **slope profiling**. Pitty (1968) has devised a simple apparatus called a **pantometer** for measuring slopes. It is simple to construct, from four pieces of wood, a protractor and a spirit level. It can be operated by one person and the fact that the slope is divided into equal lengths (between the legs of the pantometer) means that the sampled angles are objectively chosen. Slope angles are simply read from the protractor (fig. 6.13).

Of course, slope maps may also be prepared from contour maps—though only a large-scale map (1:25000 or larger) is worth contemplating because contours of at least 25-feet intervals are required to give meaningful results. Even so the angles calculated do not often agree with those measured in the field. This is mainly the

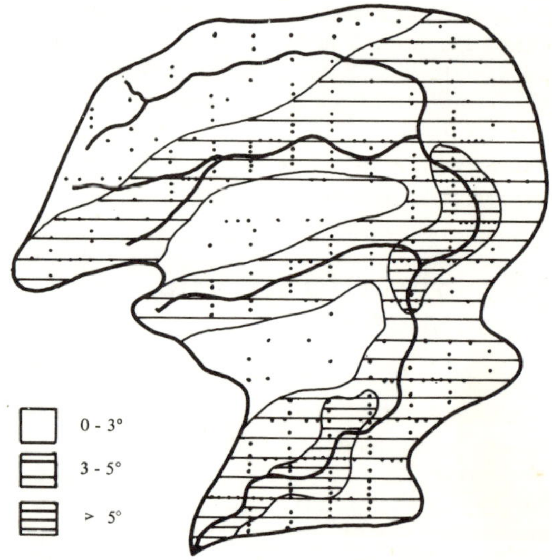

Fig. 6.14 A slope map prepared by counting the intersections of contours with a grid of lines (shown as dots). By counting the intersections and using a conversion factor an angular measurement can be obtained for each square although the correspondence with field measurements is not good

effect of reduced scale—many relief features not being represented by contours, though they affect angular measurements in the field. It should be borne in mind that the Ordnance Survey contours are merely form lines (that is they are not mapped in by levelling at that precise altitude) in three cases out of four. Several methods of slope map construction are described by Monkhouse and Wilkinson (1963, pages 97–112). One is shown in fig. 6.14. A further method is suggested in work to which we will refer when dealing with the effect of aspect on slopes. As well as being basic data to the geomorphologist and, as demonstrated by chapter 3, to the climatologist, information on slopes is of great military and engineering value. Notable investigations have been carried out in Australia into *land systems*—repeated patterns of slopes with strongly associated soils and vegetation features which will help resource planning in such a large and uncharted country. Land systems investigations in Britain have had a military bias, for instance the investigation of slope and soil moisture properties as a guide to their suitability for military transport, including tanks, if conventional combat were to be waged.

6B 1 Analysis of slope data

Because the human eye is bad at estimating slopes correctly we have advocated instrumental measurement of slope angles, best done actually standing on the slope in question. Even so, the eye can still be misled with the data produced. Since the task of drawing a straight or curved line to fit the surveyed points resembles that of drawing a line by eye in regression analysis, it follows that a more objective way of describing slopes once measured is to fit equations to the data, relating altitude to distance. This inevitably smooths the slope, a feature which may be undesirable. The relationship may be simply linear or correspond with some of the more complicated equations shown in chapter 1. Such equations also give a more objective way of detecting differences between slopes; for instance, between those on different rocks or on opposite sides of valleys. However, the smoothing may remove those actual irregularities which are the centre of interest. With the addition of data from the third dimension, **trend surface analysis** will provide equations which idealise the whole set of slopes in an area rather than selected profiles. Again the equations may be simply linear (being the equation of a tilting flat surface, over and below which actual variations of altitude are minimised) or undulating.

A more common means of dealing with slope data, this time with the sample of angles measured, perhaps during the preparation of a morphological map, is to investigate their *statistical distribution*. The data is plotted, as in fig. 615 with angle of slope on the abscissa and the frequency of its occurrence on the ordinate. There are differing opinions on how the frequency data should be derived. Obviously if 50 out of 100 slope angles measured are 5° this represents a frequency of 50 per cent. However, if the sample had been gathered almost totally from those slopes of easy access in the drainage basin this would be a biased sample. Slope angles should either be gathered randomly or they should be

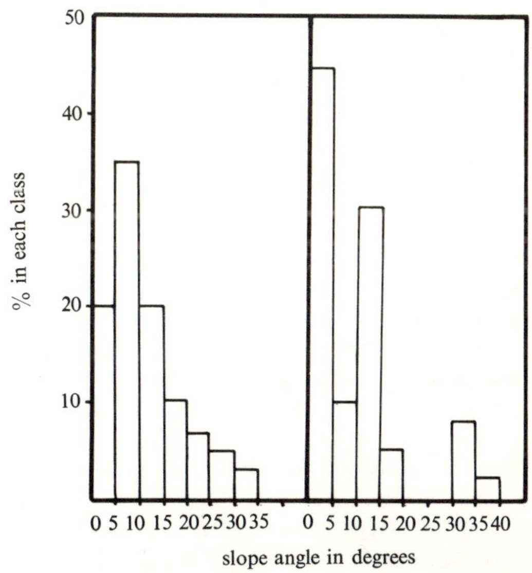

Fig. 6.15 Slope frequency histograms for two drainage basins

weighted according to their area in the basin—in which case 5° slopes would have to cover $26 km^2$ in a $52 km^2$ basin to obtain a frequency of 50 per cent. Beware too that the area a slope covers is more than that shown on a flat map. Using the mean slope angles derived from a series of such histograms for different basins, or different rock types, comparison can only be achieved objectively by **analysis of variance**. This assesses whether the means are significantly different. Using analysis of variance the mean angle of slope on different rock types has been found to differ significantly in some areas.

In some cases the histogram of slope on different rock types has been found to differ significantly in some areas.

In some cases the histogram of slope angles has more than one peak. In such cases the angles occurring most frequently are called the **characteristic angles** and may be found to be typical of one rock type, one location or of one type of landform relic from processes (such as glaciation) not now active. In urban areas the slopes of man-made waste, such as slag heaps, demonstrate typical angles. It has been suggested that characteristic angles result from a typical *angle of repose* of particles, produced by weathering or even industrial process in the case of slag heaps or with other artificial particles (see plate 6.4). The concept of repose angles will be familiar to those who have tried to pile their spoons high with materials of different size and shape during a meal! Consider, for instance, the difference between castor sugar and icing sugar, coffee and cornflakes. However, slopes do not often attain their form merely by the piling up of material produced by weathering. It is to the processes of slope development and their measurement that we now turn.

LABORATORY WORK

Place two pieces of board flat on the laboratory table, edge to edge. Place rock chippings across the line of contact between the two boards and build up a pile of soil. If they are now pulled gradually apart the pile of material will be split in two and the innermost slopes should be characteristic of this grade of material. Perform the experiment several times on each grade of material to obtain a scatter of values. Are the means of these values significantly different between various materials?

How do the angular measurements of these slopes differ from those found in the field? Why?

Plate 6.4 Artificial slopes. These two heaps of roadstone are of different grades. Is there a difference in slope angle or are we in a bad position to judge?

6B 2 Slope development

There are three important factors which influence the slopes we measure in the drainage basin of today:

a) The original form of the slope. The earth's surface has relief and hence slopes because a variety of internal, *endogenetic* processes have raised parts of it to considerable elevations above sea level. The initial slopes caused by such movement will depend on the rapidity of the uplift and the material which is being raised. It is important to remember that imperceptible uplift is locally continuing and that exogenetic processes of weathering and erosion are not withheld from the land surface during this. Many of the older theories of slope development envisaged separate periods of uplift, followed by slope and river development.

b) Changes in erosion processes over time. Both the weathering and transport of materials on slopes are affected by climate. In considering slopes throughout most of the British Isles it is important to realise the influence of Quaternary glaciation and periglaciation. The slopes may be mantled by material deposited there by glaciers and present a form quite unlike the bedrock profile they hide. The valley floors may be filled by a rock sludge derived by freezing and thawing on the rock outcrops and even gentle slopes above.

c) The activity of the stream at the base of the slope is important. It removes the material conveyed to it from the surrounding slopes. It can move around the bottom of the valley and undercut its banks. Thus the slipping of material on slopes and the removal of sediment from stream banks are but parts of a continuum involved in drainage basin dynamics (see plates 6.5 and 6.6). Historically it is likely that the stream has been much wider than it is now during the melting of glaciers or snow fields.

d) Man has had a considerable influence on slope development through agricultural and industrial activity.

Plate 6.5 shows the effects of soil creep on slopes oversteepened by the action of the stream below. Plate 6.6 shows bank material (including sediment from upslope) being removed by the stream (mainly in large blocks which collapse away after under cutting)

Because of the complexities introduced by the above factors, many workers have adopted more controllable conditions in which to study slope development; either with laboratory experiments, or by studying man-made slopes like slag heaps, or by developing mathematical models which assume controlled conditions. Whatever method is used, there are two basic processes affecting slopes: the *production of material* by weathering of bedrock or the comminution of already weathered material and the *transport* of such material downslope. In the early models of slope development a great division occurred between those who said that slopes *retreated* from the valley bottom maintaining a constant angle, and those who said that slope angle progressively declined. In fact, both situations occur and may be thought of as those dominated by

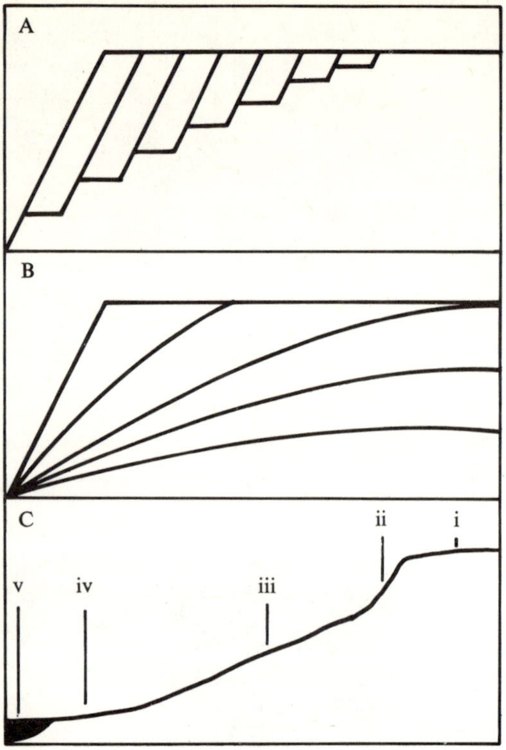

Fig. 6.16 Concepts of slope development: (A) Parallel retreat; (B) Downwasting; (C) Mixed processes acting on a typical temperate slope form. The processes are: i) subsoil weathering, ii) fall and slide, iii) mass movement, surface and subsurface water flows, iv) deposition and some surface water action, v) channel processes (an important influence on slope development)

the weathering of bedrock (parallel retreat) and the transport of waste (decline). Both kinds of slope development are shown in fig. 6.16, together with a typical profile found in the British Isles which indicates the presence of both on one slope. It is instructive to measure a series of profiles in the field and superimpose them graphically.

Because of the known influence of the two basic processes on slope form there have been several investigations into the different angles shown by slopes on opposite sides of valleys. Such *asymmetrical valleys* in the Chilterns have steeper west- and south-facing slopes which were possibly developed because of the greater number of freezing and thawing cycles which they would have experienced during the Pleistocene periglacial conditions, leading to breakdown and transport of the soft chalk to the stream below. The fact that the valleys are now dry and that periglaciation has ceased, emphasises the importance of climatic changes on slope development. In Yorkshire it has been suggested that west-facing slopes above 1000 feet are steeper because of the influence of prevailing winds on snow cover, whilst below 1000 feet north-facing slopes tend to be steeper because of differences in sunshine (again both in the Pleistocene). Present-day studies of slope development are badly needed to throw light on asymmetry. As shown in chapter 3, present-day climate also varies with slope aspect and should not be neglected. Certainly climatic explanations of assymmetry often seem more realistic than the many structural (geological) ones which were formerly ventured.

6B 3 Slope processes

Although *weathering* is the source of the particles we find on slopes the measurement of weathering will not be fully dealt with here. Chorley deals with the role of water in both chemical and mechanical weathering, *Water, Earth and Man* (pp. 135–55). Since solution is a very active process in soil profiles the removal of matter dissolved in soil and groundwater must not be neglected in a study of slope development. As regards mechanical comminution on slopes this avails itself of similar methods to those used for such processes in streams. Usually, however, the particles are much finer. For particle size analyses in the finer ranges, sieves cannot be used and apparatus is used which times the settling of the particles in a column of water. Since the size of a particle has a theoretical relationship with its settling velocity the amount of settling after set intervals of time can be used to calculate the distribution of grain sizes in the total sample (see chapter 4). One expects very fine material to move downslope in suspension, either through minute pores in the soil or by surface routes.

Surface movement or *wash* of fine material is obviously important on slopes which experience heavy rainfall, or are damp enough, to produce overland flow. It is difficult to measure the amount of material removed in this way but some sort of *trough* is clearly the most direct way. Let into the soil profile, the trough collects material washed into it from the slope above. Obviously it is easy to disturb the soil when inserting the trough in a way likely to encourage the upslope soil particles to fall into it.

FIELDWORK

Rain-splash erosion can be demonstrated and measured by surrounding a control plot of bare earth with clean white paper, well secured. After heavy rain it can be carefully removed, dried and the material splashed on from the control plot brushed off for weighing. The experiment can be set up in the garden or the park.

Both wash and rain-splash are important where slopes lack the protective cover of vegetation. It has been discovered that in South Wales, where atmospheric pollution at the height of industrialisation removed vegetation from steep slopes, wash has removed almost half a metre of soil from these slopes in under 200 years. Such historical movements on slopes can be measured if buried soil profiles are discovered or if the present-day ones appear truncated of their upper horizons. Vegetation both protects the soil from rain-splash with its leaves and binds the upper horizons against wash with its roots. It is scarcely surprising therefore that the study of these processes is well developed in the U.S.A. and New Zealand; countries with severe erosion problems caused by unprecedented rates of forest clearance at the hands of early settlers. This country also is not immune during land clearance for building or road making.

Gullying occurs when, during extreme conditions of runoff, the stream network is extended by the development of channels down what have been previously mere local depressions in the slope contributing to the nearby permanent channel. Again they occur mainly in the absence of a vegetation cover but can also be found in forests and in cropped fields. Obviously, certain agricultural practices, like ploughing up and down a slope, will induce gullying. During the period between storms of intense runoff, gullies tend to fill with material moving downslope into them by splash, wash or *creep*.

Creep is the slowest of the forms of *mass-movement* active on slopes. Mass-movements are not directly the result of the transporting power of water on the slope but of the downslope pull of gravity on a mass of soil or rock debris. Soil creep can be demonstrated by the accumulation of soil on the upslope side of field boundaries, or the curved trunks of some trees on slopes, or the miniature terraces found in sloping pastures (often exaggerated as a result of trampling by animals). Measurement of soil creep is obviously a long business requiring precision. Unfortunately, all the methods mentioned by Kirkby (1967) require disturbance of the soil. In the first case a pit is dug into the soil on a slope and along the wall which is running directly downslope, a series of pegs, a few centimetres long and made of coarse wire, are carefully inserted—so that one end is just visible. Their positions are then measured relative to some fixed points. Further pegs driven hard into the floor of the pits so as to be based in bedrock are suitable fixed points. The face with the pegs inserted is proofed with paper and the pits refilled. Some months later the pit is re-excavated, working carefully up to the paper which is then removed to reveal the ends of the pegs. Their movement or otherwise can be assessed by further measurements relative to the fixed points. Another method is to insert stakes from the surface of a slope. Either their position can be surveyed by chaining and levelling from a known point, or they can be equipped with spirit levels which record their tilt during soil creep.

However, any exposed measuring device is liable to be moved by animals or vandals and a further solution suggested is the removal of a large block of soil to the laboratory where it can be set up on a similar slope. If housed in a glass-sided box the movement of pegs inserted in a similar way to those in the wall of the pit can be easily measured. Applications of water can be made and if a sprinkler is used studies can be simultaneously made of rain splash or wash. Kirkby found that expansion and contraction of soil with wetting and drying resulted in a nett downslope movement and that the amount of movement was linearly related to the amount of water. Wetting and drying were stated to be more important than freezing and thawing in soil creep.

There are, of course, mass-movements which are more rapid than soil creep such as *landslides* and *avalanches*. Landslides are usually most widespread after long periods of gentle rain which thoroughly soaks the soil profile. The water in the soil adds immensely to its weight, becomes highly pressurised and acts also as a lubricant, should a film accumulate at certain horizons. In some cases where plant roots penetrate a thin soil mass and spread out at the bedrock contact, the whole soil mass may suddenly slip off like a carpet from a floor. In November 1971 several Japanese scientists were killed during a large-scale outdoor experiment designed to learn more about the disastrous landslides which affect Japan. A steep vegetated slope was sprayed with water for several days before the experiment got out of control and the whole slope moved too rapidly for those taking measurements and film to escape death. Such steep slopes as railway cuttings are frequently affected and cause chaos. The Aberfan coal tip disaster was caused by failure of a man-made slope. While at a stable angle of repose for dry conditions, the tip formed its own rainfall catchment and became unstable when charged with water.

Avalanches are a danger to skiiers and mountaineers but seldom considered as a geomorphological process. The fact is that mechanical breakdown under alpine or glacial conditions is accelerated and occasionally material accumulates to an unstable degree whereafter heavy snow will remove it to a lower level, usually as the snow begins to melt. *Screes* may be considered as the periglacial equivalent. A cliff face, perhaps exposed by glaciers, becomes disintegrated by freezing and thawing. The resultant material builds up an impressive boulder-strewn slope below (see plate 6.7). The screes left after the Pleistocene in Britain are quite active today and experiments can be done with painted boulders placed at surveyed locations on screes Measurement of the angle of rest, compass direction, shape and size of scree blocks produces useful conclusions about the disintegration of rock and its position of rest after falling (plate 6.8). A team of workers on one slope can quickly measure a good sample of fallen blocks. Screes can be studied in the laboratory. An artificial slope is made, with roughness imparted by chicken wire. Chippings of rock are dropped from various heights and the slope angle can also be varied. Recordings are made of the eventual position of rest for a hundred or so chippings. The size and shape of chippings can also be varied. Do results begin to change as falling particles become influenced by those already accumulated?

The present phase of geomorphological process study is likely to become even more biased to slopes as two recent books (Young, 1972 and Kirkby and Carson, 1972) have shown.

6C Rates of erosion

Where have all the techniques so far described for measuring the work done in a drainage basin, both in the channel and on the slopes, been leading us? Many of them have been shown to have applied value but what is their eventual value to the geomorphologist? Along the way he has learned the precise nature of forms by measurement instead of 'eyeballing' and, by a mixture of instrumental records and statistics, he has learned the relationship between processes and these forms. In the end, however, it is the magnitude of the output of materials which stimulates him most, matched as it is,

Plate 6.7 shows a scree slope (middle distance) as yet not fixed by grass like those above it. Compare the angle of the screes. What size of material forms them? Plate 6.8 shows measurement of compass orientation (left) and inclination (right) of scree blocks

paradoxically, with a timescale of development which seems infinite compared with his own lifetime. At this stage the geomorphologist becomes something of a philosopher and indeed it was just this topic which resulted in some of the early geomorphology texts being written by theologians and thinkers rather than experimental scientists.

It can be argued that to work out the average *rate of lowering* for a drainage basin, by whatever technique, is unproductive. The geomorphologist, it may be claimed, seeks to explain the variety of forms in the basin and his preoccupation therefore must be with the localisation of erosion which produce features of note. Thus, while it is relatively easy to calculate the average rate of lowering in limestone country from the solutional loads carried by the streams, most of this solution occurs quite locally between the joints of the rock or at the soil/rock contact. The answer is that rates of lowering are calculated to express the total amount of work done in a drainage basin in terms of simple concepts. To say that several million tonnes of material are removed in a basin during one year means less to any reader than to say that this represents an equivalent of 'x' centimetres lowering over the whole basin in 1000 years. It also allows an easy term by which basins may be compared and over several basins the rate of lowering can be related to climatic or morphometric properties.

In spite of the above justification, few rates of erosion have actually been calculated. Limestone areas have been most favoured due to the pioneer work of Jean Corbel. He proposed this formula for calculating the erosion rate:

$$x = \frac{4\ \mathrm{E.t.n.}}{100}$$

where E is the total annual runoff in decimetres (of rainfall over the basin), it is the average total hardness of the spring water and n is the denominator of the fraction of the surface area of the drainage basic occupied by limestone.

Assuming a rock density of 2500 kgm $^{-3}$, x is the rate of lowering in mm/1000 years or $m^3\,km^{-2}$ annum. Using it, or a modification, rates of lowering of from 10 to 150 mm per 1000 years have been obtained in Ireland, 22 to 102 mm in the Mendip Hills in Somerset and 60 to 190 mm in the Peak District. Limestone is, however, a special case of such studies which has used a set of rather simple assumptions in calculating erosion rates. In more heterogeneous basins where both solution and mechanical comminution lead to the removal of material, the accepted method of calculating erosion rates involves continuous records of stream discharge. From these **flow duration curves** are constructed. The frequency of discharge of various magnitudes is compared with the work performed at that discharge by means of data such as that shown in fig. 6.11. These may be referred to as **sediment and solute rating curves.** If discharge 'x' occurs on 50 per cent of occasions it occurs on 182 days a year. During the period over which this flow prevails the total removal of both sediment and solutes is calculated by working out the nett volume of water flowing during the period and multiplying this figure by the average concentration of both types of load established for such a flow.

A simpler method but one which can only be used on rare occasions is to calculate the volume of material deposited in a reservoir over a period of years. The best opportunity occurs if the reservoir is drained. In 1956 the Strines Reservoir, Sheffield was drained and Young (1969) calculated that some 85 000 m^3 of sediment had accumulated on the floor of the reservoir since its construction in 1869, representing a rate of lowering of 1016 mm per 1000 years. Young compared all the estimates of rates published at the time and found that the rates were positively related to relief, median values for 'normal' relief being 46 mm per 1000 years and for steep relief 500 mm. Of course, many of the estimates for limestone areas have ignored the solids removed by streams and those for non-limestone areas have ignored the dissolved load. However, the link with relief does seem clear. Young stated that rock type seemed not to have much effect, except for unconsolidated rocks which showed very high rates of erosion. A positive relationship has been found between mean annual effective precipitation (i.e. that causing runoff) and erosion rates, reaching a peak around 250 to 350 mm of annual effective precipitation. This is a figure typical of semi-arid areas. At precipitation figures above this, the relationship with erosion becomes negative (fig. 6.17) because of the increase in vegetation which protects the land surface from erosion.

Returning to the argument that the geomorphologist needs to know where the materials have their source, Kirkby finds that the rate of soil creep on slopes is too slow to directly provide any more than 7 per cent of the material removed in a year by the stream at the foot of those slopes. The remaining 93 per cent, he claims, results

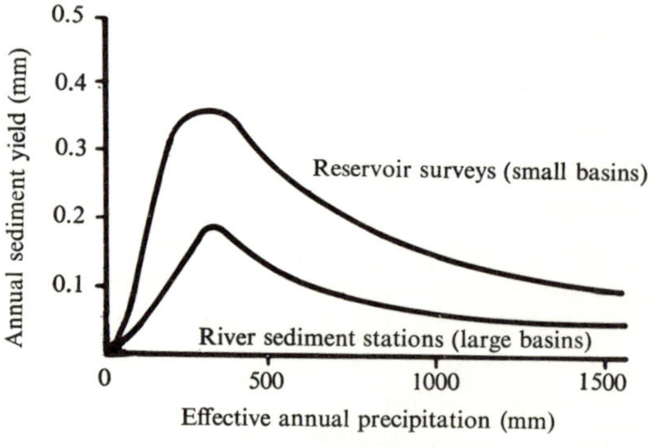

Fig. 6.17 The variation of drainage basin erosion with effective annual precipitation (from the work in the United States done by W. B. Langbein and S. A. Schumm)

from erosion of the bed and banks of the stream itself (see plate 6.6). This is easy to perceive by inspecting the slumped banks of most streams after flood. Others have, however, claimed that at least half the material removed by streams is derived from slopes, usually those on which they claim wash or gullying is active.

Tracing the eroded material back to its bedrock surface could possibly be effected using radioactive techniques, but these would be slow and involve considerable danger. However, instrumental techniques have been recently developed for *directly* measuring the rate of lowering on rock surfaces. All methods employing flow and sediment or dissolved load data may be described as *indirect*. The construction and use of an engineer's dial gauge (used for detecting very small amounts of metal wear), mounted on a rigid steel plate (the whole apparatus being called a **micro-erosion meter**) has proved that such direct methods are accurate and produce meaningful results. To provide a permanent datum, from which the dial gauge can successively record lowering by erosion, studs are driven into the rock surface in a triangular pattern around the experimental site. High and Hanna give results from the limestone floors of cave streams in Ireland which are eroding at a rate of 400 mm per 1000 years.

6C 1 Periods of rapid erosion

Those who have become thoroughly convinced that landscape development in fluvial terrains is a slow process may find this view jolted by a tour of damaged areas after a really big flood. Though much of the effect on human lives is caused by nature reasserting itself to demolish artificial structures put in its way (bridges etc.), the amounts of natural material moved by floodwaters, both on slopes and in channels, are very remarkable. The replotting of fig. 6.11 as fig. 6.18 shows how rapidly the *mass* of both solutional and solid material in streams increases (though the *concentration* of dissolved rock may decrease as fig. 6.11 shows).

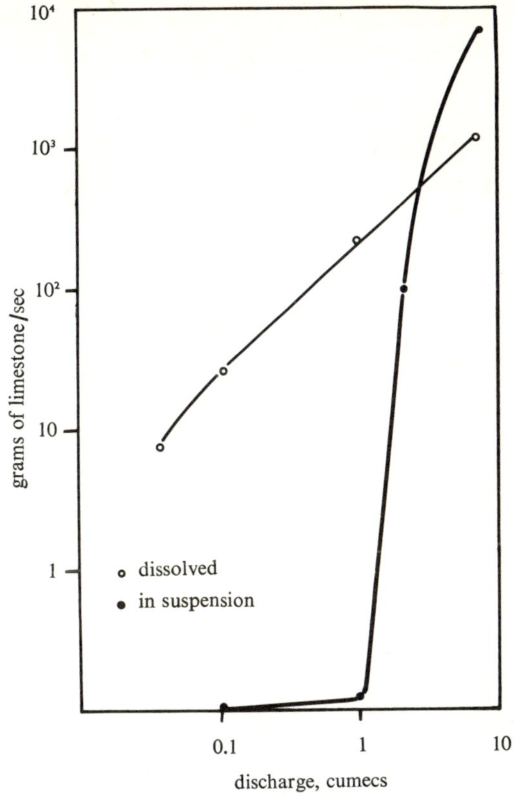

Fig. 6.18 A re-plot of Fig. 6.11 to show the total *mass* of material removed at various discharges (as opposed to its *concentration* in Fig. 6.11)

The two views of landscape processes, one stressing slow, continuous action and the other stressing the role of rare events are known as *uninformitarianism* and *catastrophism*. The merits of each were hotly disputed during the 'theological' era of geomorphology. Luckily we now have statistical means of considering the matter. By a combination of the flow duration curve, flood frequency analysis and sediment/solute rating curves we can assess the role of the rare event. The conclusion is that, in spite of the large amount of work performed by 'rare' floods, their very rarity means that 'normal' processes operating between floods of such proportions do an equal or greater amount of work because of their greater duration. The general relationships are shown in fig. 6.19. From this it is clear that events of moderate

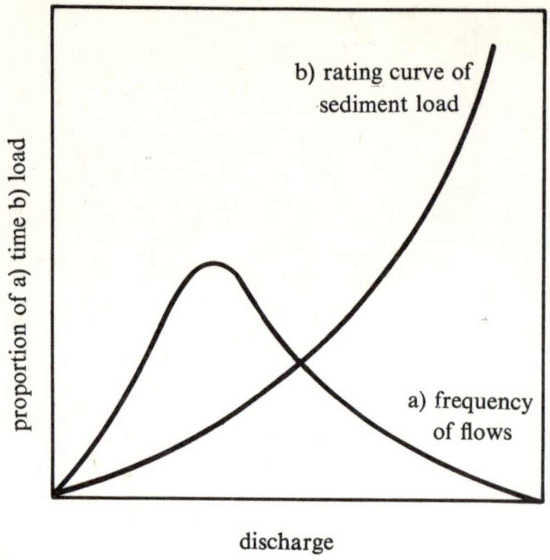

Fig. 6.19 A comparison of the transport of stream load with varying discharges and the frequency of their occurrence. Over the years, at what discharge is most load removed?

Plate 6.9 shows the Thames at bankfull discharge (Wallingford Bridge). At such levels the velocity is too great to permit boats to use the river. Plate 6.10 shows low flow at the same site, with nearly two metres of bank exposed and pleasure craft enjoying the calm conditions

frequency are the most important to the rate of erosion.

The flood event which seems to be most important for the river channel system is that which fills the channel to *bankfull*. This is an impressive site on a large river as plate 6.9 shows of the Thames at Wallingford Bridge. Plate 6.10 shows a similar view during base flow conditions. At bankfull discharge, equilibrium is most closely approached, with velocity identical throughout the channel network and sediment transport nicely balanced to maintain characteristic channel shapes. The frequency of bankfull discharge on a series of British rivers has been found to be approximately 2·2 days per year (that is 0·6 per cent on the flow duration curve, a return period of about half a year). A series of equations relating channel properties to bankfull discharge has been developed:

$$\text{width} = 1{\cdot}65 \; Q_b^{\frac{1}{2}}$$
$$\text{depth} = 0{\cdot}545 \; Q_b^{\frac{1}{3}}$$
$$\text{Area} = 0{\cdot}9 \; Q_b^{5/6}$$
$$\text{mean velocity} = 1{\cdot}112 \; Q_b^{1/6}$$

(Q_b is bankfull discharge)

The relationship between meander dimensions and bankfull discharge has already been mentioned. The scatter of points around the regression line linking width to Q_b is less than with the other relationships suggesting that rivers adjust themselves in width (where possible) more rapidly than in depth or velocity. It is possible, using the equations above, to predict the natural channel properties in reaches affected by engineering works, allowing bank strengthening or dredging to be planned realistically by river authorities.

We may conclude this section with the view that, though floods are of dominant importance in doing work in the drainage basin, this tendency is less strong beyond floods of moderate size. Thereafter our thoughts of a catastrophic explanation for the fluvial landscape may result from self-interest since when rivers escape their banks it is man who suffers most!

7 Shoreline Geomorphology

Though the coastline of Britain is within easy reach of everyone, and has been the traditional magnet for those seeking recreation since the railways were built, as a topic of serious study it has been left to those with the resources and time to cope with the large scale of coastal features and processes. We may here draw a contrast with fluvial geomorphology, where small isolated field units like mountain streams may demonstrate to the researcher in great clarity the factors operating on the larger scale elements of the fluvial system. An individual can frequently cope with the instrumentation and description of such a unit. On the coast, however, it is more likely that boat trips must be made or even diving undertaken to make the parallel measurements to those possible in the drainage basin. Many kilometres of travel may also be required of the shore-based personnel. These may be some of the reasons for the currently less widespread instrumental work on coastal processes round our shores.

Coastal geomorphological studies, therefore, tend to be in the hands of agencies like the Admiralty, harbour boards, the Unit of Coastal Sedimentation, the Nature Conservancy and the Hydraulics Research Station. Attention is still focused largely on the spectacular and sometimes unique examples of coastal activity such as Orfordness, Chesil Beach, Dungeness and Holderness. It is fair to say, however, that the incentive given to fluvial geomorphology by hydrology and its measurements may soon have a parallel in coastal studies as oceanography develops under similar environmental pressures to those behind the growth of hydrology. Coastal processes must be properly understood in a variety of engineering fields, as well as now in the broader aspects of resource planning.

Despite operational disadvantages, there are some advantages to coastal geomorphology as a theme of study, inherent in the nature of coasts. The growth of spits and the erosion of unconsolidated cliffs are perhaps two of the most thoroughly documented aspects of geomorphology; enough evidence being available in the form of old maps, charts and photographs to enable useful quantitative estimates of the rates at which processes are operating. With such rapid rates of change in areas of unconsolidated materials like sand banks, shingle bars, or cliffs of glacial material, it is easy to see why most of the work so far on coasts concerns these materials rather than coasts in hard rocks. However, since the latter tend to preserve better records of past sea-levels, in the form of raised beaches and wave-cut platforms, they receive attention from historical geomorphologists. Since these geologically recent changes of sea level, both upwards and downwards, are uniquely important as a preface to the study of today's coasts they are dealt with first.

7A The moving shoreline

7A 1 Changing sea levels

Shorelines move in two ways; one is similar to that of stream channels—by erosion and deposition (movement in plan). The other, movement of the sea level itself, is a new factor entering our studies. It, of course, brings changes in plan too.

It is now apparent from tidal records being kept on the east coast of England that sea level is rising there. Because there is also information from recent oil and natural gas exploitation about the faulted floor of the North Sea we can say that the rise of sea

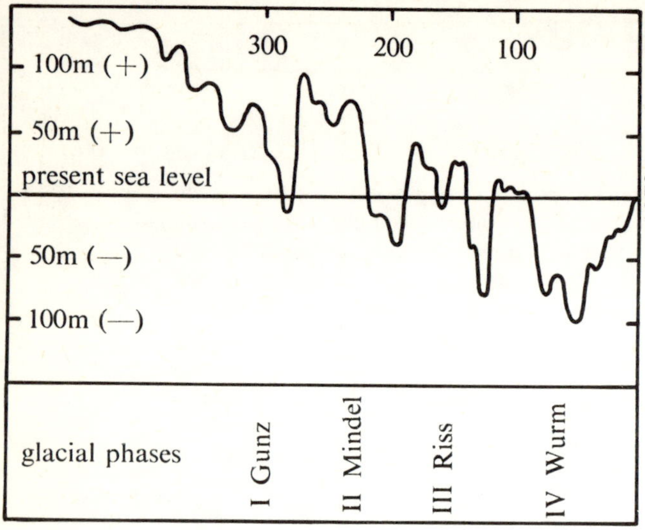

Fig. 7.1 Pleistocene sea levels relative to that of today. The time scale is in millions of years before present

level is only apparent; it is being brought about by the sinking of the land on the east coast by about 2 mm per year. These situations highlight the fact that both land and sea are capable of movement relative to one another. Since topographic survey forces us to measure these changes by reference to a surveying datum, namely sea level today, it is very difficult to be precise about sea-level changes. We usually refer to those changes in which the sea is apparently rising *relative* to the land as *positive* and those in the reverse direction as *negative*, whether their cause be movement of land or sea. It is clear that many positive changes have been enormous in the geological past—it is difficult to imagine that the sea in which our Chalk was deposited in Cretaceous times once covered the tops of what are now the Welsh mountains. The fluctuating levels of land and sea in the geological past are best demonstrated by **palaeogeographic maps** such as those in R. Kay Greswell's book (1963). However, it is with the smaller and more recent changes that we are most concerned since they effect today's coastal features.

The Pleistocene glaciation in coastal Europe is generally considered as having had four phases. The nature and terminology of these will be returned to in chapter 8. For the moment they are described as I, II, III and IV for the purposes of fig. 7.1 which shows the sea level, above or below the present one, which is thought to have prevailed during the four glacial periods and their intervening *interglacials*. Clearly there is a pattern in fig. 7.1. It is one of cyclic oscillations between positive and negative movements becoming less in *amplitude* (or range) until present sea level is reached. Despite the fluctuations caused by glacial phases there is a general trend towards a negative change. Some have used this to add weight to a theory of the earth's expansion. The nature of the evidence on which we base our conclusions about such changes are also shown in fig. 7.1. The causes of the changes are difficult to separate. With glaciation goes a submergence of land under the weight of glaciers. However, at the same time oceanic water is being added to ice sheets and sea level consequently falls. We may distinguish between the *isostatic* adjustments occurring when the land is depressed under the weight of glaciers and the *eustatic* changes of sea level which result from other causes, including that of glacier growth or decay and change in the capacity of oceans resulting from sedimentation by the world's rivers. It has been calculated that the eustatic rise of sea level which would occur if all present-day glaciers and ice sheets melted would be approximately 100m. At their present rate of decay a rise of only a millimetre or so per annum results on a world scale. Parts of Scandinavia and Scotland are still experiencing isostatic recovery of the land surface; the relative movement is therefore negligible or negative. This topic is returned to in chapter 8.

One important omission from fig. 7.1 is a prediction of future sea levels. Virtually no part of the coast of Britain is stable, unless southern England is considered on a European scale and it is likely that further changes will occur, particularly on the North Sea coast.

7A 2 Erosion and deposition

The form of the coast in plan, defined along a nominal scale, needs very little explanation since most of the terms used are in everyday use—bay, beach, headland, cliff, etc. Most of these are based on the form of the feature in plan, though cliff and beach have sectional elements to their definition. Generally, this means that different editions of maps, charts and aerial photographs are of great use, especially in areas of rapid coastal change.

Steers (1953, chapter 3) has shown what sort of historical data may be used in studies of *coastal changes*. Several Cornish churches whose records still remain have themselves been overwhelmed by sand dunes, while Dunwich in Suffolk has lost its churches to an ever-retreating line of sandy cliffs. While many such records contain mention of positions and dimensions their graphic representation on maps is often more useful. Steers shows how maps from the time of Henry VIII and Elizabeth I point to the recent growth of Orfordness. Fig. 7.2 shows how the deposition of shingle had already begun to affect the prosperity of the Medieval port of Orford in the time of Henry VIII. Several other ports in East Anglia have been even worse affected by single deposition, for example Benacre, Thorpe and Blakeney. A study of any particular section of coast could, therefore, well begin in the local library or museum. Of course there are large problems associated with comparisons between maps of such great age and modern maps. Though the scales may be standardised by redrawing, the accuracy of antique surveys was not great and the symbols open to misinterpretation. In more recent times successive editions of maps by accredited surveyors can be used to make actual measurements of coastal changes. Aerial photographs have proved very valuable, particularly of intertidal features on inaccessible beaches and mudflats. The earliest photographs taken aerially of classic coastal features are dated 1924. Carr (1965) uses a mixture of ground survey and aerial photography to plot the growth of Orfordness between 1945 and 1970 (see fig. 7.3). Whilst dealing with surveying it is worth stating that many map-makers have tended to 'lose' errors in their surveying by adjusting the coast line to compensate; the date and author of any map is clearly significant information to depict on the map itself.

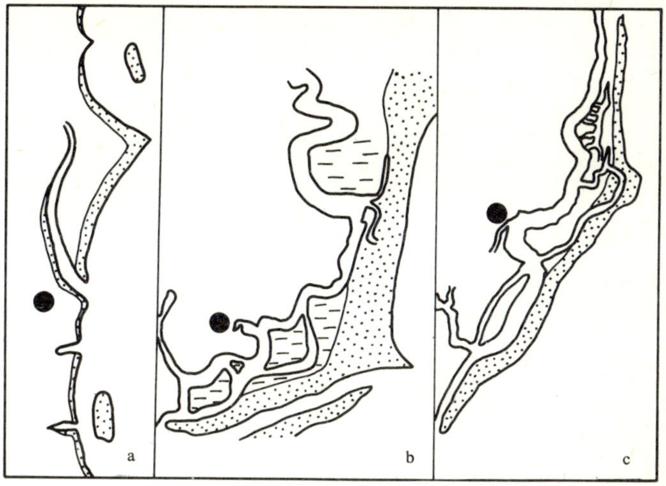

Fig. 7.2 Use of pre-Ordnance Survey maps to compare the shape of Orfordness, Suffolk, in the reigns of (a) Henry VIII, (b) Elizabeth I, and (c) according to the work of Norden, a seventeenth century cartographer. Shingle accumulations are stippled. (after Steers)

CLASSWORK

Most members of the class will have visited the coast for a holiday or seen postcards or photographs. What parts of the coast of Britain have been eroding or depositing? Where are sea walls and other defences strongest?

If possible, old and new Ordnance Survey maps of those areas which are experiencing change should be obtained (fig. 7.4). Look, too, at the Admiralty Charts made for our coastal waters. These are produced by the Hydrographic Department of the Admiralty and are updated at regular intervals, especially in major estuaries. How do sand and mudbanks change?

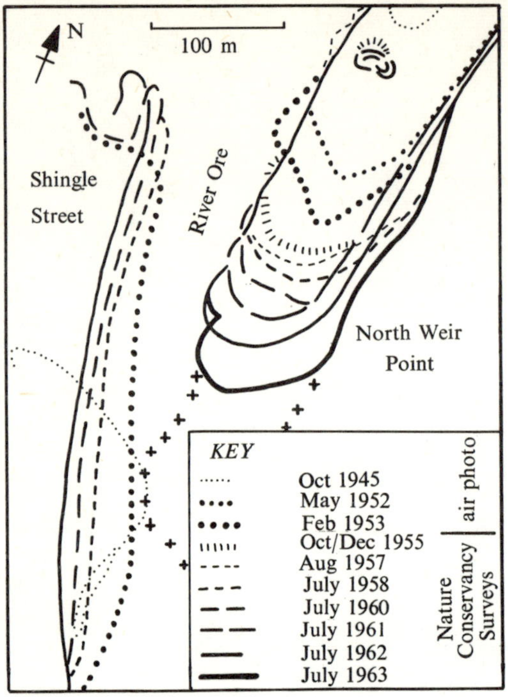

Fig. 7.3 Use of air-photo and ground survey to plot the growth of Orfordness in recent years. The most recent survey to be published, showing the 1970 position cannot be included in full on this diagram, passing as it does behind the Key! It is shown as a row of crosses. (after Carr)

The form of a *beach profile* is dependent on several variables, one of which is the nature of the waves which effect the beach, steep storm waves from onshore winds causing erosion of the foreshore and low flat swells, often the result of offshore winds, building it up. Repeated surveys can be made with conventional surveyors' methods or by using the slope pantometer described in chapter 6 and tying in the survey to a reference point. Such stable reference points may be hard to find on a coast but such buildings as coastguard stations, Martello towers and look-out beacons are obvious choices for locating beach transects. They may also have bench marks from which to to do levelling. Doornkamp and King (1971, page 216) describe the statistical procedures involved in determining that erosion of the beach at Marsden Bay, County Durham, is associated with onshore winds and that fill is associated with offshore winds. In another example (page 220) the two authors plot the relationship between beach slope in Marsden Bay and the steepness of waves. For the results of 40 measurements they calculate a correlation coefficient of $+0{\cdot}52$, meaning a significant positive relationship between the two. There is also a significant relationship between wave period and beach slope, with a coefficient of $+0{\cdot}48$. King (in a collection of essays edited by Steers, 1971) concludes that, because the sand at Marsden Bay is coarse, percolation of the backwash is rapid. Thus, with increasing swash, material is piled up at a steep gradient which is not reduced by a proportional increase in backwash. Beach profiles on finer materials are less steep owing to a greater ratio of backwash to swash. (Wave steepness, period, swash and backwash are defined in section 7B.)

A great number of beach surveys have been made to ascertain the standard form. However, no universal set of relationships has yet been obtained between the size of particles and the force of waves on a beach. However, the analogy with the modern definition of graded river profiles is worth following up. Beaches obviously have several measurable variables subject to dynamic equilibrium relationships. The variable under modification at any particular times is likely to be governed by chance.

In dealing with *cliffed coastlines* it must be borne in mind that many of today's cliffs were formerly valley-side slopes before the general rise of sea level following deglaciation. Thus, the upper parts of the cliff have almost certainly developed by processes with which we are familiar from the drainage basin and are continuing to develop in this way. They are typically convex to the sky and show signs of soil creep. The important part played by the sea is that of an exaggerated stream channel, continually removing the products of weathering, creep and slide from above. Thus, the concavity associated with depostional facets of slope is absent. Very

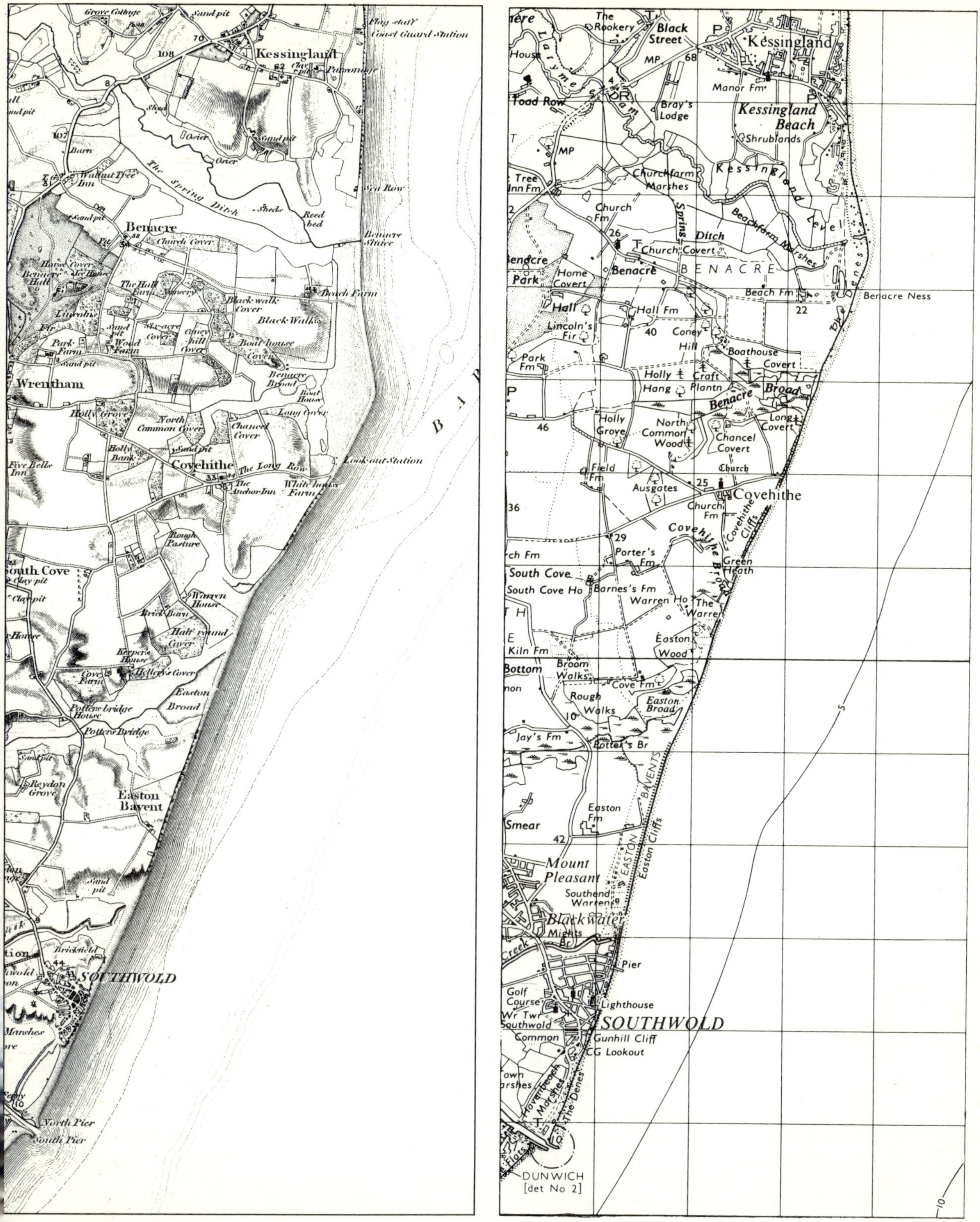

(Crown Copyright. Reproduced by permission, HMSO)

Fig. 7.4 Use of the first edition of the one-inch Ordnance Survey map and that of 1969 to compare the coast of 1837 and 1966

often the sea undercuts a considerable height of cliff and stability is re-established when this overhang collapses. Gradually by continued cliff recession and removal of the debris by the waves, a *wave-cut platform* is cut which extends seaward from the base of the cliff. Cliffed coasts provide good opportunities for slope profiling since headlands can be sketched or photographed, standing out as they do in silhouette, and angular measurements made later.

FIELDWORK

In spite of a common desire to sunbathe or swim on a visit to the coast, a fine day can accomplish the collection of a lot of data on coastlines. Equipment should include simple instruments for measuring slope angles. Where a bay occurs with a central beach and cliffed headlands, a study of both forms is possible. Of how many slope elements is the beach or cliff composed? What is the tidal variation? Does low tide reveal a gently shelving beach or a wave-cut platform? For those near the coast, regular measurements of the beach will reveal if an equilibrium form exists. A photographic record is useful too.

In a study of the retreat of Chalk cliffs, V. J. May (1971) has made use of borough council surveys in Sussex to calculate the changes in cliff line during the last century. He concludes that cliff-top retreat between Seaford and Beachy Head has averaged 0·42m annually between 1872 and 1962. At one point nearly half the loss of the cliff between 1950 and 1962 occurred in the winters of 1959–60 and 1960–61. There seems to be no explanation for this in terms of storms but it may be that undercutting during previous severe winters was stabilised by collapse.

One of the most spectacular cliff-line changes ever recorded was not wholly the result of wave action; groundwater was partly to blame. It occurred at Christmas in 1839, between Axmouth (Devon) and Lyme Regis. Chalk is underlain by sands, and water percolating through the chalk accumulates in the sands because of the clay below, making an extremely unstable stratum. Autumn 1893 was very wet and storm waves had lashed the shore. During one night a great movement occurred seawards, carrying several fields and cottages with it. A chasm over 100m wide opened between an 'island' and the new cliff line, although the sea did not invade the chasm. The story of the event is excitingly recounted, from contemporary sources, by Arber (in Steers, ed., 1971,). The former sea cliffs were pushed further seawards and much reduced in size.

We have demonstrated the importance of particle size and shape in studies of soil and fluvial sediments. We now turn our scale of study to the material on the shoreline for insights to the processes of erosion and deposition. An illustration from Chesil Beach, Dorset, will show the value of measuring the size and shapes of beach material (Carr, 1969). The pebble beach extends over 20 km, increasing progressively in height from west to east. *Longshore drift* operates from west to east. It is a common process caused by the swash of waves breaking at an angle to the beach (in this case from the southwest) while backwash occurs normal to the beach. It can be easily

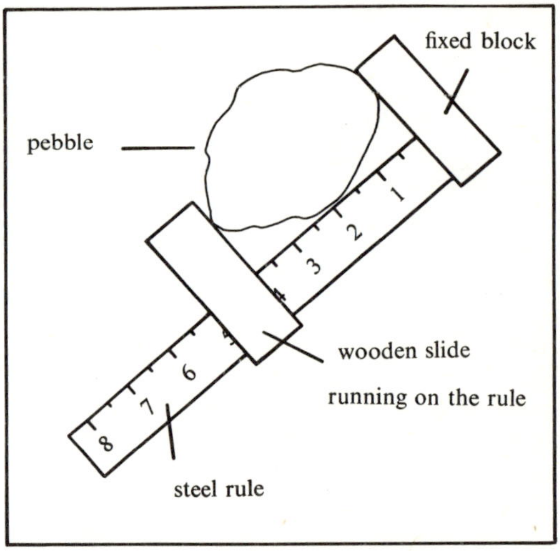

Fig. 7.5 A 'pebble-ometer', an alternative to callipers for measuring pebble diameters

demonstrated using brightly-coloured floats. Since new material is not being added in significant quantities from either end of the beach, nor from offshore bars, the beach may be treated as a closed system. Carr measured the longest and shortest dimensions of 500 pebbles at each of 23 lines of section along the beach (fig. 7.5) using calipers. He also measured their weights. Since nearly all the pebbles on the beach are flint or chert, the effect on size and shape of lithological differences may be discounted and all pebbles were considered as being members of one population. It was found that the mean value of the long diameter in each sample increased steadily eastwards. Fig. 7.5 also shows a steady eastward increase in the standard deviation of long diameters. Clearly a grading process is going on but there are various explanations. One is that the wave height differs progressively along the beach. Carr prefers the differential transport of large and small pebbles, the large ones often moving more rapidly than the small ones once movement commences, owing to the fact that they are less impeded by other beach material as they move along. In other words, the easiest thing to roll across a surface of golf balls is not another golf ball but a football. Another conclusion from the work was that long diameter, short diameter and weight keep broadly constant proportions during transport, irrespective of size and angularity changes brought about by abrasion. Abrasion on beaches is commonly rapid and highly-rounded pebbles and boulders result (plate 7.1).

FIELDWORK

Try to make the 'pebble-ometer' shown in fig. 7.6. Chesil Beach is a classic example but one is likely to find size gradings on any beach, especially where waves move along the shore. Try to demonstrate this with floats. Does pebble size change down the beach? If there is beach material above the high-tide mark how does its size differ? It is likely that it was deposited there during storms. Does a storm move *more* material

Plate 7.1 Large boulders, many well-rounded, scattered on a wave cut platform (seen upper, right) on the west coast of Ireland

Fig. 7.6 Sediment studies on Chesil Beach, Dorset. The change in the mean long diameter and its standard deviation from west to east along the high water mark is shown in the graph below. (after Carr)

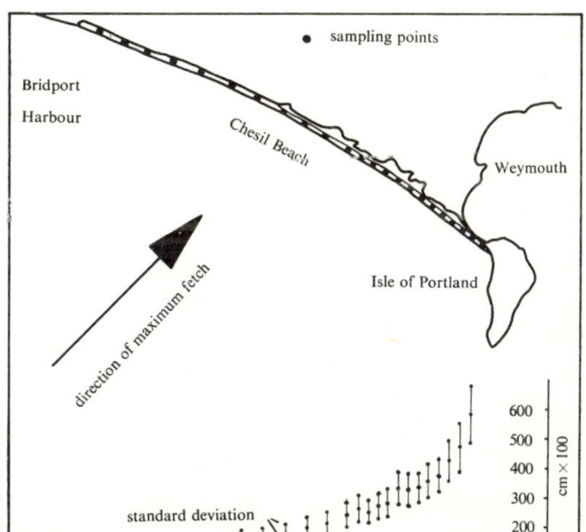

or *larger* material? Is more than one rock type represented in the beach material? How rounded is the material? (See chapter 9 for help with this.)

7B Shoreline dynamics

Just as in the study of drainage basins, measurement of energy flows and materials enters coastal studies. The scale difficulties of *measuring* coastal processes have already been mentioned. It will often be necessary, therefore, that the reader considers the *observation* of coastal processes to be his main direction of study, using shoreline deposits as indicators (as demonstrated by the Chesil Beach study above).

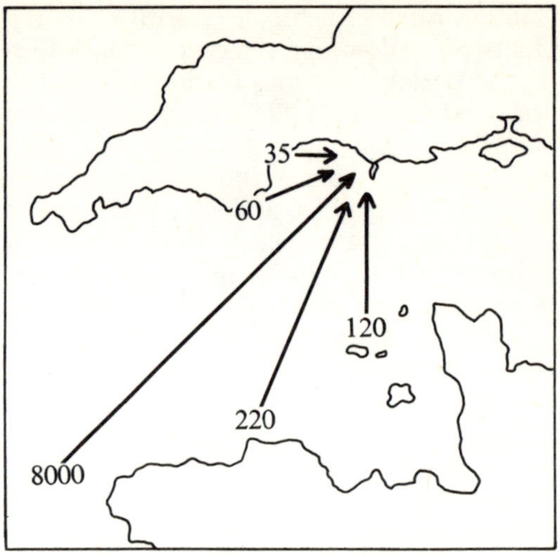

Fig. 7.7 Direction and length of fetch for waves breaking on Chesil Beach (distances in kilometres)

7B 1 Wave forms and motions

We are familiar, from our recreational visits to the coast, with the relationship between the apparent force of the waves and the strength or direction of the wind. It will be remembered that the potential energy of rivers is imparted by gravity but that the sea operates as a base level in the conversion of this energy. Hence the major source of energy on coasts is different, being mainly in the circulation of the atmosphere and the resultant winds. Though small waves may be generated in lakes of moderate size the formation of those large waves which play a large part in coastline development requires the frictional action of turbulent wind over large stretches of sea or ocean. Because of this the direction of wind is important as well as its strength. Thus, the 'Atlantic swell' often seen on our west coasts owes its size to the operation of westerly winds across a large expanse of ocean. The available distance for wave generation by winds is usually referred to as *fetch*. At any coastal location the direction of maximum fetch is an important variable to assess before coastal studies begin (fig. 7.7). Records of wind direction and force should be collected for those coastal meteorological stations whose reports contribute to the shipping forecasts on radio or the Daily Weather Report issued by the Met. Office. Some coastal data is therefore easily available at home, certainly enough to make an elementary *wind rose* (see fig. 3.7).

As with wave forms encountered in physics, measurement of sea waves involves two basic variables, wave *height* and *period*. The period is the time (which is in turn proportional to the distance, or *wave length*) between successive wave crests, from a stationary position such as a pier or anchored vessel. Wave *steepness* is the ratio of wave height to wave length. Theoretically this value cannot be more than about 0·14 or the wave *breaks*. Most wave steepness values are below 0·02. The effect of fetch is demonstrated by the average Atlantic wave height (3 m) which, however, occurs only two per cent of the time in the sheltered Irish Sea. Average wave period in the Irish Sea is 5 seconds compared to 15 seconds in the Atlantic.

Waves are a purely surface phenomenon in the oceans. The oscillatory movement which causes them is rapidly attenuated

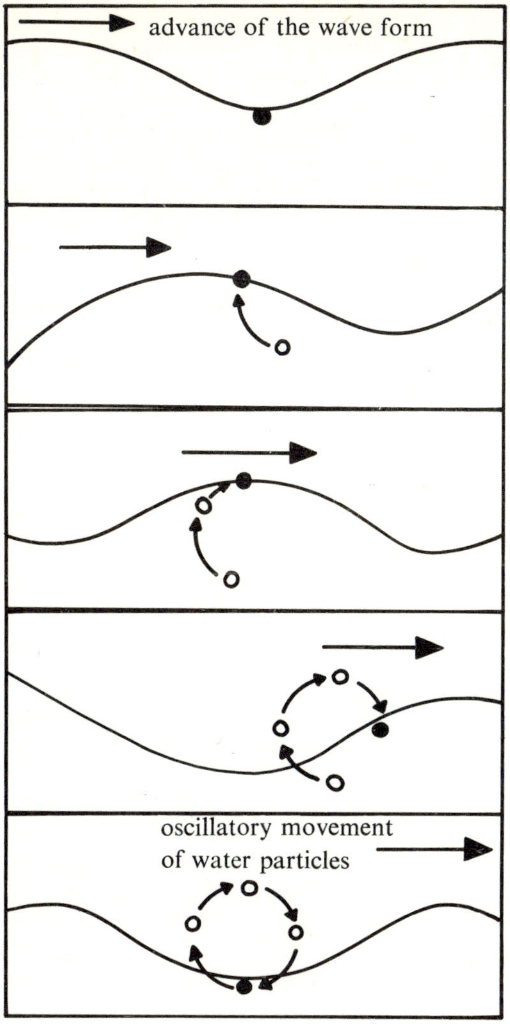

Fig. 7.8 The wave form, its movement and the oscillatory motion of water particles within it

with depth. It is important to realise that the mass of water enclosed by the wave form remains largely in its original position after the passage of the wave (fig. 7.8). Though our illustration deals with a simple or *ideal* wave, the natural situation is one of a complex superimposition of waves of various fetches, heights and lengths; bathers will be aware that at certain intervals one can expect larger waves.

There are important differences between waves in the open ocean and those which we study from the shore. As waves approach the coast, the depth of oscillatory movement (which is negligible in the ocean, compared with the great depth of water) begins to interact with the sea bed. The most interesting of the resulting changes for the coastal geomorphologist are those which result in the wave breaking on the shore. Wave period is unaffected by decreasing depth but velocity and length decline. The velocity to which we refer is that of the wave form (as we might measure it relative to our pier or boat). Since the oscillatory pattern changes to an ellipse, however, the velocity of revolution of each water particle increases; the cause of breaking is therefore the wave form being 'overtaken' by the water of which it is composed. Breaking may occur very gradually (which delights those with surf boards!) to produce *spilling breakers*, or may start and finish quickly near the beach, giving *plunging breakers*. Clearly, the energy imparted to waves depends on the height from which they break, the way in which they break and the wave period. Fig. 7.9 shows a high plunging breaker (*destructive wave*) and a flatter, spilling breaker (*constructive wave*). Waves of periods above thirteen per minute are also considered destructive.

Another feature of waves entering the shallower coastal area is that in the deeper parts, for example in bays, the velocity of the wave form does not decrease so rapidly as along the rest of the wave and *refraction* occurs. This can be observed from headlands as the wave crest in the bay appears to be ahead of its counterpart off the headland. It results in waves breaking round the flanks of the headland which might otherwise be protected.

After waves break they result in two basic components which affect the shore; a *swash* up the beach and a *backwash* as the same water returns. The distance between the breaking point of the wave and the highest point reached by the swash is known as the *run-up*. On a beach of constant slope angle and roughness the length and velocity of run-up can be used to index wave energy. Run-up will be familiar to seaside visitors who are forced to run up

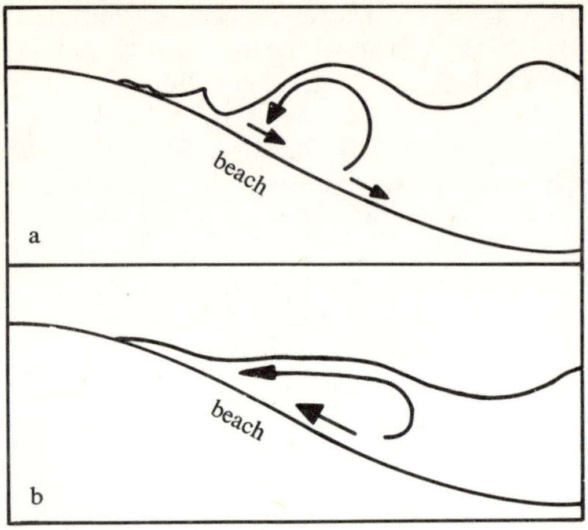

Fig. 7.9 Wave shapes: (a) destructive, (b) constructive

the beach themselves by the occasional wave which threatens to swamp their best shoes! Much of the volume of swash water fails to return via the beach surface but percolates through the beach material, returning seawards through the spaces between the beach particles just as the groundwater component of river discharge behaves on a larger scale. The different depths of swash and backwash can be simply measured on a graduated stake driven hard into the beach in the run-up zone. Obviously the volume of water which percolates into beach material is less effective at transporting sediment than that which returns down the surface of the beach, whose work potential is similar to that of a surface stream or river. Bathers will be familiar with the sting of sand and gravel, in suspension or rolling down the beach, during the backwash of a wave. Where waves encounter solid rock cliffs or man-made sea defence walls it is possible, through the entrapment of pockets of air in fissures, for very high pressures to be created during the breaking of a wave. However, they last for a mere fraction of a second and only develop if the wave breaks exactly on the cliff. Where this does not occur the wave is likely to be *reflected* and *interfere* with incoming waves. Much of the design of sea walls hinges around adequate wave reflection and interference.

Wave heights and periods close to shore can be measured at stable structures like piers. Further out at sea it requires a firmly anchored boat to take measurements. Velocity and sediment concentration measurements may be made from boats and although involving far more problems than stream measurements of the same variables both may also be performed by hand-held equipment in the swash zone. Perhaps it is the difficulties in taking such measurements that has forced processes of coastal development to be studied from their results rather than from measurements of the work performed.

FIELDWORK

Make a study of waves. If visibility is good, observe them as they near the shore. Does refraction occur? Does this reveal anything about the offshore zone not visible from the shore? What sort of waves are breaking, what is their period and how far do they run up the beach? If there are cliffs do the waves break on the cliff?

If observations of wave height are possible against the tide board of a pier these figures should be compared with those of run-up on the shore. Which way has the wind blown in the last week and which way is it blowing now?

The results together with those from previous coastal exercises should be kept as in table 7.1. Such recordings should be made at several locations on the beach, over a period of time at one location on the beach or, preferably, at several beaches.

7B 2 Currents and tides

Although sea *level* is our surveying datum, a level sea is of course an average condition. In reality there are build-ups of water around the coast at a variety of scales. In zones of higher waves there tends to be a larger mass of water near the coastline and it becomes dispersed laterally as currents (*longshore currents*) along the coast. Where such currents converge and flow seawards

Table 7.1 Coastal fieldwork sheet; an example

LOCATION	
DATE	
OPERATOR	

(A) **Wave Data**

Wind direction/ Force ___________
Tide rising/Falling ___________
Breakers plunging/Spilling ___________

Estimate No.	Period (per min.)	Length of run-up	Approx. height	Approx. steepness
1				
2				
3				
4				
5				
Average				

(B) **Beach Profile Data**

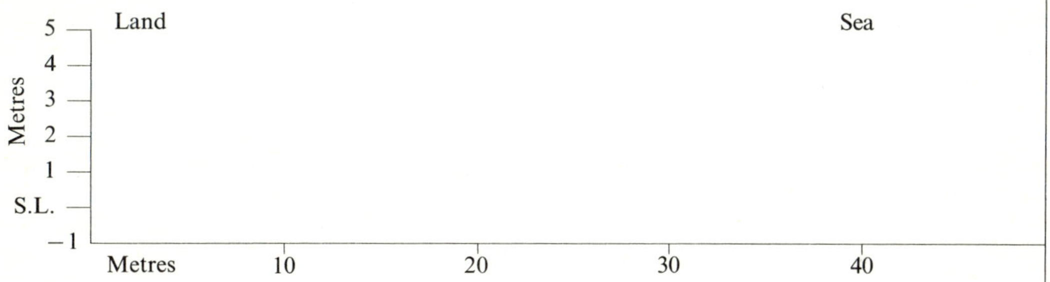

(C) **Beach Sediment Data**

No.	Long diam.	Short diam.	Rounded or Angular
1	mm X	mm	
2	mm X	mm	
3	mm X	mm	
4	mm X	mm	
5	mm X	mm	
6	mm X	mm	
7	mm X		
8	mm X		

Average long diam.	
Average short diam.	
Dominant shape	

again the *rip current* feared by bathers develops. Another cause of longshire movement is the *drift* caused by refracted waves which run up the beach obliquely but whose backwash occurs normal to the beach.

Another cause of unequal water levels, resulting in currents, is tidal movement. Tides themselves are important in the energy relationships of coastal processes; twice a day in most places the action of waves is moved landwards across a beach or wave-cut platform and returns. This gives a vertical and horizontal range of erosive activity not present on a diurnal scale in rivers. Tidal currents, unlike waves, do not diminish greatly with depth and are very important in determining the supply and removal of material to and from the coastal zone where it is worked by waves. The *ocean currents* marked on most atlases are important on the surface but have little geomorphological significance. In general the surface properties of all currents are well documented because of their effect on shipping. Much more information is required about bottom currents.

One method of measuring the direction and velocity of bottom currents which has been successfully tried is the *Woodhead sea bed drifter*. It consists of a polythene saucer, pierced by holes and attached to a solid polyvinyl rod (fig. 7.10). The rod is weighted with copper to give slight negative buoyancy.

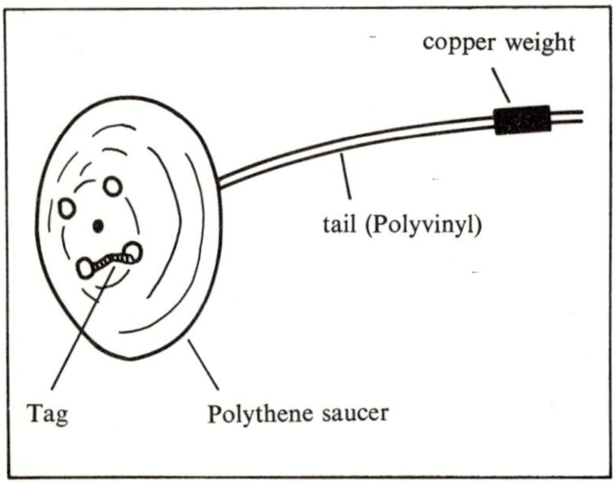

Fig. 7.10 The Woodhead seabed drifter

After release at sea in the area of interest (the precise point of release can be ascertained by normal techniques of maritime navigation) the drifters move along the sea bed at a velocity virtually that of the current. Their recovery by fishermen or people on the beach is made easy by their bright red colour and a reward is offered in a message attached to the drifter for the accurate completion of time and place of the finding.

7B 3 Beach material and its movement

One of the most accepted ways of studying coastal processes is by tracing the movement of silt, sand or shingle by tagging it in some way. Some work on this began as far back as 1907 when the Government established a Royal Commission on Coastal Erosion. Most progress has been made since 1954, largely through the activities of another official body, the Nature Conservancy. A variety of materials and methods has been used and the advantages and disadvantages of each are listed by Kidson and Carr (in Steers, ed., 1971).

Perhaps the simplest method is the use of bright coloured waterproof paints. Unfortunately paint can only be successfully applied to the coarser sizes of sediment; the application of paint to sand is both practically impossible and would alter the natural specific gravity of the particles. Though marine paints resist the abrasion suffered during transportation in sea-water the most resistant coating available is one of oven-hardened epoxy-resin, now available from hardware merchants. The application of dyes to particles overcomes the viscosity and density problems associated with paint. It is, therefore, possible to coat coarse sands with fluorescent dyes provided small amounts of a fixative are added. Fluorescence allows quick recovery with the naked eye in daylight and an even more distinctive reaction is given under an ultraviolet lamp during darkness; the latter means collecting samples for return to the laboratory and this is not always possible. The cheapest method is to make use of the distinctive colouring of 'erratic' pebbles brought from another locality and not found naturally on the beach in question, for example granite pebbles on a flint beach.

In experiments at Bridgwater Bay, Somerset, Kidson and Carr experimented with the use of fireclay bricks because the rapid abrasion of the soft local beach materials prevented the use of coloured coatings. The bricks were almost exactly the same density as the local rocks and their perforated nature allowed a galvanised wire tag, recording the original site of the brick, to be attached to each one. All the bricks were started off in a line at right-angles to the direction of the beach between low tide and the top of the beach. Recovery rates of bricks moved from the line by the first series of tides were high and the most important discovery was that movement was extremely slow; after six years the furthest-travelled marker had moved only 2·25 km from the line. It was also discovered that some of the bricks became incorporated in the beach ridges, holding up their movement. Though the three sizes of brick used moved differentially, the fact that each became frequently arrested meant that a lateral size grading could not occur since the slower, steady movers could easily overhaul those which started quickly but became trapped.

The Bridgwater Bay results were very different to those performed on Orfordness by Kidson, Carr and Smith (1958). Here another technique for tagging shingle was employed, that of coating each particle with radioactive barium 140-lanthanum 140. The isotope emits high-energy gamma rays enabling it to be detected at distances of about 250 mm in water or if buried at a similar depth on the shingle beach. Detection is by Geiger counter, towed behind a boat or by shore-based personnel on a hand-pulled sledge. The short half-life (see chapter 9) of the substance is essential to eliminate the health risks of continued radioactivity. Though Orfordness has clearly been built up by littoral drifting of the shingle from north to south the first weeks of the experiment showed that the reverse movement was possible when winds blew from a southerly quarter. The fact that one pebble was recovered 2 km from the point at which all were inserted (see fig. 7.11) after only 4 weeks, demonstrates that movement here was many times faster than at Bridgwater Bay. The drift of pebbles to the north was arrested and reversed after the onset of northerly winds. Another point of interest, from a further experiment at Orfordness, is that virtually no movement of shingle from offshore occurs on to the

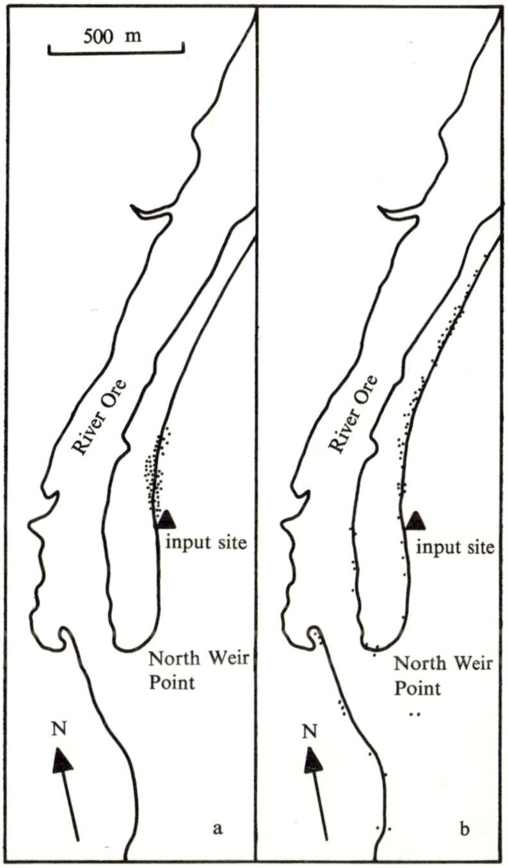

Fig. 7.11 Movements of shingle on Orfordness as revealed by the radioactive tracer experiments of Kidson, Carr and Smith. (a) Shows recovery sites after 2 days, (b) after one month.

beach, though it is equally clear that material only moves across the estuary of the River Ore at some distance offshore before being washed up again on the Shingle Street beach.

7B 4 Models

In coastal studies, more than other aspects of geomorphology, the use of scaled-down hardware models of field conditions has been both necessary and successful. The problems of coastal measurements have been mentioned above. It must not be surmised, however, that all one requires in order to study coastal processes in a model is a large tank, some sand, water and a large paddle to produce waves and currents! Though the advantage of model studies is their ease of monitoring and inspection during operation, coupled with the ability to be able to control conditions, a major drawback is in the precision required to *scale down* all the dimensions in the model. If any dimension is not properly scaled erroneous results may be obtained. A list of scale factors applying in the British Transport Docks Board model of the Humber Estuary demonstrates the detail and precision needed in scaling:

horizontal scale	1:720
vertical scale	1:72
velocity scale	1:8·5
time scale	1:85
discharge scale	1:439,870
volume scale	1:37,000,000
tidal period	8 min. 46 secs.

A wide range of artificial and natural materials is used to simulate the natural sands, silts and gravels found on the coast in question. Electrical apparatus controls the pumping of water in and out of the model to simulate tides. Automatic sensing is also used to record the movement of waves and tides and to survey the changes on the 'sea' bed.

For many years now the centre of hardware coastal models in Britain has been at the Hydraulics Research Station, Wallingford, Berkshire. The Station not only studies the basic processes of wave motion and its effects on sediment but also undertakes applied studies of specific stretches of coast. As an example of the former activity the Station has used natural scale waves and shingle in a Pulsating Wave Tunnel to predict the wave conditions necessary to move shingle of a particular size and known relative density. They found that other variables such as the shape of the particles did not affect the wave conditions at which movement is initiated. Two examples demonstrate their conclusions:

a) In 4·5 m of water a wave of 6 second period would need to be 2 m high to move shingle of 35 mm diameter and above.
b) A wave of 8 second period, 3·5 m high in 20 m of water can initiate the movement of 6·4 mm diameter shingle.

A working diagram was developed for engineers to allow calculation of such facts for other sizes of wave and shingle, assuming that the relative density of the shingle is 2650 kg m^{-3}, a common value in nature. A useful field exercise would be to relate wave height or period to the size of artificial object which can be moved. Bricks or brightly coloured marbles could be used and compared with local natural particles.

As an example of the applied work possible with scale models at the Hydraulics Research Station, the problem posed by the Teignmouth Harbour Commissioners is worth quoting. Existing shipping, of from 500 to 1000 tonnes displacement, is to be replaced by larger vessels but both the entry to the harbour and the quays themselves are not deep enough to allow this. The Pole Sands have been shown, by fluorescent tracer study in the field, to circulate in the estuary of the Teign (fig. 7.12). The predominantly southwesterly winds sweep material from the Ness beach across the estuary and it is augmented by coarser material swept out from the Teign on an ebb tide. Though the Ness Pole grows considerably as a spit during southwesterly winds, it only requires short spells of easterly weather to break it down again and drive much of the material ashore on

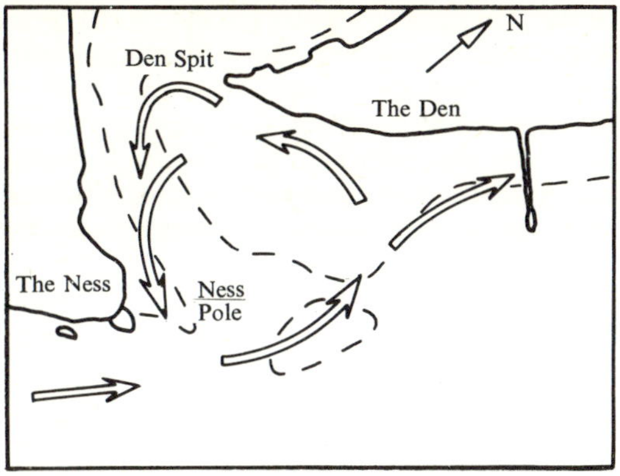

Fig. 7.12 The circulation of material in Teignmouth Harbour, based on work by the Hydraulics Research Station in the hardware model shown in Plate 7.2.

Plate 7.2 The scale model constructed by the Hydraulics Research Station in order to study remedial action to stop the silting of Teignmouth Harbour, Devon.
(Crown Copyright. Reproduced by permission, HMSO)

the Den Spit. A scale model was set up at a horizontal scale of 1:2000, a vertical scale of 1:50 and employing sintered ash to represent the pea-sized gravel found in nature (plate 7.2). Southwesterly and easterly waves were faithfully reproduced and it was possible to observe a most realistic pattern of bar formation and movement. The influence of three remedial designs to alleviate shoaling was studied in the model. The one chosen was a curved wall projecting south-south-east from the Den Spit. As well as helping to cure shoaling of the harbour entrance this scheme had important side effects. It led to replenishment of the Den beach during easterly winds, an important point for a tourist centre, and it also led to an improvement in the dispersal of sewage. It is highly likely that increased awareness of pollution in our coastal seas will lead to an upsurge of work on wave and current motion, par-

ticularly if the use of desalination plants means that they become a source of water supply or if diminishing food stocks focus our appetites heavily upon sea foods.

Successful scale models of coastal processes are possible in schools. The Northgate Grammar School for Boys at Ipswich has for some time been conducting experiments in a 3-metre-square tank made of lead-lined wood, with one plate glass side for inspection of sections. Shorelines are simulated with mixtures of kaolin and sharp sand while waves are produced in 50 to 100 millimetre depths of water by a paddle worked by hand or by electric motor. By varying the motor by means of a rheostat and the pitch of the paddle by means of adjustable nuts, constructive and destructive waves can be generated. Estuarine features are producible by means of 'streams' created by siphon action from a tank 'inland'; a similar but longer tank is used at the school for experiments in stream channel processes.

7C Storms and tidal surges

Just as in fluvial geomorphology, the 'rare' event on the coast is impressive for the amount of work which natural agents accomplish in a short spell of extreme conditions. Historical records tell us of the 1334 storms in the North Sea, which are reputed to have formed the Zuider Zee and Frisian Islands of Holland. Other major floods from the sea occurred in 1421 and 1855. The most disastrous tidal surge in living memory occurred in 1953. 18000 people lost their lives in the Low Countries. The flood is the topic of a paper by Steers (in Steers, ed., 1971) in which he states that the inundation of the English coast from the Tees to Dover would have been far worse had the rivers too been in flood and the tide been at its highest level. As it was almost 1000 square kilometres were flooded and 307 people lost their lives. The near-hurricane force winds from the north which focused a surge of water in the narrowing North Sea were caused by a depression (of 968 millibars at its centre on 31 January 1953) over the Sea itself. The very presence of the low pressure over the sea led to much higher water levels. Similar to normal tidal progression, the flooding occurred later in the south than in the north. At King's Lynn in the Wash the surge was 2·5 m high, the greatest for many years, perhaps centuries. Though there was disruption of industry and transport, as well as damage to agricultural land through salt water, the damage to natural features on the coast was mainly local, small breaches often admitting large volumes of water; for example, although Orfordness was temporarily breached at Slaughden, it was quickly reformed. Steers concludes that, wherever possible, man-made efforts to help the natural accumulation of beach dunes, shingle ridges and salt marsh are a crucial item of policy to protect the east coast of England from the sea. A full documentary report on the North Sea floods of 1953 can be found in *Geography* for July 1953.

As mentioned earlier, the North Sea is known to be moving positively, relative to our east coast. In the region of the Thames the movement is 1–2 mm per year, which makes the Thames itself tidal to Teddington, instead of to the area of London Bridge, as it was in Roman times. Clearly the extensive preparations of sandbags and evacuation procedures which preceded the autumn high tides of 1971 were wise even though they were not needed and studies of flooding in the Thames Estuary carried out on models by the Hydraulics Research Station are being used to plan more permanent protection. National co-ordination is clearly required when, as Steers says, the installation of groynes to catch protective shingle by one coastal borough may well deprive the neighbouring resort and render it more liable to flooding. He quotes the example of Great Yarmouth gaining at the expense of Gorleston and Lowestoft, and Pakefield losing material as a result of Lowestoft harbour works. Valentin (in Steers, ed., 1971) describes the struggle against coastal erosion at Holderness. A geological map reveals just how much of the east coast of England is similarly composed of low cliffs of weak glacial deposits or of salt marshes and dunes. He calculates that it would cost more than £10000000 to protect the unguarded parts of Holderness. Meanwhile the annual loss of cliff-top land is estimated at just over 1000. Thus, the defences would, if built, have to last 10000 years to be economically viable. Clearly, there is room in coastal studies for the geographer, the civil engineer and the economist (see Brunn, also in Steers' collection of essays).

8 The Historical Dimension

'Every drop of the Thames is a liquid 'istory' (saying to transatlantic visitors attributed to John Burns).

Our interest in researching the *history* of natural systems depends on the time taken for them to adjust to changing external conditions—their *relaxation time*. Where such adjustment is rapid there will be few remnants of the past and little interest in the historical dimension. It has been stressed elsewhere in this book, however, that the landscape is slow enough in adjusting to environmental changes to warrant extreme care in the interpretation of, for example, valley-side slopes or sea cliffs in terms of only today's processes. Historical geomorphology is therefore a useful and valid branch of the subject. Rates of change in most natural systems including physiographic ones are controlled by *feedback* relationships in which a change in element A causes change in element B which then effects element A again (fig. 8.1). In the case of a preponderance of *negative feedback*, change in a certain direction brings adjustment in the causative factors so as to curtail further change (fig. 8.1b). A model of the elements of the stream channel erosion system reveals how such balancing relationships caused by negative feedbacks, lead to the conservative steady behaviour of the landscape (fig. 8.2). Thus we return to the concept of a dynamic but steady state outlined in chapter 1.

It is because of the residual effect of historical processes in the natural landscape that geomorphologists, more than meteorologists or hydrologists have adopted techniques of studying the past. Geomorphology has an important interface with geology for this reason. Change in climatological system is a basic cause of change in landscape; changes in climate are brought about comparatively rapidly by more common positive feedbacks as in the case of glacier growth (see section 8D). The

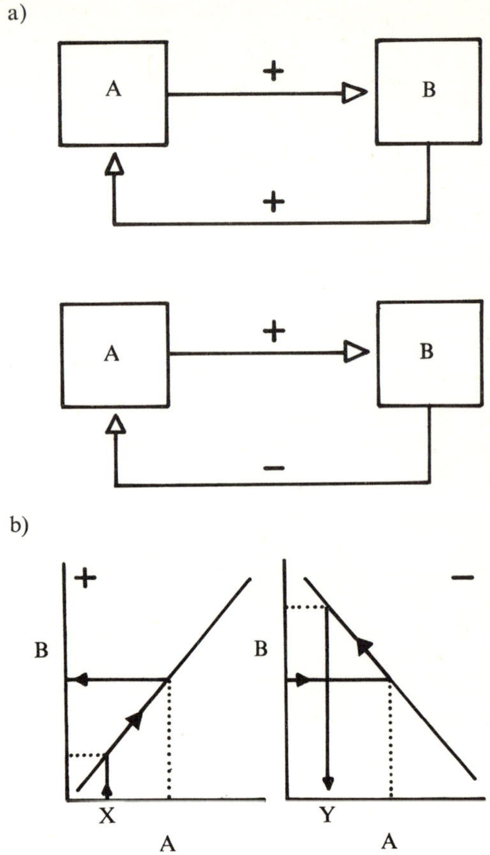

Fig. 8.1 Feedback relationships. In (a) the upper diagram shows a positive feedback, an increase in *A* resulting in an increase in *B* which in turn increases *A*. The lower diagram shows how *B* may negatively influence *A*, resulting in negative feedback.
In (b) the same effects are graphed. Enter the left hand graph at 'x' value of *A*. Increase in the value of *A* means following the arrows on to the *B* axis to record the positive effect on values of *B* (dotted lines show the original value of *B* and the new value of *A*. If we enter the new value of *B* on the right hand graph and further increase *B*, the result is a decrease in *A*.

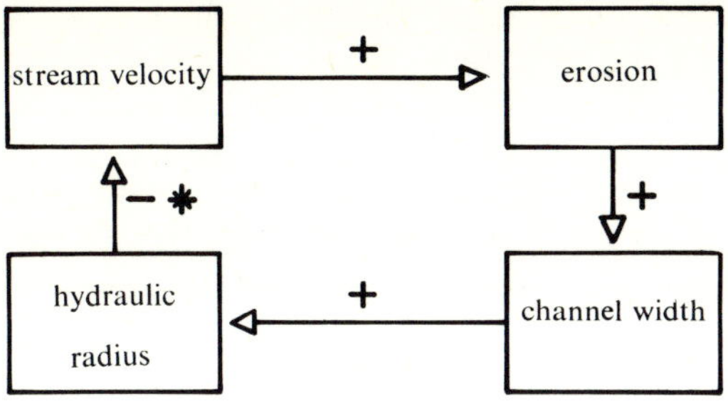

Fig. 8.2 Feedback in the stream channel system. The important negative link which keeps a dynamic balance between processes is marked by an asterisk.

climatologist is also therefore forced to consider landscape as a source of historical information. The speedier relaxation time of climatological systems leaves him little residual evidence in today's climate. We do not now experience rainfall in response to Tertiary convection, nor sunshine associated with Pleistocene high pressure!

Table 8.1 The whole geological time scale. Quaternary is the distance between the two lines (see table 8.2)

Years B.P.			
	TERTIARY	Pliocene Miocene Oligocene Eocene	Alpine orogeny
100 M	MESOZOIC	Cretaceous	
		Jurassic	
200 M		Triassic	
		Permian	
	UPPER PALAEOZOIC		Armorican orogeny
300 M		Carboniferous	
		Devonian	
400 M	LOWER PALAEOZOIC	Silurian	Caledonian orogeny
		Ordovician	
500 M		Cambrian	
600 M ?		Precambrian	Charnian orogeny

8A Climatic changes: when and how?

Some 250 million years have elapsed since the Permian and Carboniferous periods when such rocks as the Carboniferous Limestone and Coal Measures were deposited (table 8.1). At the end of their deposition, *orogenesis* raised up the familiar ranges of Carboniferous and older rocks in the south-west of England, South Wales, and the Pennines. While this orogenesis was going on there is clear evidence from the southern hemisphere that a widespread glaciation was occurring. Fossil boulder clay or *tillite* has been found in a stratigraphical position similar to our limestone and coal. In Britain it is clear from the surface features and geology of an area such as the Mendip Hills in Somerset that the Triassic period, which followed the Carboniferous, was one in which the newly-uplifted hills were eroded under arid or semi-arid conditions. A cover of Coal Measures which had overlain the uplifted Mendips must have been stripped off by such arid erosion to expose the Carboniferous Limestone because boulders and pebbles of the latter occur in the Triassic conglomerate deposited around the edge of the hills. The *matrix* or cementing material of the conglomerate has a red colouration, typical of desert conditions, in which oxidation of iron minerals (such as pyrite in the Coal Measure rocks) to haematite is common. In the more characteristic sandy Triassic rocks of the Midlands and Cheshire the red colouration is pronounced, one reason for the rock being called the New Red Sandstone. The grains of the sandstone are very well rounded and dune features have been discovered from the distinctive way in which the beds of sandstone rest upon one another. In Charnwood Forest the sandstones may be seen to occupy steep-sided gorges in the older rocks. One characteristic of desert areas is a landscape guided strictly by *structural* features; joints and bedding planes are eroded by sand blasting. Since the debris is removed by wind, steep, structurally-controlled slopes develop.

The rich resources of salt in the Triassic of Cheshire are clear indications of the presence there of large salt lakes during the Triassic. Gypsum bands are also common in arid deposits. It is probable that much of the Midlands resembled a great 'Dead Sea', while on the land reptiles such as *Thecodontes* have left fossil bones and teeth in Triassic cave deposits.

It is possible that the climatic fluctuations which caused both the aridity in the northern hemisphere and the earlier glaciation in the southern hemisphere were not caused by movements in climatic belts but by moving continents. This topic can be followed up under the heading of *continental drift* in geology texts. How can continental drift be explained?

Between the Permocarboniferous and the Pleistocene, which began some two million years ago, climate on a world scale was fairly stable. Lower Tertiary fossils do show a poleward extension of tropical conditions, probably bringing very warm, moist conditions to Britain. An important evolutionary step brought a cover of grasses to much of the world, whose importance in binding mineral soils has already been mentioned.

A downward trend of temperature (see fig. 8.3) during the later Tertiary signified the onset of greater climatic instability during the Pleistocene. The higher latitudes became very cool and there was a general equatorial movement of climatic belts. Wetter, '*pluvial*' conditions came to the deserts of North Africa, as evidenced by stream channel features and a large body of groundwater which still feeds springs in some areas; arid conditions too moved towards the equator, fossil sand dunes being found in many parts of the northern tropics which now support quite dense vegetation.

The onset of glaciation can be attributed to a variety of factors. Positive feedbacks involved in the process are shown in fig. 8.4. There is still no sure explanation of the basic independent causative factors, which must lie outside the terrestrial climatic

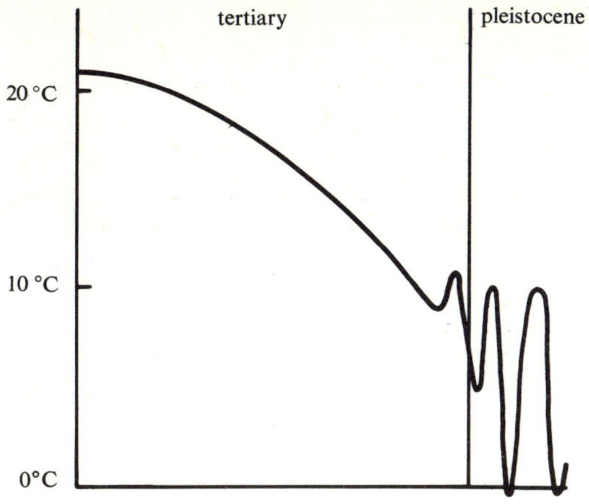

Fig. 8.3 The fall of mean annual temperature during the late Tertiary era as a prelude to Pleistocene oscillations

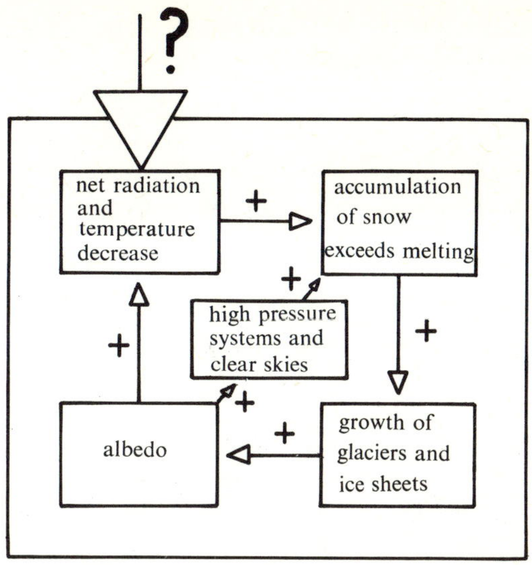

Fig. 8.4 The growth of glaciers and ice sheets is a process which, once initiated, is dominated by positive feedbacks

system in the earth's orbital movements or in variations of sunspot activity. Radiation changes associated with volcanic activity on Earth are a possible cause. Today, as the world's fossil fuels are burned up, the increasing obscurity of the atmosphere is said by some scientists to be causing a future extension of glaciers and ice caps. Others state that the reduction of outgoing radiation will make the earth warm up! While dealing with man's responsibility for future climates it is worth mentioning the postulation that the position of the earth's axis could well be changed (so as to shift the climatic belts) by using huge nuclear explosions.

Four main glacial phases are recognised in lowland Europe, although only three caused extensive glaciation of the British Isles. Between the glacial phases the warmer interglacials allowed renewed development of river systems and the growth of a large flora and fauna which eventually included man. During the period since the last glacial phase, climate has not ceased to oscillate. The changes have been extreme at the scale of human sensitivity to climate but their slow passage and man's short lifespan make them less noticeable. They are detectable in part by means of the records of man's activity. The 'optimum' period of climate following the last glaciation is thought to have lasted from 2000 to 4000 B.C. Temperatures were several degrees above those of today. There followed a general decline, interrupted by a mild, dry period from A.D. 1000 to 1250. From then a wetter stormier period lasted until 1400, followed by colder conditions which culminated in the 'Little Ice Age' (a bad term since there was no widespread glaciation) from 1600 to 1850. Alpine valley glaciers advanced a short distance three times during this period. The period since then has been one in which instrumental records of weather have been kept. Though there is little evidence for present-day cyclic changes in these records, they show general retreat of glaciers and a stronger northward movement of warm Atlantic water into Arctic latitudes. We are told that during the last few years Britain has experienced drier, warmer autumns and cooler, later springs due to an increasing 'continental' influence. Can you verify this? What are the views of elderly people you know, particularly those who carefully observe weather through gardening or farming?

8B The present as the key to the past?

The theme of this book is measurement, mainly of active processes. The reader may be wondering how the processes of the past can possibly be studied and clearly some adaptation is needed for our 'Let's find out' principle to enable its application to glaciation and so on. We may describe three ways of incorporating the study of process into historical geomorphology and climatology:

a) Processes thought to have once affected the area of interest can be studied in another area where those processes are still active. The example of glaciologists visiting Greenland, Norway, or the Alps is obvious. We may call this the **geographical approach**.

b) The processes may be replicated under controlled conditions in the laboratory, either by using a scaled-down version of natural processes (like a refrigerator for studying freeze/thaw processes) or by using analogous processes (such as the use of a kaolin glacier to study the flow of real ice). This may be termed the **experimental approach**.

c) We may infer past processes from the effects they had on morphological systems. This **inferential approach** is based on results we may obtain about relationships between processes and morphology today. For example, the enormous discharge of rivers during the late glacial stage may be inferred from the wavelength of the valley meanders they left, using the relationships derived between present-day discharges and channel wavelengths (Dury, 1970). This, of course, assumes that the valley meanders were caused by a river filling the valley floor. Dury concludes that precipitation may have been 100% higher than at present.

We have already presented the non-geologist with a geological timetable (table 8.1). We now want to concentrate, however, on the uppermost portions of this table since the detectable landscape elements remaining today are largely not older than the Tertiary period. Therefore table 8.2 has been drawn; this time the duration of each subdivision of time cannot be made proportional to the space it occupies in the table, as was the case with table 8.1. Is there any more appropriate scale, for example a logarithmic time scale? Refer, if necessary to the introduction to this book. There is room for experimentation, using the age figures given in the table. Readers should note that time in the Quaternary era may be expressed in years *Before Christ* or *Before Present* (B.C. and B.P. respectively). The latter system has grown up with the technique of radioactive dating (section 8F). We begin the story so to speak, as the sea which deposited the Chalk in Britain was receding and rivers became initiated on an eastward-dipping land surface. The easterly flow element can be seen in several of the major rivers of Britain even today—look at the Tyne, Tees, Trent, Thames and the rivers of west Dorset. How much of their total length is composed of eastward-flowing elements?

Before describing the landscape changes which occurred during the Tertiary era we must first describe the kinds of evidence on which geomorphologists base their interpretations of historical systems in both the Tertiary and Quaternary eras. We may delimit *landform variables* and *sedimentary* variables. The landform variables are:

a) The site of the landform. Is it away from or above the main lines of operation for present-day processes?

b) The shape of the landform. Is it regular and is the regularity unrelated to present-day processes in the area?

The sedimentary variables are more numerous and sediments are usually less misleading as indicators of past conditions:

a) Are the sediments stratified or are the individual particles orientated in a certain direction? Is there a smooth transition into neighbouring deposits of a different age or type?

b) What rocks and minerals are contained? Are they local or far-travelled?

c) What size are the sedimentary particles? Are they well mixed or all the same size?

d) What shape are the sedimentary particles?

Table 8.2

Age, B.P. in Million years	Period/ Glacial Phase	Rock type and Landform Development	Man
·005	Climatic optimum	The Post Glacial (see table 8.4)	Neolithic Mesolithic Upper Palaeozoic
	Würm glacial	Deposition of Newer Drift with fresh glacial topography of drumlins etc. Four separate glacial phases.	
·12	Interglacial	Called the Ipswichian. Further Thames terraces and the 7·5 metre raised beach.	
·23	Riss glaciation	Great Chalky Boulder Clay deposited. Called Gipping glacial from E. Anglian examples.	Middle Palaeozoic
·36	The Great Interglacial	Called the Hoxnian after discovery of deposit at Hoxne, Suffolk. Thames terrace graded to sea level (and consequent beaches at 45 metres).	Lower Palaeozoic (at some stage Homo sapiens appears to replace Homo erectus)
·67	Mindel glaciation	Lower Chalky Boulder Clay deposited. Called Lowestoft glacial from E. Anglian examples.	
·78	Interglacial	Cromer Forest Bed in Norfolk indicates warmer conditions 80 metre bench cut.	
·9	Gunz glaciation	Weybourne Crag } like the Red Crag these are now locally exposed in East Anglia	
	Interglacial	Norwich Crag } like the Red Crag these are now locally exposed in East Anglia; sea cuts another bench	
	Donau glaciation	Glaciation of Alpine borders only. Deposition of Red Crag.	
1·8	PLIOCENE	Sea level high. Cuts the 180 metre 'Calabrian' bench. Deposits Coralline Crag.	
	MIOCENE	Fluvial planation. Alpine folding. Mainly affects the newer rocks of S. E. England. Volcanoes and faulting in North.	
	OLIGOCENE	Deposition of rocks in Hampshire Basin, Isle of Wight and locally in Devon	
	EOCENE	Development of eastward flowing streams on Chalk surface. Planation. Further planation as marine transgression deposits London Clay in South-East England.	

8B 1 The Tertiary physiographic system

As mentioned above, the Tertiary era is about as far back as we can reasonably go to find evidence of historical systems in present-day landforms. This is mainly because of the reasonable geological stability, particularly since the Alpine earth movements which occurred during the mid-Tertiary. Erosion has played a proportionally greater part in the British Isles since the Tertiary period than has the widespread deposition of thick sedimentary rock strata which qualifies our area over much of the rest of the known geological column. Yet considerable controversy surrounds all interpretations of the Tertiary physiography, mainly on the topic of the effects of the Alpine earth movements. There is controversy about their relationship with reduction of the land to a surface of low relief by rivers, or similar planation achieved by transgressions of the sea. Professor Wooldridge spoke of, 'this great lost interval in the geological story'. To both fluvial and marine erosion have been ascribed the *erosion surfaces* which exist up to considerable altitudes in Britain. It

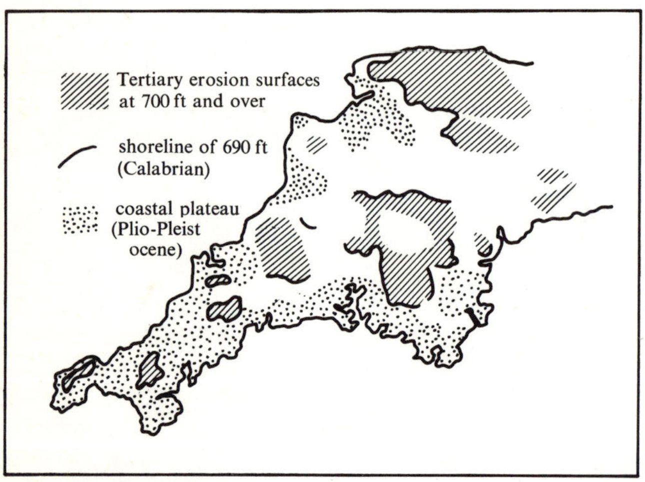

Fig. 8.5 A generalised distribution of erosion surfaces in southwest England (modified from Gregory)

certainly is a remarkable feature of upland Britain that much of the land between the slopes bordering present-day drainage basins can be demonstrated, both on maps and in the field, to be flat or gently undulating (see fig. 8.5). The fact that these areas may be shown to accord in altitude over wide areas, despite crossing important geological boundaries, has led to their interpretation as remnants of peneplains, the final 'old age' features of Davis' cyclic model of erosion. Since the coast is never far away in Britain there has also been a great willingness to attribute the features to coastal erosion, likening them to extensive versions of today's wave-cut platforms, cut when sea levels were higher. In all work on erosion surfaces we are left with largely no recourse to the basic elements of our historical methodology: landforms and sediments. Though the typical landform is a flattened or gently undulating surface we cannot be sure that such a surface could have developed in the Tertiary or, if it did, have been unmodified by earth movements and glacial or fluvial erosion since its formation. However, before considering some other conceptual problems associated with interpreting the Tertiary physiographic regime, let us consider the methods which have been applied to the study of erosion surfaces.

There are several ways in which we may quantify our first impression of a map showing summits, erosion surfaces in the form of spurs and flattened interfluves which accord in height (fig. 8.6). Firstly, the altitude of all the summits, interfluves and spurs on a map sheet can be merely tabulated or the altitude of each intersection on a superimposed grid recorded and an **altimetric frequency** diagram drawn (fig. 8.6c). As in the case of diagrams of slope frequency, the frequency of each altitude class may either be calculated from its proportion of the total number of intersections whose altitude is recorded, or with a much lengthier analysis the proportion of the total area which it occupies. A further modification is to be more objective about the sampling procedure and gather altitudes from random points on the map. For an

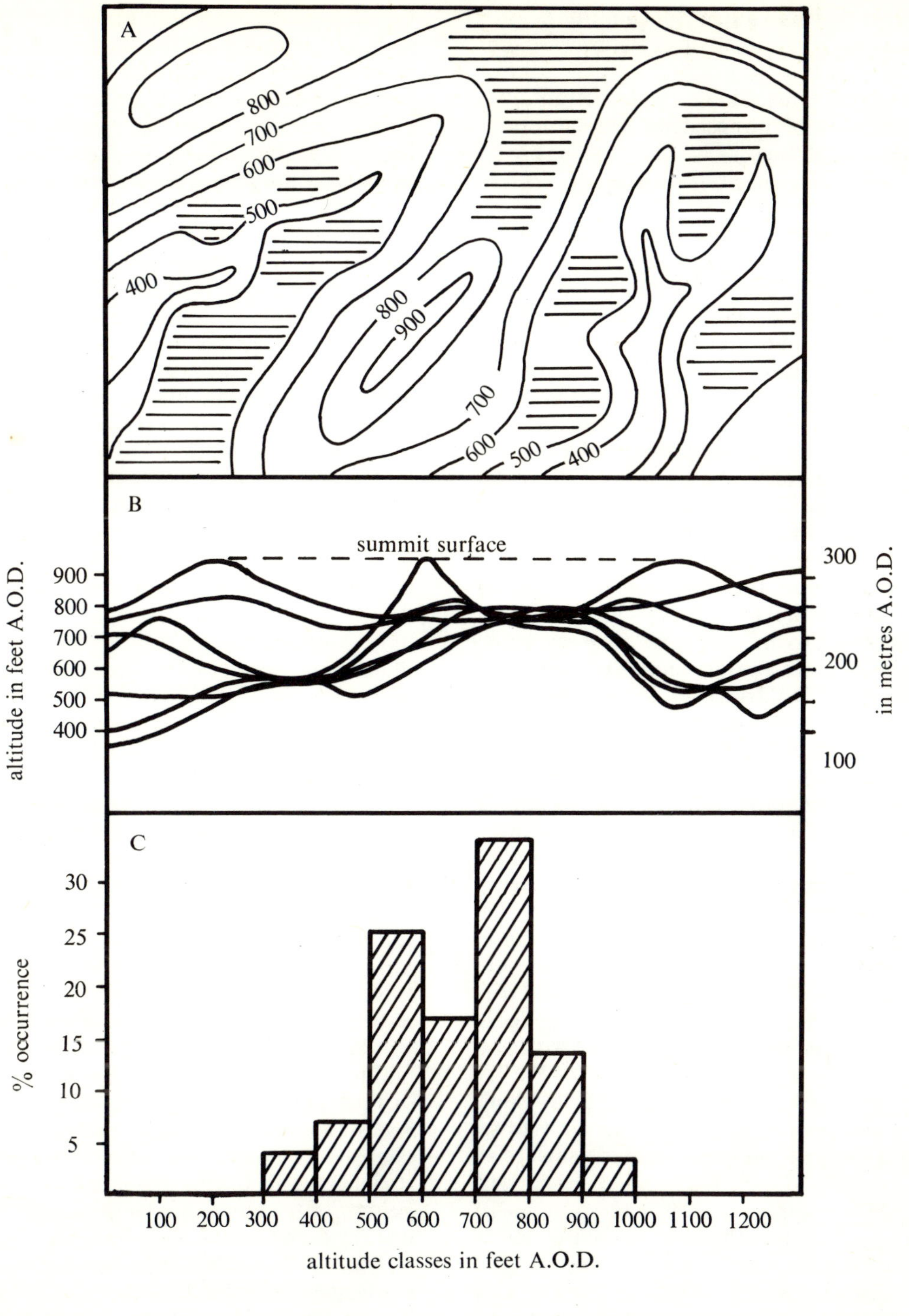

Fig. 8.6 Methods of mapwork analysis to delimit erosion surfaces. The map (A) has been shaded where large areas of flat land occur between contours. These are further highlighted by the superimposed profiles in (B) and the distribution of altitudinal frequencies is summarised by (C).

excellent visual guide to summit heights and the altitudinal accordance of flat areas a series of profiles may be drawn across the area along cross-section lines and then superimposed on the same diagram, (fig. 8.6b).

MAPWORK

Look at 1:63360 or 1:25000 maps of one of the following areas: Dartmoor, Exmoor, the Mendips, the Peak District, Craven, and the Lake District.

Are there obvious areas of relatively flat land at high altitudes? At what altitude do they lie? Draw superimposed profiles across the area, possibly one each (as a class) on tracing paper to the same scale for easy superimposition. If the flat areas are erosion surfaces do they cross on to different rocks at the same altitudes? From your profiles decide whether the surfaces are joined by sudden steep slopes, resembling sea cliffs, or by a gradual transition. Are the long profiles of the rivers in the region polycyclic?

Trend-surface analysis has been used to detect the mathematical form of a best-fit erosion surface computed through altitude data points drawn from areas of accordant summits and flat upland surfaces. King (1969) took over 300 points from 1:63360 maps of the Askrigg and Alston blocks in Yorkshire and Northumberland. She recorded spot heights of summits and the altitudes of closed summit contours. These Z coordinate values were matched to X and Y coordinates ('eastings' and 'northings') according to their position on a grid. Computation of the surface which best fitted the altitudes revealed a quadratic surface to fit the Alston block, whereas a cubic surface was more appropriate to the summits of Askrigg. King went on to study the deviations from the computed surface represented by altitudes which rose above it (**positive residuals**) and those which fell below it (**negative residuals**). The positive residuals were said to represent *monadnocks* (peaks remaining intact above a peneplain) while the downwarped negative residuals were said to have localised drainage into the present river system of the Yorkshire Dales.

Trend-surfaces are excellent statistical devices but it should be remembered that they only statistically represent real surfaces, compressing variations as a regression line does a scatter of points: so we are back to the problem of adequately recognising and interpreting erosion surfaces in the field. Professor E. H. Brown's epic study of Wales represents the field study of three major plateau levels in Wales together with the summit surface which caps the monadnocks. Brown favours a subaerial (i.e. fluvial) origin for these surfaces; the summit may well be an exhumed surface formed in the Permo-Trias while the three plateaux are unwarped and were formed after Alpine folding. Brown located the surfaces with mapwork but then viewed each remnant in the field, in profile and actually on the surface, making field sketches and taking photographs. He was able to see how the plateaux cut across very different lithologies and geological structures without marked change of altitude. The zone of junction between a plateau and the one below was described as typical of a subaerial origin; at the edge of the major surface element it became dissected into monadnocks standing above the lower surface. Further away the lower surface became dominant and full transition had been made. By studying the direction of the major routes of drainage across each surface Brown describes how, in response to successively lower base levels the rivers of the newer, lower surface would have 'gutted' the upper surface from within. (The use of the mathematical form of river profiles to define their former base levels has been described in chapter 6.) In contrast, the junctions between the marine surfaces which Brown delimited at altitudes below 700 ft (230 m) are simple, resembling degraded cliff-lines.

The topic is clearly as complicated as it is controversial. We may summarise the findings of a wide variety of workers as follows.

At the end of the Cretaceous deposition of Chalk, uplift occurred and the eastward-flowing drainage achieved some planation before the transgression of the Eocene sea which trimmed the subaerial surface. There is some evidence of this wave-trimmed surface in southeast England but it tends to dip under the Eocene and Oligocene rocks. Folding affected the southeast in the late Oligocene and Miocene. After it, planation occurred, possibly as many as three surfaces being formed. An important one is that formed at between 635 and 745 feet (210 and 245 m) on the Chalk in southeast England. It is variously mantled with a deposit of clay-with-flints and large silicified stones called *sarsens* (especially on the Marlborough Downs). Both these rocks indicate the tropical weathering whch has also been deduced from other evidence for the Miocene period. Deep weathering under such conditions has been suggested as the reason behind *tors* in the Pennines and Dartmoor; glaciation and periglaciation in the Pleistocene having a large amount of loose, weathered mantle to remove, leaving smooth rock (see fig. 8.20). Below the Miocene surface comes a series of levels attributed by nearly all workers to marine action along the coasts of those seas which deposited Pliocene and early Pleistocene *crag* deposits in East Anglia and the southeast before glaciation began. At lower altitudes still are the platforms and *raised benches* more easily attributable to interglacial levels of the sea.

We may summarise the stronger objections to the above conclusions as a series of questions:

a) Can surfaces form in the ways described? Is there time for subaerial peneplanation and can it be widespread? Can marine planation be widely effective? Today's wave-cut platforms are never very wide (plate 7.1).
b) Can the surfaces delimited be contemporary over the whole land, in view of the continuation of earth movements to the present day, or are we reading too much into an association of accordant altitudes which have come about by chance? So much of the interpretation of the age of these surfaces depends on their accordant altitude. Bear in mind that uniformly-spaced streams, bordered by uniformly-developing slopes will tend to produce interfluves at a similar altitude.
c) Since the surfaces cut across a variety of rock types, eroding at measurably different rates, why should surfaces still accord in height?

8B 2 The Quaternary physiographic system

After the torture of making conclusions about the Tertiary with so little unequivocal data, the fresher evidence for elements of the Quaternary system comes as a welcome change. Three subsystems may be said to dominate, *glaciation* (ice-sheets and valley glaciers), *periglaciation* and *sea-level changes*. Compared to the *Pleistocene* activity the *Recent* period of fluvial action seems relatively insignificant. It should be remembered too that between the cold phases of the Pleistocene a warm, moist climate prevailed in the interglacials. Shorter, milder phases are called *interstadials*. The glacial phases are given Alpine names (Riss, Würm, etc) but there are English equivalents (see table 8.2).

Much of the landscape of Britain is glacial and most of the rocks at the surface over many parts of the British Isles are glacially-derived. The study of glaciation has benefited from the attentions of geologists, geomorphologists, botanists, zoologists and anthropologists/archaeologists, their combined work making the Pleistocene physiographic system almost as well documented as today's. We begin, however, with the work of physicists in determining how glaciers move and therefore erode and deposit.

Glaciers form by snow *accumulation*. Glaciation occurs during periods in which annual accumulation exceeds *ablation* or loss by melting. It may take from three to thirty years for the snout of a valley glacier to move down-valley in response to excess accumulation. A period of some 5000 years may be necessary for the Antarctic or Greenland ice sheets to respond. Even during a glacial period there is some ablation and *melt water* is seasonally abundant.

Usually it eventually reaches the base of the glacier where it may receive additions in the case of temperate glaciers from *pressure melting* of the ice. (*Temperate* glaciers are those whose surface is at around freezing point while at depth the temperature falls. *Polar* glaciers have extremely low temperatures at the surface but are warmer, although still below freezing, at depth.)

Glacier ice is formed by the compaction of snow. As soon as the snow settles, with a density of less than $100\,kg\,m^{-3}$, the conversion of ice begins. At a density of around $550\,kg\,m^{-3}$ the snow becomes *firn*. This is the highest density to which the original grains of snow can be compacted without deformation, melting and recrystallisation. Further increases in density then occur up to $900\,kg\,m^{-3}$. The density of pure ice is 917 $kg\,m^{-3}$ (Sharp 1960) but this is rarely attained, except in individual crystals. Even when so compacted, the ice retains layering indicative of the major seasonal additions of snow to the surface. Since they entomb pollen, volcanic ash or other time markers, ice cores obtained by deep drilling have played a large part in determining the origin and development of glaciers and ice sheets.

In its movement, glacier ice behaves unlike both water (a Newtonian viscous material) in which the rate of yield (strain) increases regularly with increasing stress, and a perfectly plastic material which yields infinitely beyond a certain threshold of stress; ice yields at a gradually increasing rate as stress increases (fig. 8.7). The mechanisms by which this relationship is manifested in the down-valley progress of glaciers is important to the study of glacial erosion. There has long been argument over the erosive or protective action of glaciers. It can be resolved by the study of glacier movements which may be divided into *basal slipping* and *internal flow*. Basal slipping (fig. 8.8) involves the intact movement of the basal ice over the substratum, producing erosion. In fact only 10–20 per cent of the movement of thick, temperate glaciers (such as those which affected Britain) is accountable to basal slipping. Measurements of ice movement using sur-

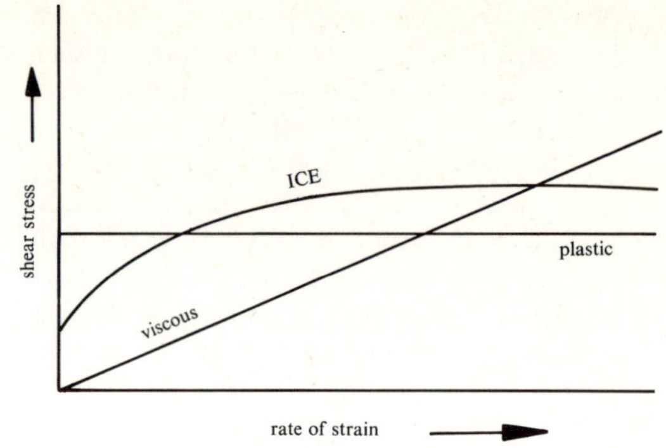

Fig. 8.7 Ice as a material. The relationship between applied stress and the rate of strain in response

Fig. 8.8 The two major components of glacier flow, basal slipping and internal flow

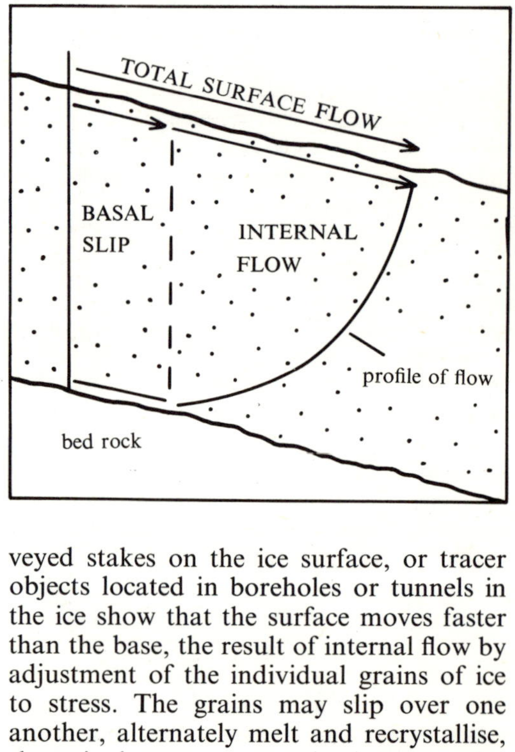

veyed stakes on the ice surface, or tracer objects located in boreholes or tunnels in the ice show that the surface moves faster than the base, the result of internal flow by adjustment of the individual grains of ice to stress. The grains may slip over one another, alternately melt and recrystallise, shear in layers or move by intragranular gliding (fig. 8.9). The latter is most likely. Although basal sliding is the most important in eroding the glacial valley the nature of the internal flow is important too. Nye (1952) has shown how glacier flow is affected by valley profile and width. In steep or narrow reaches *compressive flow* occurs.

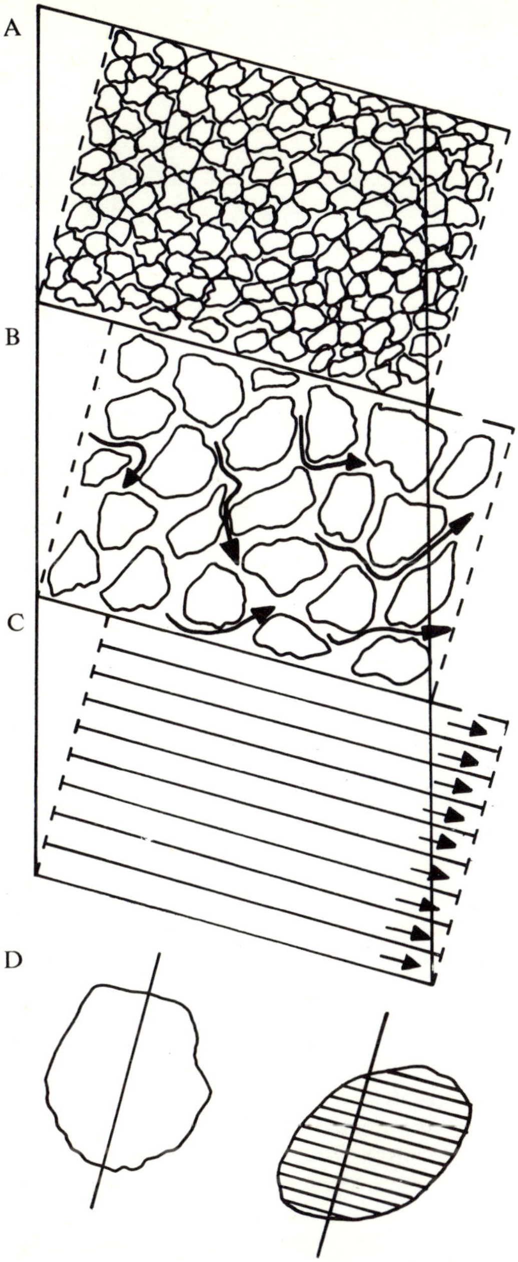

Fig. 8.9 Mechanisms of the internal flow of glaciers: (A) Intergranular movement, resembling that of beans in a bag; (B) phase changes (ice—water—ice) and recrystallisation; (C) internal slip planes and (D) intragranular slip in the individual grain

Internal flow moves material upward from the zone of basal sliding and therefore accelerates erossion. In wide or flatter reaches where the the reverse, *extending flow*, occurs, movement is tangential to the valley floor and sides resulting in less entrapment and erosion of material.

LABWORK

How do various materials flow? It may seem messy to suggest experiments with syrup, blancmange, jam and so on but they all flow. Geographers tend to be ignorant of materials and their properties—try to learn about *shear strength* and so on.

Lewis and Miller (1955) have suggested that a model glacier can be made from blancmange, or kaolin. The former is cheapest but the latter produces better results.

A plaster of paris valley (U-shaped and rugged) is produced at a scale of 1:1 000 to 10 000 with a corrie hollow at its head. It is tilted at five degrees or so and a mixture of one part water to two parts kaolin is emptied into the corrie. Flow should be measurable with matchsticks on the surface of the 'glacier'. Annual layers of fresh 'snow' can be added using dusts such as powdered charcoal—these will be incorporated and moved down-valley.

We must not, therefore, imagine the glacier plugging its way down valleys, crushing and grinding bedrock. The recorded rates of movement are in the order of a few centimetres per day, requiring expert surveys to detect. In some cases the flow is faster and can be detected by photography; on ice falls it may be 3 to 6 metres per day. When *surges* are propagated by increased accumulation at the head of a valley glacier their effect may be cumulative and the snout may be pushed down-valley at a rate of 100 metres or more per day.

8C Our glacial inheritance?

In the British Isles during the Pleistocene we must picture a series of glaciers and ice sheets encountering a preglacial landscape, deeply weathered in the Tertiary and further weathered by periglaciation before direct glaciation. This mantle of weathered rock became entrapped in basal ice which therefore became highly abrasive to fresh rock. *Plucking* of fresh rock may have been aided by *pressure release*, deep-seated rock bulging upwards as its overburden rock, of density $2500 kg\ m^{-3}$, is replaced by ice of density $900 kg\ m^{-3}$. Obstructions of small size or lying parallel to ice movement were scratched, producing the typical glacial *striae* on rocks, or rounded off up-glacier to form a *roche moutonnée* in the case of larger masses.

8C 1 How much erosion?

Perhaps the most typical form associated with glacial erosion is the 'U-shaped' valley **cross profile.** It is perhaps due to the most frequent occurrence of entrapped debris at the side of glaciers; much of the preweathered mantle is there and periglaciation above the ice itself deposits rock on the sides of glaciers, aiding erosion. Perhaps because of its widespread qualitative recognition the characteristic profile has not been often measured geometrically. A parabola seems to give the best approximation, with an equation of the form:

$$y = ax^n$$

in which y = height above the valley floor of any point on its sides

x = distance of the point from the centre of the valley in similar units

n = an exponent, in the case of a true parabola n = 2

a = a coefficient which describes the steepness of the valley sides

By assuming that the shape is a parabola and that n = 2, the value of *a* can be gathered from trial and error on a cross section of a valley drawn from maps (see fig. 8.10). Doornkamp and King (page 277)

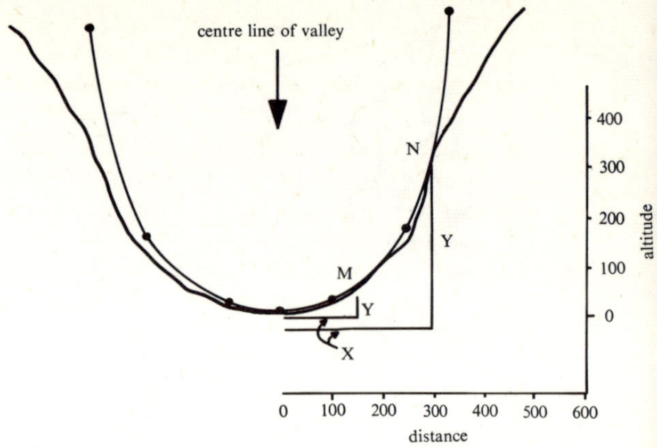

Fig. 8.10 A parabola fitted to a 'U' shaped valley. Two points (M and N) are selected on the valley side. Using the distance and height (X and Y) for each point the equation $y=aX^2$ is solved. Values for a of 0·16 and 0·11 are obtained. A parabola is drawn taking the mean value of these two solutions (a=0·135).

obtain the following equations by this method:

$y = 0{\cdot}08x^2$ for two valleys in North Wales

$y = 0{\cdot}04x^2$ for Bishopdale in Yorkshire

$y = 0{\cdot}1x^2$ for Grisedale in the Lake District

Those who attempt this exercise should remember that the value of *a* derived from your cross-section drawing should always be divided by the vertical exaggeration you have used. Vertical exaggeration steepens valley sides artificially.

The **longitudinal profile** of a glacial valley is very typical. Its marked irregularity at every scale of inspection contrasts it with the river long profile. This irregularity has been ascribed by Professor King to the preponderance of positive feedback relationships in glacial erosion. Preglacial hollows tended to be deeply weathered and became deeply scoured by glaciers which, as they deepened and showed compressive flow, eroded even more strongly and encouraged pressure release at considerable depths into the fresh bedrock. Hence the growth of hollows into deep rock basins, now often filled by lakes. Increases in erosion by streams of water, however, tend to be compensated by negative feedbacks. For instance, erosion of bed and banks by increased velocity of streamflow tends to increase the hydraulic radius, reduce velocity and cause deposition.

MAPWORK

Choose a valley in a glaciated area. What is the coefficient of steepness of the valley sides, using the parabola model? Does the value vary at other points in the valley? If the long profile of the valley is irregular is there a geological reason? Has the stream in the valley graded a profile? Are there hanging valleys? The cross profile should be examined at points where the long profile shows steps. Is the parabola a useful model? Draw a flow diagram indicating feedbacks in glacial erosion.

Two further features of glacial valleys are *corries* and *hanging valleys*. It is undeniable that corries and cirques are associated with glacial valleys but whether the head of the glacier plays much part in the development of the typical form (whose height is approximately three times its length) is debatable. It seems obvious that the corrie is the development of a sheltered hollow or preglacial valley-head which focuses snow accumulation during the onset of glaciation. Snow sapping of the corrie floor and freeze-thaw action on the headwalls are other processes which must operate. The headwalls are exposed by the *bergschrund*, a large crevasse which occurs between the glacier and rock. The smooth, cup-like shape of corries has been thought to signify erosion by rotational sliding of ice in the hollow but this is now a less favoured hypothesis. It has been claimed that many of the typical corrie features in our mountain areas were occupied by ice relatively late in the Pleistocene when only the corries themselves were glaciated. Reconstruction of the snow line and hence temperatures and wind directions of the time may be made by collection of data on the height and aspect of corries from maps. A map of corrie features can be found in the *Atlas of Great Britain* (1963). There is a strong preference for snow accumulation on the sheltered sides of mountains and thus most late Glacial corrie glaciers face northeast.

The amount by which unglaciated tributary valleys 'hang' above glacial valleys has been used to calculate the amount of erosion by glaciation. This exercise can also be done from maps. Where post-glacial erosion has graded the lower part of the tributary to the present main valley floor the upper part of the tributary profile may be extrapolated, as in the case of river profiles used in Tertiary reconstruction. Of other methods used to calculate the amount of glacial erosion perhaps the earliest was the cutting of a three-millimetre-deep cross in bedrock before the Dachstein glacier advanced over it in 1856. The surface was bare and polished after its retreat, with no sign of the cross. Sediment sampling in glacial meltwater streams shows them to carry loads in excess of normal rivers (being frequently over 1000 mg/l); their solutional loads are also high. It is difficult, however, to determine how much of the material was eroded by glaciation and is not merely preglacially-weathered material eroded by the fast-flowing turbulent meltwater streams. Erosion by meltwater, often under considerable pressure when beneath the ice, is responsible for a type of channel found in glaciated areas which appears to have no present-day drainage. Such channels are frequently completely out of context to the present channel network under fluvial conditions, cutting spurs, vertically descending valley sides or having a 'humped' profile, with an uphill gradient for part of their length. They were formerly interpreted as the overflow routes of glacial lakes, dammed by the ice. However, ice is not very impermeable to water. Particularly during stagnation and decay, when glacial lakes would have been formed, the glacier takes on the appearance of a karst limestone landscape with streams, typically descending sink holes or flowing through ice caves. It is such streams which have carved channels with scant regard for topography or geology where they have intersected the land surface below the ice.

Some efforts have been made to assess the amount of erosion accomplished by the ice sheets which covered much of Britain, particularly during the maximum extent of the ice. The ice sheet, unlike the valley glacier, is little modified in its flow by preglacial topography. The scarp slope of the Chilterns may be divided into a northern, glaciated section in which the steep slope itself has been pushed back, east of the Chalk boundary, and a southern, unglaciated half where the slope and the geological boundary of the Chalk coincide. It is possible to calculate the volume of Chalk eroded by glaciation from comparing topographical and geological maps. The volume of the Riss boulder clay in East Anglia has been calculated as about 400 cubic kilometres; this ties in neatly with the volume of Chalk eroded.

By a variety of methods, estimates of the rate of glacial erosion per year range from 0·05 to 2·8 millimetres. It is likely that glacial erosion proceeds at between *ten and twenty times* the rate of normal fluvial activity.

8C 2 Forms of deposition

As table 8.2 shows, there have been four major glacial phases in Europe, although the first is not represented in Britain by more than a series of sedimentary rocks containing arctic-type fossils. The next two, though they saw the maximum extent of the ice (fig. 8.11), do not exhibit fresh glacial features. Rather they have left behind low-lying land, undulating and dissected by river action during and since the final *Würm* glacial advances. Such is the major part of East Anglia, the south Midlands and the east Midlands; it is called the area of *Older Drift*. Drift is a word which is unspecific to glacially deposited material. What we mean by the Older Drift is a thick mantle of *boulder clay* or *till* deposition. The *Newer Drift* areas to the north and west show the freshest glacial morphology and the least-disturbed glacial deposits, having been covered by ice during the last 100,000 years (in one or all of the phases of the Würm glaciation). The Würm consisted of four glacial advances from the ice centres of the north, over progressively smaller areas.

The fullest depositional sequence of the earlier stages of the Pleistocene is found in East Anglia. There, the Pliocene and Pleistocene crag deposits show a shelly fossil content indicative of a progressively cooling climate until the Weybourne Crag which shows evidence of the earliest glacial phase of Europe, the Gunz. This had been preceded in the Alps by the Donau glaciation. It was succeeded by the Mindel and Riss glacials, the most extensive in Britain where they are called respectively the *Lowestoft* and *Gipping* glaciations after classic sites in East Anglia. Their tills may be separated where they overlap by colour as well as more sophisticated techniques described below. Only a small classic site at Hoxne provides depositional evidence of the very long interglacial which separated the Lowestoft and Gipping glaciations in time. This indicates one problem with Pleistocene deposits—their isolated occurrence. Quite often they are discovered by chance such as those of the interglacial deposit which followed the Gipping phase, called the *Ispwichian*, after discovery of the deposits during sewage works construction near Ipswich. The glacial history of East Anglia is well described in the Geological Survey Memoir for the area.

As mentioned above, the deposits of the Würm glacial phases further north are fresh, allowing both the morphological and sedimentological studies stressed in our introduction to historical geomorphology to be followed with more success.

We have already dealt with the typical forms associated with glacial erosion. We have also mentioned that the most typical feature of widespread glacial deposition is an undulating surface of till, featureless except where preglacial landscape shows through or post-glacial modification has been rapid. Before dealing with areas in which there is a fresher topography with a

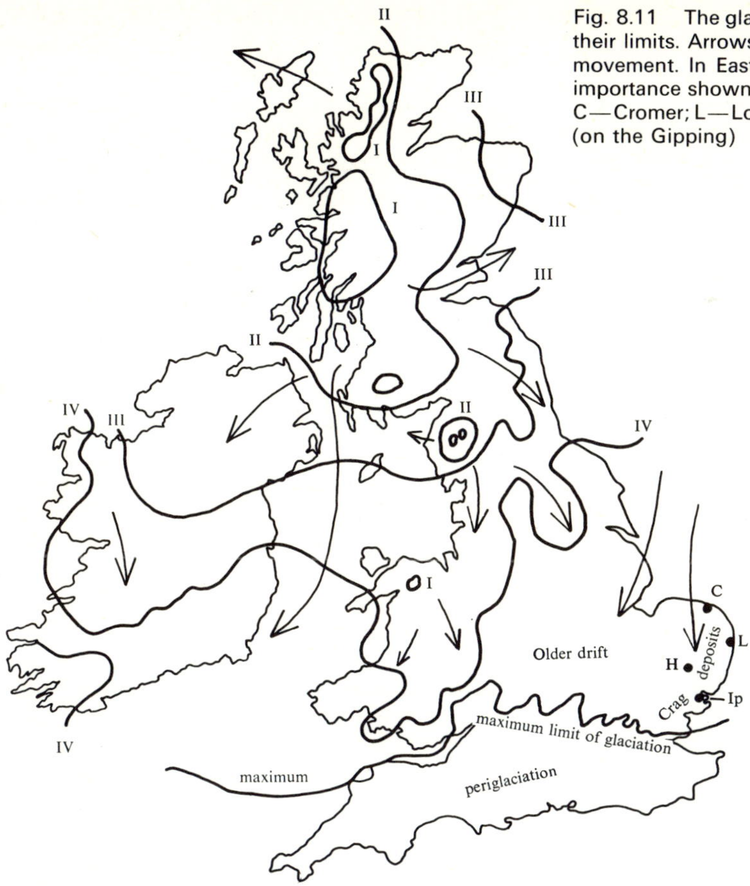

Fig. 8.11 The glacial phases in the British Isles and their limits. Arrows mark the major directions of ice movement. In East Anglia those sites of geological importance shown are:
C—Cromer; L—Lowestoft; H—Hoxne; Ip—Ipswich (on the Gipping)

range of glacial depositional features we must stress two points:

a) We will be referring to the Newer Drift areas, glaciated in the Würm.
b) The morphological features are based upon a nominal scale ('drumlin', 'esker' and so on) and are liable to different interpretation by different workers. The 'classic' examples of the features we shall mention are illustrated in such texts as Monkhouse and Holmes, and defined in Charlesworth's book on glaciation. A more recent definitive volume is that by Embleton and King (1969).

In recently-glaciated areas certain regularities of outline characterise the boulder clay landscape. The most obvious case is that of *drumlins*. These small, rounded, boulder clay hills, occurring in swarms and producing a 'basket-of-eggs' topography, are undoubtedly formed by the pressure of moving ice. The upglacier (or ice-sheet) end is broader and higher, the form tapering to the distal end. Charlesworth says the shape is one offering minimum resistance to flow by hindering the formation of vortices in the rear which would act as a drag on the moving body; in other words it is streamlined. Chorley (1959) has likened the shape to that of airfoils and birds' eggs. The egg is still plastic while passing down the oviduct and becomes characteristically shaped in response to the external forces during this passage. It has been found that in birds laying eggs which are of large size compared to the bird itself the tapering is most pronounced. Charlesworth noticed

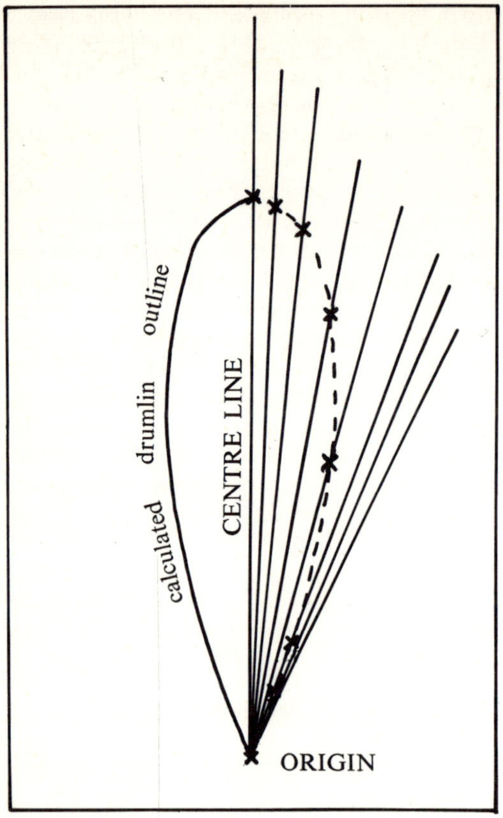

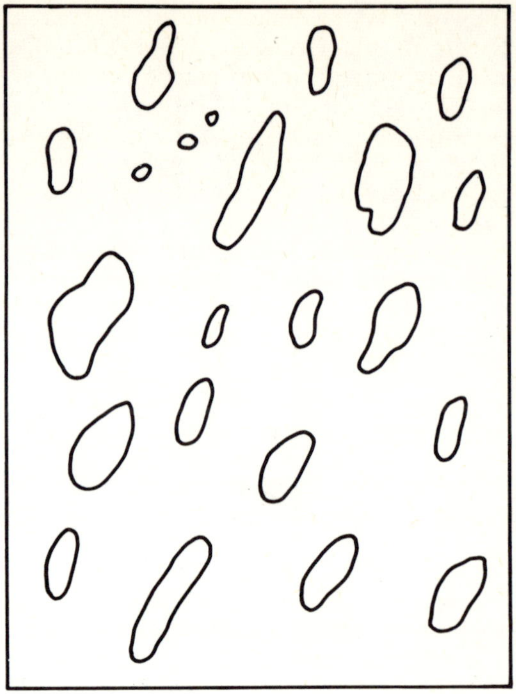

Fig. 8.13 The shape of some closed contours (indicating drumlin forms) taken from maps of the Solway area

Fig. 8.12 The lemniscate loop fitted to drumlin form. Firstly k is calculated; this value is used in the expression $r = \sqrt{a^2 . \cos . k . \theta}$, where r is the distance from the origin to the outside of the curve along lines at $\theta°$ to the centre line. Each value of θ is multiplied by k before its cosine is taken. a is the area of the drumlin. The same analysis may be applied to drainage basin outlines.

that drumlins formed under conditions favouring rapid ice movement were most tapered, whereas those on rising ground (which would slow the ice) were broader and shorter. There are other factors too, such as the amount of material (being carried by the glacier) which is available to be 'plastered on to the landscape' to form drumlins. Chorley, seeking an objective way of comparing the shape of drumlins, experimented with the **lemniscate loop** (fig. 8.12). This is the shape generated by an equation which gives distances from an origin to the boundary of the curve at various angles from the long axis of the shape. Part of the equation, **the k factor**, may be used on its own to describe the elongation of the drumlin.

$$k = \frac{l^2}{4A}$$

where l = drumlin length and A is drumlin area.

Chorley has also experimented with the lemniscate loop and the k factor in describing the shape of drainage basins.

A simpler **elongation ratio**, for use when the lemniscate loop cannot be drawn, consists of dividing the maximum length by the maximum width of the drumlin. Though concise and accurate measurements can only be obtained by field survey, much useful work may be done from 1:25000 maps. Sheets of the Ordnance Survey 1:25000 map showing drumlins include NX 25, 26 and 27, NY 24 and 43 (Eden Valley and Solway Lowlands) and SD 77 and 87 (the head of Ribblesdale, Yorkshire). Some mapped drumlin shapes revealed by contours are shown in fig. 8.13.

Also derivable from maps are the *density* of drumlins in different areas (are they more dense where they are smaller?) and their *orientation*, which is a good indicator of the direction of ice flow.

MAP AND FIELDWORK

1 Use maps of drumlins to assess their size, shape, density and orientation. Refer to texts and the *Atlas of Great Britain* for the ice dispersion centres responsible for the drumlin field you have chosen. If the contours which indicate drumlins were not present that would be the relief in the area? In other words, where are likely locations for drumlin deposition?
2 In the field are there drumlins which do not show up on the map? Why not? Do the drumlins consist of clay or rock or a mixture? How has the present-day drainage adapted to the presence of the drumlins? Can you discover examples of drumlins 'drowned' by the post-glacial rise of sea level?

Terminal moraines are also of use in assessing the direction, and limits of glaciation. Unfortunately they are seldom as fresh in our landscape as they are in those Alpine and Scandinavian valleys which have experienced glacial advances fairly recently. In such cases they are discrete cross-valley ridges of boulder clay, each marking a stage during retreat. However, Würm terminal moraines are less easy to recognise, especially in areas where confusion arises with other mouldlike features left by stagnant ice. Too often the terminal moraines have been eroded back from the centre of the valley by post-glacial river action. Terminal moraines are most characteristic of glaciers but ice sheets too can produce them, for example the Cromer Ridge in north Norfolk, thought to mark a readvance of the Lowestoft ice across the Wash. King (1960) gives some examples of valley terminal moraines in Yorkshire.

Fluvioglacial deposition, by meltwater streams, occasionally produces striking forms called *eskers*; long, narrow ridges, often interlinking after the fashion of a braided river channel or delta. The reason for this is that most fluvioglacial deposition occurs, delta-like, as the streams lessen velocity on emergence from under the ice. The fact that eskers may be found crossing small hills and spurs is further testimony to the similarity in behaviour of subglacial streams and deep karst limestone circulations, both being under enough hydrostatic head to flow uphill. Other fluvioglacial features found hugging the sides of glaciated valleys are *kame terraces* but again their recognition is difficult without recourse to a sedimentological approach.

8C 3 Sediment fabric

A common dream of glacial geomorphologists is said to be that of owning a large mechanical excavator for the purpose of getting into glacial deposits at some depth! In many cases this is the only way to obtain undisturbed samples of material, the outer layers and surface being much modified by subsequent processes. However, there are often usable, natural exposures of glacial sediments in river banks and the opening of quarries or cuttings by engineers is always watched closely by geomorphologists and geologists. For example, the M5 motorway cuttings at Clevedon, Somerset, have yielded fresh evidence about glacial history in that area. There are instruments available to the researcher for obtaining samples but in the case of various types of *auger* which screw into the ground these inevitably mean disturbance of the deposit. In many cases the digging of a soil pit (plate 8.1) yields usable evidence. The use of seismic instruments, in which a shock, transmitted to a steel plate on the surface of the deposit using a hammer, is differentially reflected by the clays, sands and gravels below, has

Plate 8.1 Digging a soil pit in order to recover pebbles for roundness studies. This is a fluvioglacial deposit, suggested by pebbles as round as the one being held.

made geological profiles easy and quick to produce. This is important since most geological maps in Britain were compiled quite a while ago when many different types of glacial deposits were labelled as 'drift'. The maps in the *Atlas of Great Britain* are the most recent of glacial deposits.

Glacial deposits are seldom seen at their junction with bedrock. If they were, the junction would clearly be unlike those over much of the rest of the geological column where sedimentary rocks either grade from one to another in a *conformable* junction or are angularly juxtaposed along an *unconformity*. Glacial deposits are unconformable but are also mixed and contorted during deposition in a way which is not possible in water deposition. For the same reason glacial deposits may be found at all altitudes at which glaciers flowed, mantling even steep slopes. Only fluvioglacial deposits and those formed in glacial lakes show stratification. Fluvioglacial cobbles are typically *imbricated*, each lying on the one below like roof tiles, with the slope of the 'roof' being upstream. This is the typical position in which the flatter fluvial cobbles come to rest on a stream bed. Fluvioglacial deposits also show channel-like features and lenses of sand mixed with the cobbles (fig. 8.14). Glacial lake deposits can be of great value in dating glacial phases. Not only do the fine lake clays preserve pollen grains (see section 8.F1) but the layers of sediment, or *varves*, each represent an annual cycle of deposition consisting of coarser material from the spring snow melt, with a subsequent gradation to extremely fine material during the restricted inflow of meltwater in winter.

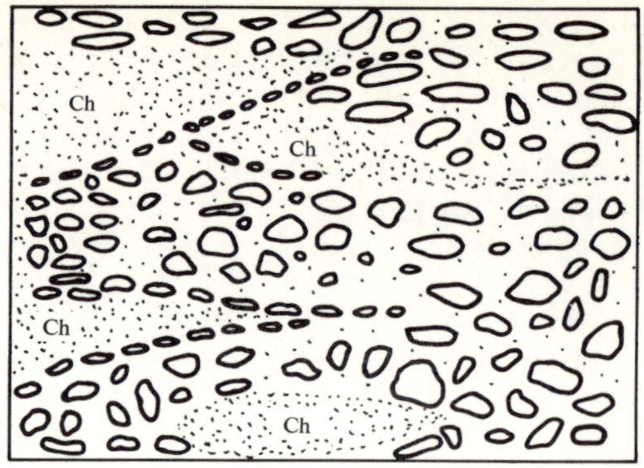

Fig. 8.14 A typical exposure of fluvioglacial material showing mainly rounded pebbles with enclosed, sandy channels (Ch)

The orientation of stones in boulder clay has been used as a guide to the direction of ice flow for over 100 years. For the same reason that drumlins are characteristically shaped, a pebble tends to become orientated parallel to the direction of movement. It has, however, been found that a smaller number of pebbles do orientate themselves at right-angles to ice movement but usually not in enough numbers to mislead a result if a large sample is taken. Fifty pebbles are the minimum sample allowable and should, ideally, be measured for orientation whilst *in situ*, although each pebble can be removed, the longest axis chosen and then replaced. This is obviously easier on banks or quarry faces from which pebbles or cobbles project. It is best to choose only those pebbles with a ratio of long to short axis lengths of 2:1. A knitting needle may be used to lie along the longest axis of the pebble *in situ* (the needle has the advantage of being capable of forcing into the boulder clay matrix alongside the measured pebble which can then be removed for other measurements) but the needle must be non-ferrous in order not to invalidate the compass readings of its orientation. At the same time it is worthwhile to measure the dip of the pebble or needle with a clinometer or Abney level (plate 8.2). Both measurements are recorded on field sheets. After measurement of orientation and dip the pebble may be removed and can be used for the size and shape analyses described below. Clearly this **till fabric analysis** is teamwork! Full details of data collection and processing in till fabric analysis are given by Andrews (1971).

The results of orientation measurements are plotted on a **rose diagram** (fig. 8.15) of compass bearings, usually divided into 20° classes. Obviously, a pebble pointing northwest also points southeast at its other end; both directions are plotted, making the diagram symmetrical. In the regional context only one of the two can be the direction of ice movement. For example, J. J. Donner and R. G. West found a northwest/southeast orientation for pebbles in the Lowestoft boulder clay of East Anglia. Only northwest is a possible direction for an ice source. The Gipping boulder clay in the area has a north/south orientation, indicating ice movement from the north.

Frequently no clear, preferred direction emerges from till fabric analyses. This is most often the result when the sample is gathered from near the surface, especially on slopes where mass movement either in periglacial or recent conditions has led to

Plate 8.2 Till fabric analysis involves taking both orientation and dip measurements. The latter is seen here being done with an Abney Level whilst the pebble is in situ.

Fig. 8.15 Till fabric orientation data plotted as a rose diagram. The direction lines are drawn from the centre to lengths proportional to the number of pebbles in that direction class. The distal ends of all these lines are then joined to make the rose. Ice movement was from the northeast.

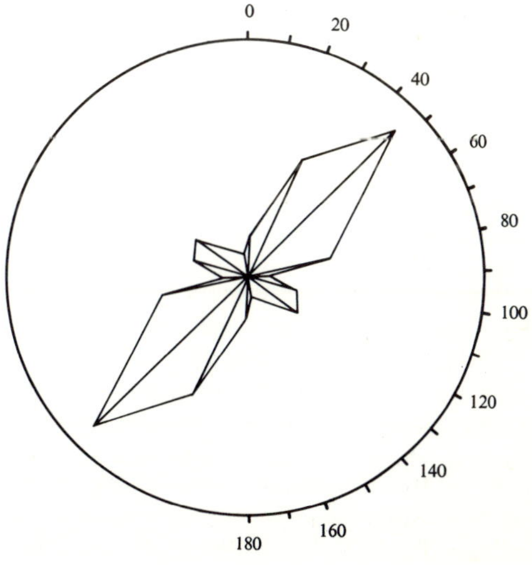

redeposition of pebbles. Mass movement deposits show a random orientation. It is sometimes possible to confirm this origin by the dip measurements which in a case of mass movement should show a similar direction to the slope from which the samples were taken.

As well as sharing the geologist's taste for relentless field study, the glacial geomorphologist must know a good deal of geology to be able to identify the rocks from a variety of source areas which may be eroded, mixed and deposited by the ice. One of the most immediate clues given by glacial deposits to their directional origin is their content of *erratics* which may be prolific enough to impart a characteristic colour to the boulder clay. Erratic particles range in size from small crystals of one mineral to rock masses large enough to dig a quarry in. Where bedrock is exposed and the boulder clay matrix has been weathered away, erratic blocks, even if composed of the same rock as the area of deposition, present a rough, littered appearance. The shapes of such blocks are usually those of quarried blocks of the same rock because of their erosion from the bedrock by the plucking process described above. Where composed of different rock type to the bedrock they are spectacularly clear as at Norber Scar in Yorkshire where Silurian boulders lie on Carboniferous Limestone, or south of the Brecon Beacons where Old Red Sandstone similarly contrasts with light grey limestone. The protection that these erratics have given the limestone bedrock in post-glacial times has allowed the formation of a *pedestal* beneath each one. The relative height of this pedestal to the surrounding limestone has been used to calculate the amount of post-glacial solutional weathering in such areas.

To be really useful in analyses of ice movement, erratic material should come from a relatively unique area close to the centre of ice dispersion. For this reason two of the most diagnostic rocks in northern Britain are those of Shap Granite from the Lake District, found in boulder clay in the Solway Lowlands and those of a rock called riebeckite from the island of Ailsa Craig, found in east and southeast Ireland and the north and west coasts of Wales. Porphyries from Scandinavia are found in Yorkshire and East Anglia, revealing the passage of ice sheets across the area now covered by the North Sea (remember the low glacial sea levels would have exposed the continental shelf). A distribution map of erratics is shown on page 19 of the *Atlas of Great Britain* (1963).

There has recently been renewed interest in the glaciation of southern Britain following a controversial theory that the 'Blue Stones' of the Stonehenge prehistoric site are glacial erratics. Archaeologists had been prepared to believe that they reached Wiltshire from their outcrop in Pembrokeshire by means of human effort in floating and rolling them eastwards. The recent study suggests that there is plenty of other erratic evidence between Pembrokeshire and Wiltshire to suggest that they moved there entrapped in glaciers. Morphological evidence, too, is used in the form of the 'U' shape of many valleys in Wiltshire and the 'subglacial' gorge-like features in the Bristol area.

One should beware of conclusions based on erratic fluids. Many of the early workers who enlisted help in searching for far-travelled erratic rocks found it better to pay men to look for, rather than find, erratics! If paid to find erratics, these rocks were brought into the area by friends. The less mischievous but similarly complicating factor of the transport of building or road materials may interfere with present-day investigations.

FIELDWORK

1 Using the drumlins studied in the previous exercise obtain a sample of pebbles for till fabric analysis. If digging is involved obtain all necessary permission before beginning. Use a ready-made section if possible; in such a case 'clean up' the section carefully and measure only undisturbed pebbles. Produce an orientation rose diagram and compare the predominant orientation with that of the drumlins themselves.

2 Look at landforms which are not obviously drumlins. Use techniques already described and back them up with size and shape analyses below.

We have already dealt with *size and shape analyses* designed to aid our interpretation of fluvial action. The methods used were those which assume the processes of particle morphogenesis to be similar throughout the river system; they are more tuned to detecting *changes over time or distance*. In this section we are hoping to use size and shape parameters to distinguish between *different processes* on the basis of their historical effect upon sediments. We are attempting to distinguish characteristic *sedimentary environments* from the size and shape of particles. Size analyses are still performed with calipers and sieves but the distribution of sizes in the deposit becomes more important. Inman (1952) gives details of the simple parameters of **mean, dispersion, skew** and **kurtosis** (see fig. 8.16 and table 8.3) which define how well *sorted* a deposit is. These parameters are easy to derive from cumulative graph plots of the size analyses. Sorting means preferential transport and eventual deposition of a particular size range of sediment to the exclusion of others. It is most typical of wind-blown deposits but also characterises fluvial sediments. Size distribution of sorted material show high values of kurtosis and often considerable skew (both terms are explained diagrammatically in Toyne and Newby, page 36). Glacial transport, however, shows very little power to sort sediments. A wide range of material, characterised by low values of kurtosis predominates in boulder clays. Put in other statistical terms, glacial deposits tend to show larger standard deviations

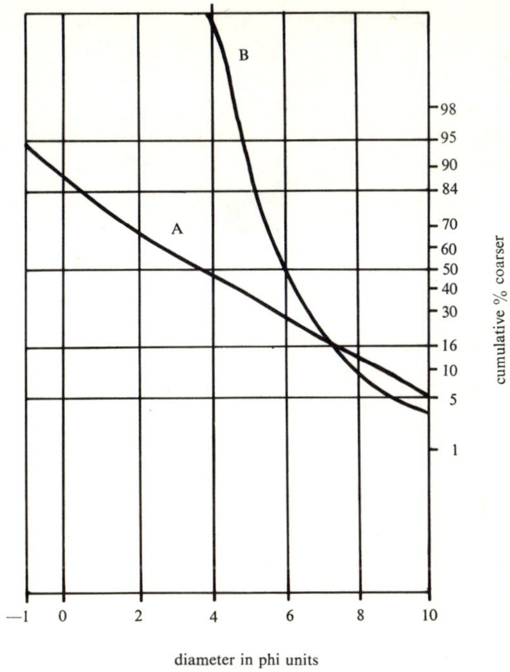

Fig. 8.16 Size distribution of two sediment samples, A and B. The percentage scale is a probability scale and the normal distribution would plot as a straight line. Sample A is most nearly normal and is less well sorted than B, a conclusion verified in Table 8.3

Table 8.3. Calculation of sedimentary size parameters arcing phi values derived from Fig. 8.16. The phi value of which 50% of tho sample is coarser is signified by $\phi 50$—and so on. Sample A is in fact a glacial fill and B a wind-blown loess

PARAMETER	CALCULATION	SAMPLE A	SAMPLE B
Mean Diameter	$\frac{1}{2}(\phi_{16} + \phi_{84})$	3·8 (ϕ)	6·4 (ϕ)
Dispersion (sorting)	$\frac{1}{2}(\phi_{84} - \phi_{16})$	3·50 (ϕ)	0·90 (ϕ)
Skewness	$\frac{\text{mean} - \phi_{50}}{\text{dispersion}}$	0·23	0·44
Kurtosis	$\frac{\frac{1}{2}(\phi_{95} - \phi_{5}) - \text{dispersion}}{\text{dispersion}}$	0·58	2·01

(dispersion) of size, or a larger coefficient of variation. These characteristics can be demonstrated by sizing the material used for till fabric analysis. This would obviously be one way in which to distinguish an esker from a drumlin more reliably than by morphological differences alone.

In our analyses of the shape of fluvial sediment during transport we used the index of Lees which measures the sharpness of particle corners and their individual mechanical stability. This is why Lees' index is a good one for measuring abrasion and attrition. To distinguish simply between deposits of fluvial (fluvioglacial) or marine and glacial deposits a fairly coarse measurement of shape may be made. The Frenchman, Cailleux, provided two indices based upon measuring the three diameters (long, intermediate and short) of pebbles, defined as a, b and c respectively. His roundness index includes the measurement of the radius of curvature of a circle (r) which would fit into the sharpest corner of the pebble (fig. 8.17).

$$\text{roundness} = 2r/a \times 1000$$

$$\text{flatness} = \frac{(a-b)\times 100}{c}$$

In a study by King and Buckley (1968) of pebbles from various sites in the Arctic they found flatness to be diagnostic of various rock types. Significant differences were found between the flatness values of gneiss and limestone pebbles in a variety of deposits. The deposits themselves, however, were not distinguishable by flatness of pebbles, but by using the index of roundness significant differences were discovered between glacial, fluvioglacial and marine deposits:

moraines	138
solifluction (periglacial deposits)	185
kames	238
eskers	332
beaches	398

As well as being unrelated to geology, roundness also appears to have no relationship with pebble size. Thus it is left as being a clear indicator of sedimentary environments and appears to be yet another method

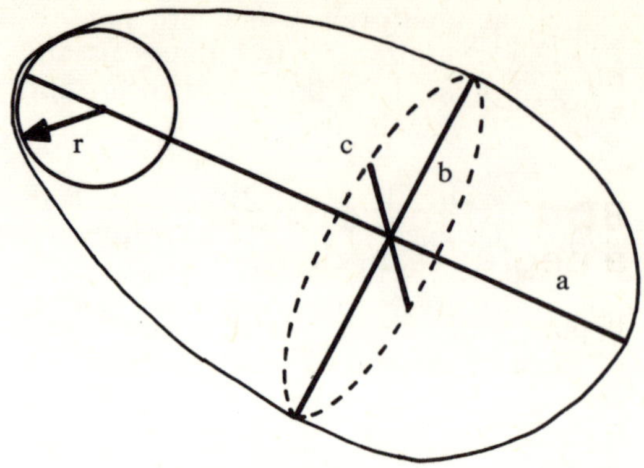

Fig. 8.17 Measurements for the calculation of Cailleux roundness and flatness indices

by which measurements can distinguish between the postulated origins of landforms. The morphological approach may well leave these in doubt.

8D Beyond the glaciers?

The effects of periglaciation may be largely ignored by those who take it to mean 'almost glaciation'. While periglacial processes do operate close to glacial ones, both in space and time, they are quite distinct in mechanism and effect. Periglaciation is dominated by the effects of freezing and thawing on both solid rocks and aggregates of smaller particles, like soil. In the case of solid rock the effect is to *weather* the rock but in the case of aggregates the effect is one of rearrangement of particles, often producing a distinct *patterned ground* and in some cases leading to mass movement of a type known as *solifluction*. Although Kirkby has ascribed the major role in slow soil movements on slopes today to wetting and drying there are many slopes, especially in the south of Britain, which owe some of their current profile to the residual effects of periglaciation during the Pleistocene. Apart from the relict effects of periglaciation in the present-day British landscape, the processes associated with freezing and thawing are important because of their effects on current human activity over large areas, for example in Arctic Canada and the Soviet Union. Even in our own islands, engineers and road builders, as well as agriculturalists, pay much attention to the mechanics and effects of freeze/thaw.

8D 1 Freeze-thaw: a neglected topic

If pure water freezes inside a pore typical of those in many rocks the resultant pressure is some ten times that necessary to fracture the rock. However, the water within pores is seldom pure and it could only have reached such a location because the pore has openings which connect to other pores. The matter is even more complicated in an aggregate such as soil; a theoretical approach in freeze/thaw investigations therefore becomes secondary to direct experimentation. Wiman (1963) studied the effects of freeze/thaw in two freezing-rooms equipped with observation panels. He was keen to demonstrate the difference between two types of periglacial climate, the *Icelandic* (with a 24-hour temperature cycle varying between −7 °C and +6 °C) and the *Siberian* (where less numerous cycles occur but the variation is between −30 °C and +15 °C). Several pieces of dried, weighed and sized rock were subjected to both sorts of 'climate' for 36 days. In those cases where no further water was added, no weathering occurred; where small amounts of water were added during the freeze/thaw cycles weathering was most effective in the Icelandic conditions.

To assess the variable effects of freeze/thaw on different rock types Wiman measured their porosity by drying specimens in an oven for several days, weighing, immersing in water for a week and then reweighing. The most porous rocks generally showed the most weathering (expressed as the weight of weathered fragments as a percentage of the weight of the original lump). Periglacial weathering of rock may therefore be claimed to be most effective in moist areas with frequent alternations of temperature above and below freezing point; in such areas it is most effective on porous rocks. Similar experiments can be performed by use of a domestic fridge although only soft rocks like Chalk will weather.

The major result of the 9 per cent expansion when water freezes in soil is *frost-heave*. This still occurs during fairly hard frosts in lowland Britain today (when temperatures reach −4 °C to −6 °C); maximum heave is generally between 10 mm and 20 mm. In describing methods of measuring frost-heave in the field James (1971) differentiates between *needle ice* and *ice lenses*. Most commonly in Britain today needles of ice grow perpendicularly to the soil surface around the crumbs of soil and stones. Ice lenses grow parallel to the surface and usually further down in the soil. The growth of both types of ice is by migration of soil moisture towards the ice. The amount of heave produced at the surface can be measured by several methods, including simple precision survey down to the soil surface (from a reference level such as a tight thread between two stakes). The use of a wooden frame driven into the soil in which small rods, with their bases on the soil, slide up and down through slots in the horizontal members of the frame, has also yielded valuable results. Both methods will show measurable results during a typical winter. Amounts of movement can be related to the grain size of the soil and thus its porosity; results are different to those obtained for rock weathering since in the most porous soils ice needles and wedges do not form, heave being therefore a more typical phenomenon in fine-grain soils. The Road Research Laboratory has shown that frost-heave liable to damage roads in Britain can be prevented by adequate drainage of the compacted earth beneath the road. Moisture is clearly the most important variable.

One effect of frost heave is that the larger particles in an aggregate are gradually lifted to the surface. The process appears to be due to the finer material falling, as the heaved soil relaxes during thaw, into voids left during the freeze beneath the coarser fragments. Thus after each cycle the stones appear to have risen. Once on the surface, stones are moved into the geometric distributions typical of patterned ground.

Circles, nets, polygons, steps and *stripes* of stones are the dominant forms, some being found on mountain sides in temperate areas such as Britain but the most impressive types being found in those Arctic and sub-Arctic areas of *permafrost.* The exact mechanism of patterned ground formation is still not clearly understood but seems to involve some function of differential movement in response to freeze/thaw by particles of different sizes. The gradient of the surface is an important factor. Circles, nets and polygons are restricted to flat or nearly flat surfaces, elongated forms occur on slopes of 2°–6°, and stripes in particular are found on slopes of 6°–30°.

Though permanently frozen ground, or permafrost, does not affect Britain now, relict features of the Pleistocene clearly show its evidence in central and eastern England. For example in East Anglia there are large scale polygons, almost 10m across. They are best demonstrated by vegetation patterns on aerial photographs, heather growing on the sandy polygons while grass covers the chalky centres. Permafrost is not only distinguished by the excellence of its patterned ground but by the effects of an annual thaw at the surface which forms a muddy *active layer.* It is during the wetting season that solifluction leads to movement of debris downslope. During refreezing at the surface the active layer is subject to compression which produces *involutions*, contortions of the finer grained material (fig. 8.18). Involutions can sometimes be observed in roadside cuttings through soil or weathered material in Britain, being most clear where layering of different sized or different coloured particles has become contorted. Another ice form typical of permafrost is the *ice wedge* which differs from the needle or lense of ice in that it forms by polygonal cracking of the surface. The cracks admit water which freezes into wedges which, when they finally melt, are replaced by fine material or stones lying vertically. Such features are also easy to spot in quarries or cuttings (fig. 8.19).

Those features of permafrost found in southern Britain are usually ascribed to the

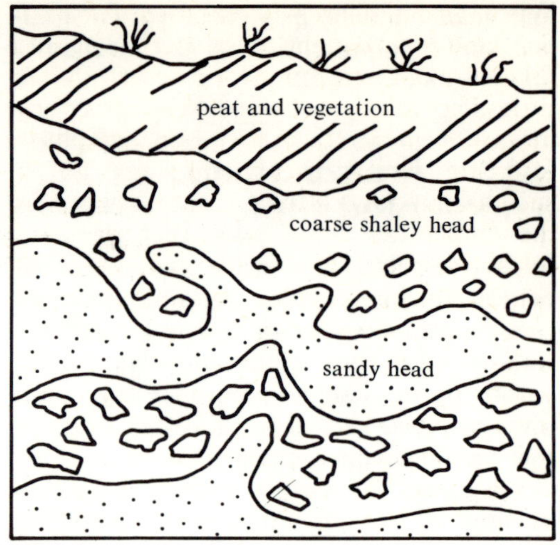

Fig. 8.18 Involutions in head deposits

Fig. 8.19 A fossil ice-wedge feature

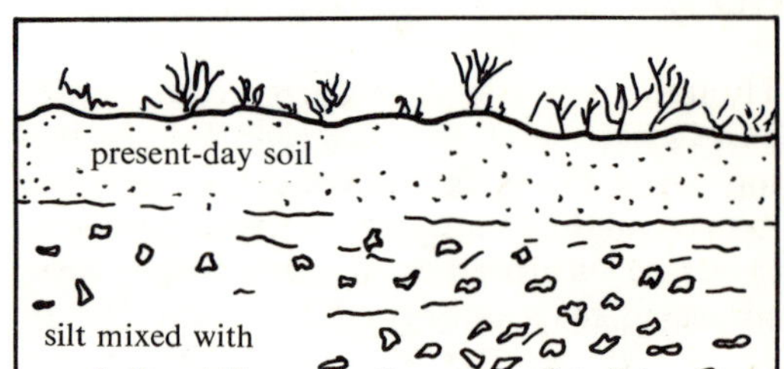

Würm glacial phase whereas those found in Scotland are thought to be later. The late Glacial periglacial phase in Scotland had little effect in terms of slope development by mass movement. It was a long enough phase to create areas of patterned ground but in such a humid, maritime climate, as soon as the temperature drops below that necessary for active periglaciation, the landscape becomes quickly covered by a blanket of permanent snow or ice which prevents further freeze/thaw cycles. However, in the south, during the longer and more continental phase of periglaciation during the Würm, the effects of periglaciation on slope development are clearly shown, especially in those central and eastern areas in which permafrost led to extensive solifluction. In the southwest solifluction occurred but to a lesser extent.

8D 2 The flowing soil

Though river action in periglacial regimes is restricted to occasional river floods during rapid thaws the landscape is far from 'frozen' in the usual sense of the word. There is active weathering of rocks and a very suitable environment for transport of weathering products downslope. Meltwater from ground ice and snow patches saturates the active layer, being prevented from deeper circulation by the permafrost. Vegetation is scanty and the lack of binding roots facilitates mass movement on even gentle slopes. There are three main types of mass movement in periglacial regions: gravitational deposition of weathered rock (as in screes), slow creep of aggregates downslope by alternate freeze and thaw, and the most characteristic movements of the sludgy active layer referred to as *solifluction* (flowing soil). The effects of solifluction on the landscape in Britain was first recognised in 1839 when the term '*head*' was applied to accumulation of rubbly waste mantling slopes and valley bottom in southwest England. Further east, on the Chalk, the word 'coombe rock' was applied because of the association of such deposits with the classic dry valley or coombe forms of the Chalk escarpment.

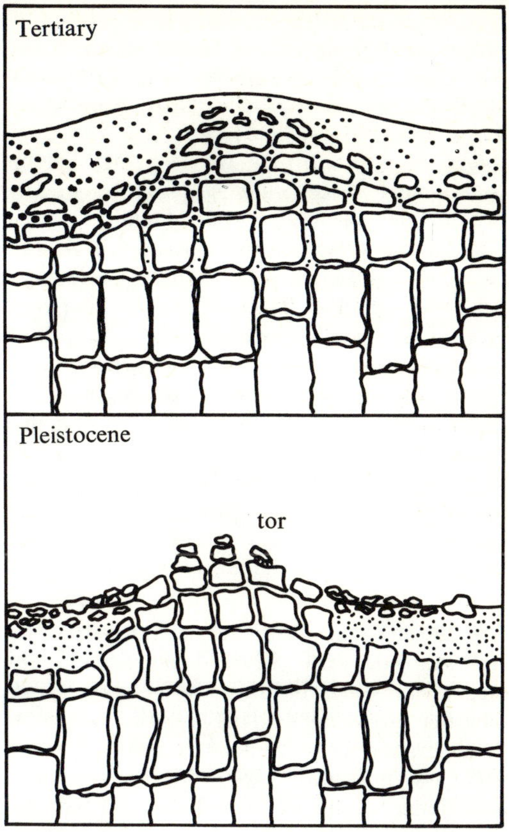

Fig. 8.20 Tor development in two phases, Tertiary deep weathering and Pleistocene mass movement

The date of the periglacial phase in the southwest has been tentatively supplied by the occurrence of two head deposits on Bodmin Moor, separated by a layer of peat. The date obtained by a radio-carbon treatment of the peat suggests that the lower head was produced during the Würm glacial advances further north while the upper head is late Glacial in age. The nature of the head deposits on Dartmoor shows how Tertiary or interglacial deep weathering under warm, moist conditions prepared a fine-grained mantle of weathered granite to the unweathered core stones (fig. 8.20). This mantle was removed by periglacial action and the core stones were further weathered by freeze/thaw to produce a coarser upper head which now overlies the finer lower head on the upper parts of the valley sides. The weathered core stones are now exposed to give the characteristic *tor* landform (Linton 1955).

Another example of Tertiary deep weathering producing geological effects which later show up in periglacial action is that of the sarsen stones on the Chalk. On Marlborough Downs (Wiltshire) these large silicified blocks, formed in the Tertiary period on the plateaux and upper slopes of the Chalk have been moved down by periglacial action to form a *rock stream* in the base of a dry valley. Periglacial action has found widespread support as the cause of dry valley formation and so such evidence from deposits is reassuring. Ollier and Thomasson have shown how the effect of aspect during periglacial phases has produced asymmetrical Chalk dry valleys in the Chilterns. On the North Downs deposits in a deep coombe have revealed fossils typical of the late Glacial and which are also indicative of seasonal periglaciation, not permafrost. The valleys thus formed produce a network of drainage lines much out of equilibrium with the fluvial system of today; this can be demonstrated by comparing the figures of drainage density for the dry valleys with drainage density of streams on the Chalk and streams on impermeable rocks. The Chalk areas of southern Britain were probably most affected by solifluction although the relatively easy weathering of the soft crumbly rock by freeze/thaw was clearly an important property too.

Apart from contributing to the formation of dry valleys periglaciation has produced many of the present-day slope deposits, now moving more slowly by wetting and drying cycles, and a valley-bottom infill in southern central Britain. Certain slope forms are also indicative of solifluction, including lobes and terraces of jumbled, contorted material. Some hollows and terraces on slopes are thought to be the result of *nivation*, the continuous freeze/thaw action which goes on around perennial snow patches lying in the moister periglacial areas.

Slope development did not only occur in periglacial phases through downslope movement of weathered material from exposed plateaux. In some cases the whole slope foundered because of the effect of periglaciation on the softer members of the local geological outcrops. In the neighbourhood of Bath, Somerset, the sands and clays sandwiched between hard limestones became very mobile under periglacial conditions and the complex hummocky slopes and unstable engineering foundations on the steep valley sides of the area today are undoubtedly the result of large scale landslips and mudflows during the Pleistocene.

One final phenomenon resulting from periglacial conditions is the widespread occurrence in Britain of wind-blown *loess* deposits. These well-sorted silts or sands were blown off the loose surface of fresh glacial deposits in Europe by strong anticyclonic wind systems accompanying glaciation. They may be built up to great thicknesses in sheltered pockets or may be merely admixtures of the present-day soil.

To summarise, the effects of periglaciation have been profound, more especially south of the maximum advance of the ice, because of the lengthy spells of periglaciation which must have been experienced. Permafrost is likely to have been present, at least in the Würm. In glaciated areas further north there was less time for periglaciation between glacial phases, though a late Glacial phase did have some effect in producing valleys in the south and patterned ground in the north. Finally it must be remembered that some minor effects of freeze/thaw still cause problems in engineering and agriculture.

FIELDWORK AND LABWORK

1 What is the effect of frosty weather on soil in parks and gardens? How would you measure frost action? There is a good opportunity to study frost weathering where soft rocks such as chalk outcrop or are quarried. Chalk too can be used for experiments in a refrigerator ice box or deep freeze. Keep on the good side of

mother or the school canteen! How important is moisture? Try soaking the chalk with water.

2 In the field try till fabric analysis on a deposit which is not obviously glacial. Highly angular fragments are typical—the length of travel is not great and the movement slow, so little rounding occurs. If a valley is studied, particularly in chalk areas, compare the depths of soil on the hillsides and plateau top with that in the valley bottom. If there are cultivated fields on both plateau and valley bottom in which fields are lumps of chalk or flints most common?

8E Since the glaciers?

We have mentioned that the British Isles have an immense glacial heritage of landforms. What degree of modification has occurred since glaciation? First we must go back to the glacial phases in order to understand the motivation of sea level changes which have still not been completed.

8E 1 Land and sea

In dealing with techniques of studying coastal geomorphology it was stressed that changes of sea level have meant that coastal processes have operated in a variety of situations both above and below today's sea level. The fact that wave energy can be focused at a variety of levels by changes involving glacial phases, tectonic movement or even ordinary diurnal tides is one reason for the extensive erosive effects of the sea.

Fig. 7.1 shows the variability of sea level during the Pleistocene glaciations. We have already dealt with the basic causes of these changes. The evidence around our present-day coasts consists of a series of raised beaches and rock platforms; sea levels below today's are demonstrated by buried river channels in estuary mud or terrestrial peats and forests which are exposed around the present-day low-tide mark.

The coastal erosion surfaces at or below 700 ft (230 m) in Britain are now assumed to be marine in origin. Their various altitudes need not bother the reader; the important point to realise is that, with cyclic variations in sea level above and below that of today, it becomes increasingly difficult to ascribe greater age to the higher surface. Sea level may 'revisit' a level abandoned thousands of years before and go on rising to cut a newer, higher bench, beach or platform. For this reason the interpretation of most of the lower features (below 60 ft or 20 m) is largely based on the inter-relationships with glacial or periglacial deposits or landforms. To briefly state some of the techniques used, the coasts of Britain may be divided into stable and unstable sections. Basically the raised shoreline features of the south coasts of Britain and Ireland are level and unaffected by either the eustatic downwarping which is influencing the North Sea area or the isostatic recovery which has affected Scotland.

A comparison of sea-level changes in a stable area and those in an area subject to isostatic recovery is shown in fig. 8.21.

Stevens and Synge (1966) dealing with the stable coasts of Devon and Cornwall, describe a series of raised rock platforms at various heights below 60 feet (20 m) which can in places be seen to grade into each other. Thus, they cannot be thought of as three distinct sea levels. They can be dated roughly as early Pleistocene because of the

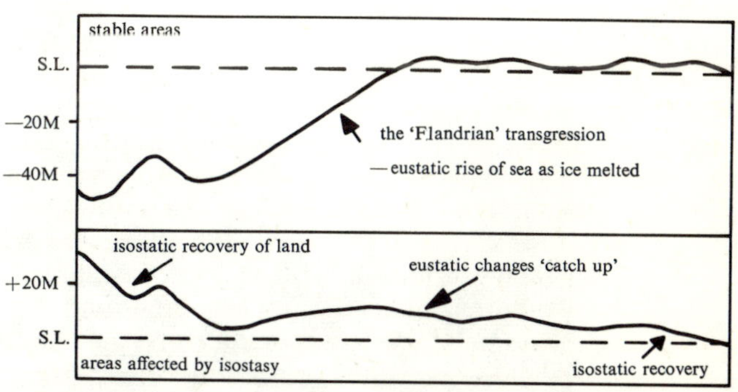

Fig. 8.21 Sea level changes in stable areas and those experiencing isostatic recovery since the last glaciation

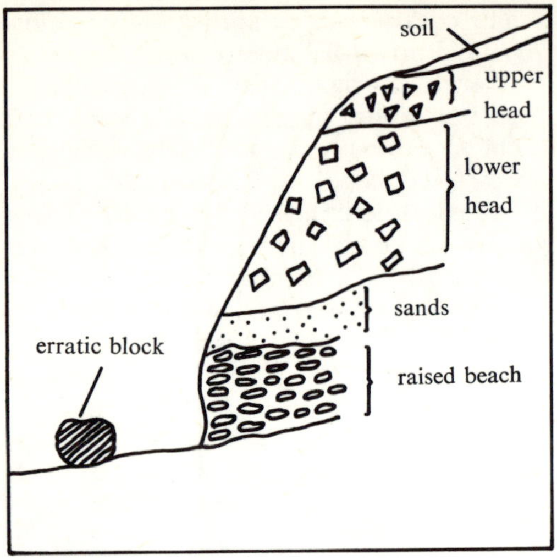

Fig. 8.22 Typical cliff profile in southwest England

Plate 8.3 A raised beach deposit shown in the lower half of a cliff near St Just, Cornwall. Above the raised beach (and distinguished from it by a clear change in angularity of the boulders) is a head deposit. The cliff is over 20 m high as revealed by the small figure at its base.

deposits which overlie them. These consist of a raised beach of beach pebbles and cobble with some sand, followed by two head deposits, the present soil profile capping the upper head (fig. 8.22). In some locations there exist large erratic boulders on the rock platforms, buried by the beach deposits and heads unless exposed by current wave action. These are thought to be associated (by the sources of the erratics) with the Mindel or Lowestoft glacial phase. This would mean that the raised beach material would almost certainly be formed in the interglacial before the Riss or Gipping glacial which itself overlaid the beach with the lower head. The upper head is most likely to be Würm, with no raised beach between it and the lower head to represent the Ipswichian interglacial; there is instead a weathered layer on top of the lower head signifying warmer and moister conditions. Clearly these techniques are based on the stratigraphy of deposits at a few sites and therefore open to misinterpretation, especially bearing in mind the variability of coastal action with rock type, exposure and wave fetch. The presence or absence of raised beach material could well depend on exposure to or shelter from waves. For example, plate 8.3 shows the great thickness of raised beach cobbles found in a sheltered Cornish Bay. However, the juxtaposition of rounded beach pebbles and head does provide a useful demonstration of particle size and shape distribution from different depositional environments (plate 8.3). Roundness measurements have also been used to differentiate between two raised beaches of different ages.

In Scotland there have clearly been far greater rises in sea level, recent coastal features being found well up to 100 ft (30 m) mainly due to submergence of the land by ice. Although a series of level, raised beaches was once described, the discovery by more exact surveying that the beaches are in fact sloping seawards suggests that the land was gently rising during the period in which they were formed. The most simple explanation for this is an isostatic one, slow recovery of the earth's crust from the overburden of ice sheet sometimes keeping pace with and sometimes exceeding the rise of sea level caused by the volumes of water added by melting ice. Generally speaking, eustatic rises were complete before isostatic recovery; the latter is still continuing (fig. 8.23). The fact that ice was present during the formation of the landward parts of a warped beach can be demonstrated by the relationship of the beach with glacial features and deposits. Thus the main raised beaches in Scotland are associated with glacial phases, not interglacial as in the south coast, and the two most important sets are correlated with advances in the Würm ice. There is also a raised beach associated with the late Glacial readvance of the ice in Scotland. The biggest difficulty with such a suite of raised shorelines is to separate the effects of isostatic and eustatic recovery from glaciation. Clearly the slope of the shoreline reveals the balance between both processes and very accurate surveying is required, together with stratigraphic analysis of the deposits and attempts to date the features by reference to glacial deposits or fossils contained in the beaches themselves.

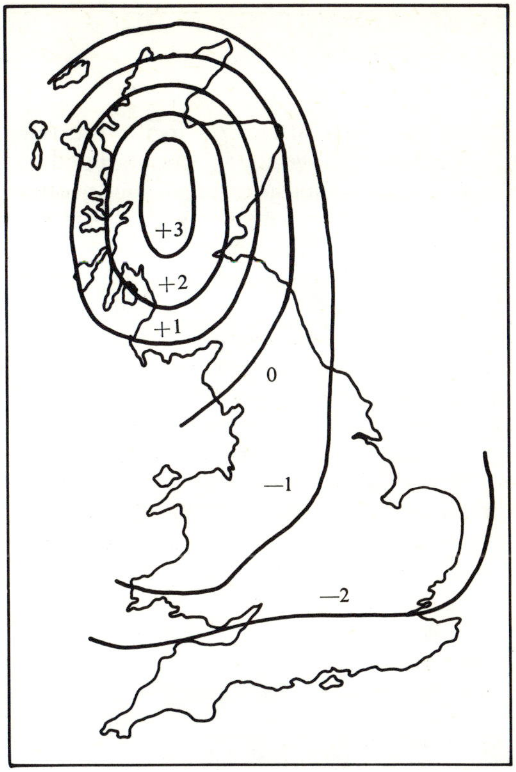

Fig. 8.23 The present rate of uplift (+) or downwarping (−) in millimetres per annum, showing the continued isostatic recovery of Scotland. Figures for Northern Sweden and Finland reach +9 mm per year.

LABWORK

Make sure you understand the two contrasting processes involved in sea-level changes by experimenting in a water-filled tank or large sink. The land surface can be represented by a compressible substance or a tin can be fixed to a bed spring. Add weights to the 'land' to represent glaciers. To preserve the analogy one should use water from that in the sink or tank to act as weights since this is what happens during glaciation. Unload the 'land' slowly. If the 'land' is marked with 'contours' the relative sea level at any stage can be recorded. What is the density of ice? Juggle with the volume required to lower the North Sea by one metre. How thick would it be and what would it weigh?

8E 2 How fossil is our landscape?

Despite the general activity in the present-day landscape suggested by our chapters on fluvial and shoreline geomorphology neither the rate of erosion and deposition, nor the morphological effects are as pronounced as during the Pleistocene. To some extent fluvial processes have not yet been re-established for long enough to destroy the evidence of Tertiary planation, periglacial slope deposits, till and so on. It is important to realise, therefore, that in studying the landforms of the present fluvial landscape we should always consider the historical dimension. The concept of the adjustment of natural systems to environmental changes via feedback relationships is a particularly useful one in this context, stressing as it does the slow relaxation time of morphological systems.

8F Can we be sure about time?

8F 1 In relative terms?

We have so far stressed the techniques used by geomorphologists to reconstruct palaeoenvironments and processes. Although a temporal sequence has been followed in this chapter, only occasional mention has been made of the methods by which the complete sequence of events has been put in time order. Where reference has been made it has been to **relative dating**, that is the establishment of the fact that a landform or deposit is older than, or younger than, its neighbour; in other words time has been measured along an ordinal scale. A major technique in this field has been largely covered by examples of the laws of **stratigraphy** which state that in most cases the uppermost of a series of deposits is the youngest. A major drawback is that, as we have seen in the case of raised beaches above, an important phase of activity may be *missing* from the randomly-occurring convenient exposures used for study.

We shall continue in this section to consider further methods of relative dating and continue our historical treatment of the Tertiary and Quaternary by stressing those techniques used to date deposits laid down in the late Glacial periods following the Würm. The more recent techniques of **absolute dating** are also discussed. They allow an actual date (time on an interval scale) to be given to a deposit based on the organic or mineral matter contained.

Both the *archaeological remains* and *pollen grains* found in Pleistocene deposits suffer from some of the disadvantages as stratigraphical analysis: there is seldom a full record of all phases in one section. This is particularly true of archaeological remains. Man is a shifting animal and very rarely does a new culture, race or tribe occupy the sites of its predecessors. Thus finds like the cave sections in France (the Dordogne Valley) and northern Spain are of enormous importance and local French terms are applied to many of the cultural phases of Stone Age man. One problem is how confident to be about the extension of the record at one site to surrounding areas. In this case the fact that the Channel was not occupied by sea again until 5000 years ago makes the extension of the European phases into Britain, where they are correlated with some of our cave sites in Derbyshire and Somerset, justifiable. Man at that stage probably occupied much of the Channel area himself. Archaeological timescales are not always fully determined, however, and controversy often rages about artifacts not immediately identifiable with a cultural phase.

Obviously archaeological evidence becomes better as we approach modern times. It is clear that the habits, homes and clothes of primitive peoples are a guide to the climatic conditions prevalent at the time. Contrast, for example, the cave dwellings of the Stone Age with the hilltop sites of the Bronze Age, now found at levels not considered warm or sheltered enough for building homes. The archaeological approach to dating geomorphological events is brought almost up to date by L. F. Curtis' discoveries of nineteenth-century china beneath slope deposits on the Yorkshire Moors.

Pollen analysis has the advantage that, in sites suitable for the preservation of grains, a fairly full record is often available; lake basins excavated by glaciers have been gradually filling with sediment and vegetative material since the Würm and as well as the stratigraphy of the deposits, the content of pollen can lead to a rapid reconstruction of environment and processes.

Pollen grains contain the male gamete of flowering plants. They are robust and produced in large numbers by wind-pollinated plants, especially trees. The shape of the grain is highly diagnostic of the plant species which produced it (see fig. 8.24). Though pollen is preserved in rocks as old as the Jurassic its main value is to help establish the palaeo-environments and relative dates for the Quaternary.

Pollen samples are obtained by cutting a cubic centimetre or so from an exposure of sediments or peat or recovering them from depth using a peat border. Often half a centimetre's depth of deposit represents fifty years of deposition. Samples are therefore taken from as many levels in the deposit as possible. Extreme care is required not to contaminate the sample with modern pollen which, as hay-fever sufferers know, is plentiful both in the field and laboratory. Crabtree (1968) describes the laboratory procedure, a complicated (and potentially dangerous) series of chemical treatments which rids the sample of most non-pollen material. What remains is mounted on a microscope slide in glycerine jelly, stained with saffranine. Viewing the slide reveals the number and species of pollen grains, the latter by reference to slides already prepared and identified by experts. Both the number and species of grains are noted

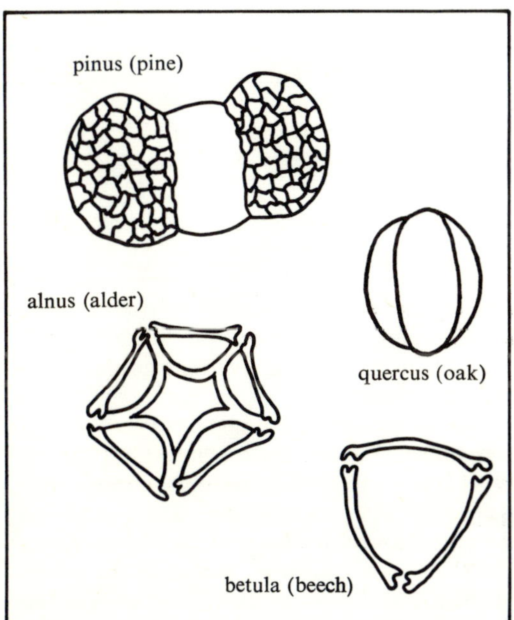

Fig. 8.24 Some species of tree pollen important in Pleistocene studies. (after Crabtree)

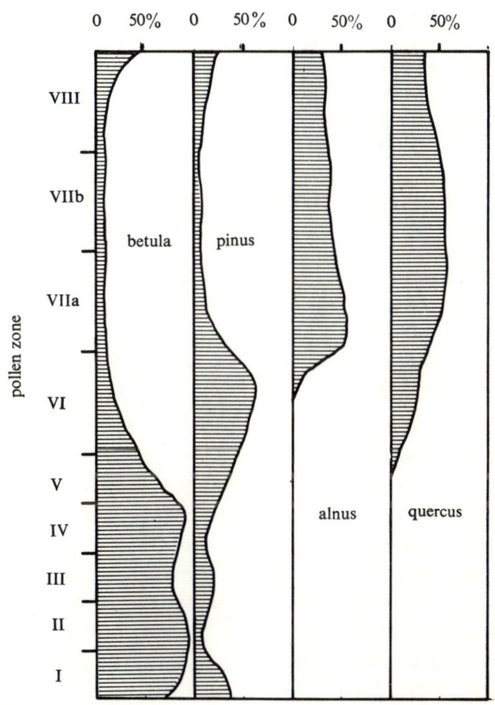

Fig. 8.25 The abundance of four types of tree pollen (shown in Fig. 9.24) during the late Glacial and Post Glacial. For an interpretation of the pollen zones see Table 8.4. (after Crabtree)

down against the position of the sample in the deposit. Between 150 and 300 grains should be counted from a representative horizon to form a representative sample. The results are drawn up as a pollen diagram (fig. 8.25). The late and post Glacial periods of the Quaternary have been divided into eight main zones based on important and widespread changes of pollen (table 8.4). It is also possible, further back in the Pleistocene, to separate the Hoxnian and Ipswichian interglacials by pollen analysis. This is important when other fossil remains, especially those of animals are more restricted in occurrence (cave animals, insects and molluscs). However, the use of animal fossils where they occur has been most useful to Pleistocene chronologies (West, 1968).

The various pollen zones have now also been absolutely dated using radioactive methods described below. Thus it becomes

Table 8.4. Phases of the Late and Post Glacial

	YEARS B.P.	POLLEN ZONE	CLIMATE	VEGETATION	MAN
POST GLACIAL					NORMAN et seq.
	1000	VIII		Alder, Birch, Oak	ANGLO SAXON
	2000				ROMAN
					IRON AGE
	3000	VII b	deterioration ↑		BRONZE AGE
	4000				NEOLITHIC
	5000		'Climatic Optimum'	Alder and Mixed Oak	
	6000	VII a			MESOLITHIC
	7000				
	8000	VI	getting warmer ↑	Hazel, Pine	
	9000	V		Hazel, Pine, Birch	
LATE GLACIAL	10000	IV		Birch	PALAEOLITHIC
		III	Cold	Tundra	
	11000				
		II	Milder	Birch forest	
	12000				
		I	Cold	Tundra scrub and mosses	

a fairly cheap method of according dates to Quaternary deposits while, with the depositional record provided by the sediments from which the pollen is sampled, it provided an insight into climate, environment and processes.

8F 2 Absolutely?

The state of decay in isotope activity is fairly specifically related to the time during which decay has been going on. The time taken for half the atoms in a radioactive system to decay is a diagnostic property of that isotope and is called the *half life*. Comparisons of the ratio of radioactivity from a particular source (for example Carbon 14) in deposits to that at present from the same source allows dating of the deposit. The most commonly used isotope for dating the Quaternary is Carbon 14, with a half life of 5730 ± 40 years (the degree of uncertainty in radioactive dating is usually expressed as $\pm$ years and the dates are given as B.P., present being 1950). Radioactive carbon is formed by cosmic ray bombardment and becomes incorporated with normal carbon in CO_2 which later becomes incorporated into organic matter. Frequent occurrences of organic matter, particularly plant remains (including the crops and timber of primitive man) in Quaternary deposits have allowed extensive radiocarbon dates to be obtained. It also occurs in the mineral carbonates of shells, stalagmites and stalactites. The expense of the analysis has made several worthy uses of the technique impossible, however, and there is still much to do.

Other useful isotope records are those of Oxygen-18 and Oxygen-16, whose ratios in ice cores and the shell of marine protozoans provide simultaneous information on the climate (hence latitude or altitude) and processes of snow (or sediment) accumulation. Isotopes with longer half-lives are used to determine the age of Tertiary material, for example potassium/argon in volcanic rocks.

There are some plants growing today which provide an absolute date and record of climate for the post-Glacial. Tree rings of such long-lived species as the Giant Redwood and Ponderosa Pine, being seasonally added according to conditions of growth, reveal cold and warm phases. The growth rate of lichens on rock surfaces in recently glaciated areas has also proved predictable enough to become a valid technique for calculating the length of time over which ice has retreated from a position covering the rock.

8G The Anthropogene?

8000 years of man influencing the landscape (and some say the climate) have left a profound mark, although it is only obvious in a few cases. Mines, quarries, cuttings, examples of man's erosive action, and slag heaps, embankments and coastal defences, examples of his deposition, show up clearly in the landscape. There are, however, more subtle influences such as the regrading of river channels to prevent flooding and the strengthening of river banks, preventing natural meanders. It is clear in many cases that a lack of attention to geomorphology has resulted in disaster for man's new landforms. Soil erosion attended the forest cutting of settlers in the United States and New Zealand. The Aberfan disaster in South Wales resulted from the failure of a steep slope flanking a slag heap during wet weather. The opening of the M4 motorway in 1971 was delayed by wash-outs on the oversteepened sides of cuttings through gravels. Increased liability to flooding has been suggested to result from drainage ditches dug during forestry operations on steep hillsides. There are endless examples. They are not restricted to recent times: the constructional activity of the prehistoric builders of Silbury Hill in Wiltshire surely merits mention. An interesting example of the interfering action of prehistoric earthworks in preventing normal chemical solution of Chalk bedrock below them has been given, amounts of half a metre in 4000 years being claimed (this being the amount of lowering on the surrounding chalk since earthwork construction).

The overall effect of man has been erosive. As Brown (1970) points out, estimates of natural erosion range from 12 to 1500 $m^3 km^2 yr$ while those in areas influenced by man vary between 1500 and 85 000 $m^3 km^2 yr$. The reader should make a list of the effects of man on the landscape in his area. Do engineers take geomorphology into account? Ask them. On a lighter note, how will the geomorphologists of the Upper Anthropogene date the phases of our Lower Anthropogene? Will we leave evidence of the Nuclear Interglacial in our diaries, films and magnetic tapes?

FIELDWORK

When confronted by a deposit of glacial material or a rock outcrop evaluate, by using a series of logical steps, the possible age of the deposit. Is it older than you? How do you know? Is it older than the Industrial Revolution? When did man evolve and inhabit the area? Is the rock older than the one above it? Try to see if the deposit contains objects or substances which would allow you to date it. Has the landform developed during the current phase of landscape evolution? What do we mean by the current phase? How long does landscape take to develop?

9 Organisation and Opportunities

'It is only through . . . application of systems analysis that considerations of the management of the natural environment can be elevated above mere *ad hoc* bookkeeping to form part of a broader scholarly discipline . . .'

R. J. Chorley and B. A. Kennedy

9A Summary and conclusions

The written conclusions to a piece of experimental work usually attempt to draw together the results of the experiments and show how they are meaningfully related to the hypotheses with which the work started. Most of the results of the exercises suggested in this book, we trust, will be in the notebooks of the reader. The authors' conclusions must therefore relate the contents of the book to the demands which we think the reader may have for sustained interest or for utility in what he has learned.

The first of these demands, that for interest, is easy to meet in a subject like physical geography; most readers will already be aware that the physical elements of climate and landscape hold a personal appeal to them, perhaps via a general exhilaration at being in the open air, perhaps in a way that cannot be immediately particularised. The second demand, that for utility or relevance, is one which geography as a whole has often found hard to meet; it is worth dwelling on here.

Most senior students and practitioners of other subjects, particularly in science and technology, have tended in the past to regard geography as an artificial discipline, borrowing from all other subjects but claiming as its *raison d'etre* the addition of the spatial (or mapable) dimension. This does not anger the economist, physicist or engineer because he has never felt threatened by geography. 'No geographer', he will claim, 'is ever likely to have more to offer to an applied problem in my field than I am; he just doesn't specialise enough.' The last point is partly true, although one would never gather so from considering the titles of theses and papers recently published in geography! The point is probably that geographers take an 'overview' of the interrelationships in science to much higher levels than do students of most subjects, specialisation coming only at a research level. The jibe that such knowledge cannot be applied is becoming more nonsensical as specialists in a coordinating scientific role become more necessary. It is no accident that two modern technological universities are soon to introduce general science courses which, if they were bolder, they might call 'physical geography'.

The way ahead for geography is probably best demonstrated by human geography which, since it became consciously quantitative, has been able to break down the traditional suspicion that the geographer is a *dilettante* ('Jack of all trades, master of none') by applying much of its work to the professional fields of urban planning, regional economics and even sociology. It is noticeable that Stamp's book *Applied Geography* contains mainly human and economic examples. Many writers and employers are now equating geography with urban or general economic studies.

What, then, of physical geography? Is it just that part of meteorology, botany, geology and hydrology concerned with maps or mapable variables? What professional niche could a physical geographer occupy? Make no mistakes about our aims in this discussion: we are not intending to examine the holding stock of physical geography to see what would sell! The value of application is that, where it is possible, it encourages in any subject a diligence and precision not generated by purely academic pursuits. Additionally it does provide some of the material for further research, both in the way of unsolved problems and monetary support. It is the authors' experience from researching and teaching physical geography that it is a subject naturally and attractively suited to application in the world and that the applied uses lend stimulation to its study.

It is to some extent fortunate for all geographers that the so-called 'environmental crisis' affecting the world today has already begun to make demands for scientists with a 'lateral thought' background. It is this type of scientist who is now required to aid the technologist in advancing society in a more careful, balanced way than before. His job is sometimes to advise on the readjustment of imbalances resulting from the hasty application of specialist knowledge, for example those resulting from the construction of the Kariba Dam. Though the ecologist may well be a much-needed scientist of the future, the geographer will also be in the front line. It is no accident that physical geography comes under the heading of 'Environmental Science' at the newer universities.

As Professor Hare has said, 'If physical geographers do not assert that one can look at man's physical environment as a scientific whole . . . then no one will, at least with adequate breadth.'

The physical geographer contributing to the work of a government department or a firm of consulting engineers or architects may well be appalled at the narrow approach which has been forced upon many specialists by the very need to apply the work. It would be little use for the geographer to tell his colleagues that they had forgotten, during their construction of, say, a block of flats, to take the change in local climate into consideration and that the balconies of all the flats would be extremely windy. He would have to quote work already done on the problem or, better, design experiments, perhaps with models, whose results would enable predictions of the effects of the building. Prediction must be mathematical. The building itself will stand or fall on a series of equations (involving the strength of beams and materials) and so must the work of the environmental scientist. Only when on a 'common wave-

Table 9.1 A few examples of the applications of physical geography and its findings

PROBLEM	FACED BY	TECHNIQUES DEALT WITH IN THIS BOOK (with page nos.)
Irrigation of farmland	Farmers and growers	61, 142
Soil warming, aspect of plots	Farmers and growers	42, 81
Shelter for crops and animals	Farmers and growers	64, 77
Air pollution	Planners and health officials	82
Climate in streets and buildings	Planners and architects	91
Central heating efficiency	Architects and engineers	92
Water pollution	Water suppliers and river authorities	163
Flood forecasting and prediction	Engineers and dam builders	145
Soil erosion	Foresters and farmers	170
Foundations for building	Engineers	171
Road building and upkeep	Engineers	174
Erosion by rivers and the sea	Estate managers, farmers, engineers	177
Sewage dispersal at sea	Local Authorities	189
Siltation of harbours and estuaries	Harbour Boards	168

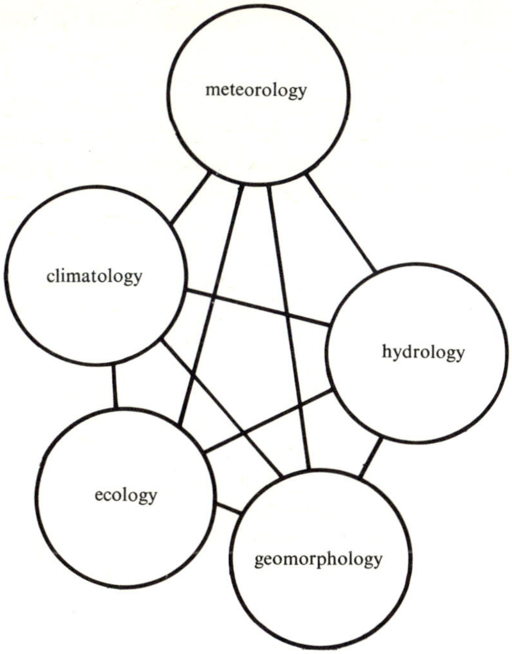

Fig. 9.1 Simple linkages of the constituent topics within physical geography

length' of techniques will the necessary teamwork between specialists and environmental scientists develop. It is by this yardstick that the techniques suggested in the book have been selected, especially that of a systems approach to research which has a useful overlap with computing and the world of business. Table 9.1 gives examples of the applications of physical geography.

It is to the basic technique, systems thinking, that we now return as a means of integrating with the specialist and of viewing physical geography as a complete and essential discipline. What seems to be suggested in the paragraphs above is that the physical geographer may well be coming to occupy the role of an environmental engineer, capable of viewing systems at a variety of scales and of knowing where and when to isolate systems for study. Just as the circuitry of the TV set, whatever make, will be familiar to the television engineer, the interconnections (or correlations) and causal relationships between various elements of climate or landscape are known to the physical geographer at whatever scale or in whatever region of the world he cares to study. To carry the analogy just one more stage, the location and repair of faults can only be carried out with this kind of knowledge; now that 'faults' have been affecting the physical environment the physical geographer may increasingly be called on to 'repair' them and prevent recurrence.

The systems approach to physical geography has recently been completely illustrated by Chorley and Kennedy (1971). It is probable that a book of the nature of *Physical Geography, a systems approach* could not have been written before because many of the connections (i.e. correlations) in the 'circuit diagram' were not known in sufficient detail. The book itself, though an invaluable reference, is rather too advanced in its scope to back up the techniques suggested here; the important thing is that it has been written. Chorley and Kennedy take for granted that the systems which they break down all fit together again to make physical geography, whereas many school or college readers of this text will not feel so confident. Fig. 9.1 shows a possible grouping of the previous seven topic chapters in isolation, arranged according to the location of each topic in relation to the earth and sky. However, it is clear that isolated study is defeating our aims in this book. Yet, in order to link all the topics in various combinations, we would need to draw 120 routes along the lines on the diagram. It is far easier to link the separate compartments by using the method described as a Venn diagram (fig. 9.2) grouping the topics according to the way they overlap and influence each other.

The student of physical geography may also wish to liken his action in considering any particular topic as similar to dialling 0 on the telephone. It is impossible to do this successfully without drawing the dial round past 9, 8, 7, 6, 5 and so on, only to then release it to return past 2, 3, 4, 5 until it is at rest. Just so is the physical geographer's attitude to environmental systems (or the chapters of this book!). This connectivity reaches its height in the soil which has a morphology, an atmosphere, a climate, a hydrology, a fauna and a flora; it is indeed

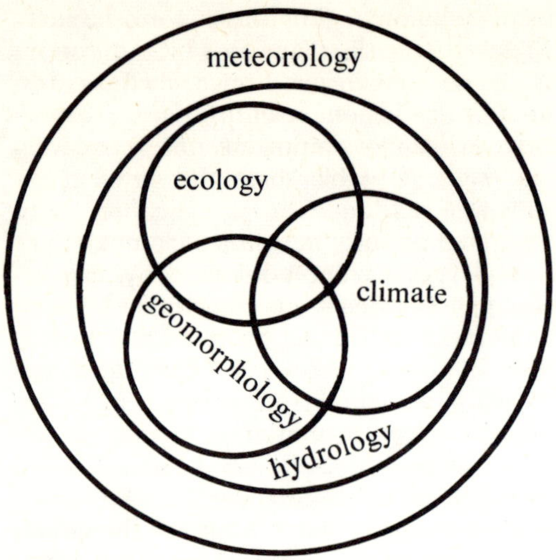

Fig. 9.2 Venn Diagram stressing the wholeness of physical geography

a microcosm of the global environment (again the scale of systems proves useful). Connectivity is also high in the actual techniques we have described. Notice how often the measurement of size, shape, slope and flow come into our work.

We need, therefore, to be convinced of a paradox. To be effective, environmental scientists must learn the complexity but practise the simplicity of things. The basic elements of dynamics: materials and forces, will always rescue simplicity for us as it did in the preceeding chapters. We may go even simpler and reduce all the world we see to energy, dissipated at various scales. This is not the place to deal with the theory of relativity but we pose fig. 9.3 as a further attempt to show physical geography as integrated science.

9B Sources of information

Most of us are destined to be in the back room of scientific discovery. Reading through this and other books the reader may well feel that everything that is to be solved has been solved; only a genius will

Fig. 9.3 The energy and scale relations of physical geography

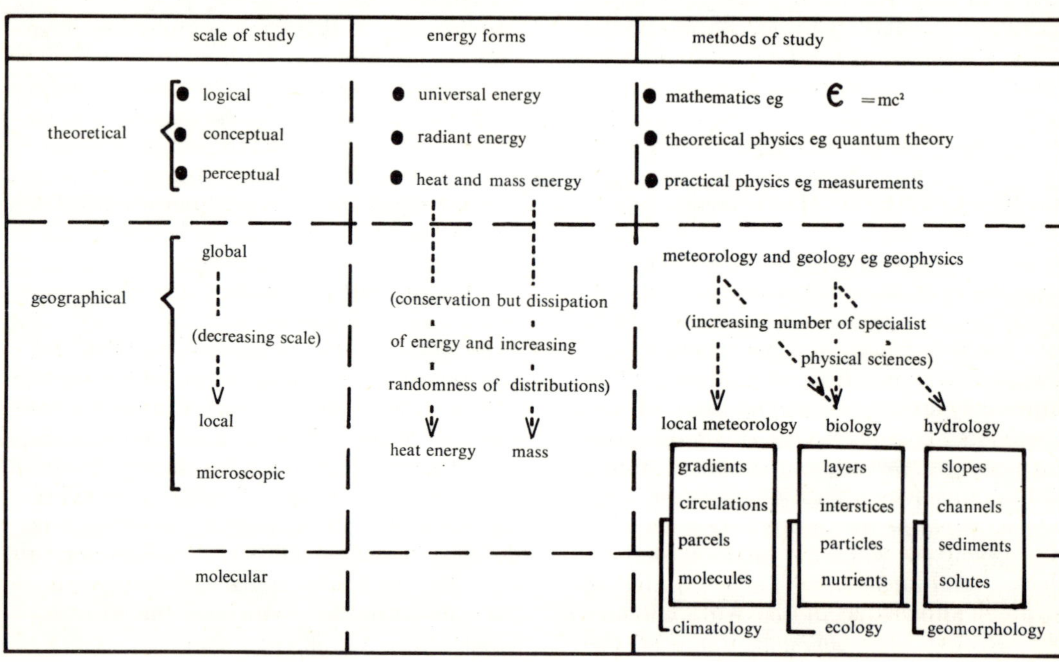

discover something new by noticing a unique truth or merely reorganising the work of others. This is partly the view of the authors and a mildly depressing one at that! However, a sustaining feature of the work we have described is that it has only just begun in an effective way.

After centuries of exploration at a global scale (and now space!), comprehensive measurements are only just beginning of river flow, sediment load, slopes, local climate and so on. While they may never reveal anything as startling as the work of Newton, Einstein and Darwin they are essential and require gifted personnel. More and more research is being done in teams and the fact that the reader is probably a member of a class at school or college has been used in the design of most of the practical work we have suggested. Much of it may need membership of the 'team' of other science subjects, especially physics, as fig. 9.3 shows. The instruments, time and dedication of your class may be only a fraction of those possessed by graduate professionals but we have tried to simplify things where possible. The work of professionals is always of interest, as evinced by the number of school parties visiting laboratories and exhibitions, and descriptions of such work have been included. However, the professional may well be jealous of your opportunities of wide ranging experiments.

9B 1 General method

As well as techniques, the basic *concepts* surrounding each topic have been discussed, especially where new frameworks for study have recently emerged (for example the concept of throughflow in hydrology) or controversy exists (as in the case of the Stonehenge 'erratics'). One of the abilities which will sort out the geniuses amongst you ('Genius does what it must; talent does what it can!', Owen Meredith) is that of being able to see which problems exist to be solved. In many cases this requires a thorough knowledge of the work already done on the topic in question and the results obtained. Obviously a likely field for further work is one in which different results or conclusions are reached by different researchers. Scientists have always thrived, like news reporters, on *controversy* and some notable breakthroughs have come as a result.

Several philosophers have claimed that a good general methodology for scientific enquiry has come from a neglected branch of learning called *heuristic*. In his book, *How to solve it,* Professor Polya gives the following guidelines to tackling a problem in science or mathematics:

1 Understanding the problem
What is the unknown? What are the data? What is the condition? He says that to draw a diagram of the problem helps during this phase. Perhaps a flow diagram?
2 Devising a plan
Have you seen the same problem in a different form? Can you use the methods or results of a related problem to help solve yours? Can you restate your problem in a simpler form?
3 Carrying out the plan
Is each step correct? Is a solution obvious? Does more data need collecting?
4 Looking back
Can you check the result? Could you have achieved a different result or the same one by a different method? What new problems have you set up? Are there any remaining problems which are solved by your effort?

We suggest that this is a useful guide for the student of physical geography, particularly the keen awareness of related problems—our justification for the existence of physical geography!

Whether or not you make startling discoveries and whether or not you are eventually able to apply your knowledge of physical geography as a kind of 'environmental engineer' we are sure that the choice, design, operation and use of simple experimental work will be tolerably exciting, will develop useful skills and will develop a critical faculty seldom aroused by digesting the 'authoritative' word of the conventional text.

Perhaps the 'last word' on general methods ought to go to Professor Mead, who has written, 'In the last analysis, the course of geographical knowledge is determined by men and women of imagination and enthusiasm.' We hope we have stirred both.

9B 2 General information

By tradition, geographers have always been in the vanguard of exploration and discovery. Whilst we may bemoan the fact that our predecessors have left us little to add to the map, the present tasks and appeal of the subject still remain exacting and encouraging. Such challenges are no less in our studies of problems and techniques than in the glamorous expeditions of old. Just as explorers must have been somewhat awestruck at the enormity of the challenges that confronted them, we would be the first to admit feeling a little daunted by the prospect of selecting and recommending but a fraction of the books and information now available to the physical geographer. What follows in fact must be considered a mere reconnaissance based upon our experience in this field; a choice aimed at furnishing the reader with a reasonably up-to-date survey by which we hope he can chart his own way through the resources and opportunities available. The aids to good navigation will be your teachers, tutors, librarians and, in the final analysis, your *own* resourcefulness!

Balchin, W. G. V. (ed.) *Geography: an outline for the intending student* (Routledge, 1970)

Balchin, W. G. V. and Richards, A. W. *Practical and experimental geography* (Methuen, 1952)

Board, C. *et al* (eds.) *Progress in Geography* 4 volumes (Edward Arnold, 1972)

Bowen, D. Q. (ed.) *A concise Physical Geography* (Hulton, 1972)

Chorley, R. J. (ed.) *Water, Earth and Man* (Methuen, 1969)

Chorley, R. J. and Haggett, P. (eds.) *Frontiers in Geographical Teaching* (Methuen, 1965)

Chorley, R. J. and Haggett, P. (eds.) *Models in Geography* (Methuen, 1967)

Chorley, R. J. and Kennedy, B. A. *Physical Geography: a systems approach* (Prentice-Hall, 1971)

Cole, J. P. and King, C. A. M. *Quantitative Geography* (John Wiley, 1968)

Cooke, R. U. and Johnson, J. H. *Trends in Geography* (Pergamon, 1969)

Dickinson, G. C. *Statistical Mapping and the Presentation of Statistics* (Edward Arnold, 1963)

Dickinson, G. C. *Maps and Air Photographs* (Edward Arnold, 1969)

Garnier, B. J. *Practical Work in Geography* (Macmillan, 1963)

Graves, N. J. (ed.) *New Movements in the Study and Teaching of Geography* (Temple Smith, 1972)

Gregory, S. *Statistical Methods and the Geographer* 2nd edn (Longmans, 1971)

Hancock, J. C. and Whiteley, P. F. *The Geographer's Vademecum* (Philip, 1968)

Harvey, D. *Explanation in Geography* (Edward Arnold, 1969)

Inc. Assoc. of Assistant Masters *The Teaching of Geography in Secondary Schools* 5th edn (Cambridge, 1967)

Long, M. (ed.) *Handbook for Geography Teachers* 5th edn (Methuen, 1964)

Long, M. and Roberson, B. S. *Teaching Geography* (Heinemann, 1966)

Martin, G. and Turner, E. *Handbook of Environmental Studies* (Blond, 1972)

Minshull, R. *The Changing Nature of Geography* (Hutchinson, 1970)

Morgan, W. B. and Pugh, J. C. (eds.) *The Field of Geography* 2 volumes (Methuen, 1971)

Monkhouse, F. J. and Wilkinson, H. R. *Maps and Diagrams: their compilation and construction* 3rd edn (Methuen, 1971)

Sauvain, P. *Practical Geography, Book IV: advanced techniques* (Hulton, 1972)

Stamp, Sir. L. D. (ed.) *A Glossary of Geographical Terms* (Longmans, 1966)

Theakstone, W. H. and Harrison, C. *The Analysis of Geographical Data* (Heinemann, 1970)

Toyne, P. and Newby, P. *Techniques in Human Geography* (Macmillan, 1971)

It may be considered a happy coincidence that the above list starts with a guide for the aspiring student of geography and ends with the companion volume of this book in the field of human geography. However, it is certainly not a coincidence that, with one exception, all of them have either been written or revised within the last decade, you may well reflect on the increasing significance of the library and librarian as a resource and aid to the physical geographer.

We do not expect that your own libraries will have all of these books, or many of the papers referred to later. Neither do we expect that you will read them all from cover to cover: we certainly have not! Clearly, the techniques of acquiring information and selecting what is relevant must be part of your repertoire. Only experience will help you with the latter, of course, and so we must confine ourselves to some libraries from which you may loan suitable books and information with the help of a librarian or your teacher. Most of the societies and organisations given below publish reports, periodicals, journals and magazines of the sort we have referred to in each chapter.

Association of Science Education, 52 Bateman Street, Cambridge

British Association for the Advancement of Science, 3 Sanctuary Buildings, 20 Great Smith Sq., London, S.W.1

British Cartographical Society, 4 Tamesa House, Chertsey Road, Shepperton, Middlesex

British Ecological Society, Nature Conservancy, Abbots Ripton, Huntingdonshire

Commonwealth Bureau of Soils, Rothamsted Experimental Station, Harpenden, Hertfordshire

Council for Environmental Education, 26 Bedford Square, London, W.C.1

Council for Nature, Zoological Gardens, Regents Park, London, N.W.1

Field Studies Council, 9 Devereux Court, London, W.C.2 and Preston Montford Field Centre, near Shrewsbury, Shropshire

Geographical Association, 343 Fulwood Road, Sheffield S10 3BP

Geologists' Association, University College, London, W.C.2

Institute of British Geographers, 1 Kensington Gore, London, S.W.7

Institute of Hydrology, Howbery Park, Wallingford, Berkshire

Meteorological Office, London Road, Bracknell, Berkshire

Ordnance Survey, Romsey Road, Maybush, Southampton SO9 4DH

Physics and the Physical Society, 47 Belgrave Sq., London, S.W.1

Royal Geographical Society, 1 Kensington Gore, London, S.W.7

Royal Meteorological Society, Cromwell House, High Street, Bracknell, Berkshire

Schools Natural Science Society, 8 Sandy Lane, Sevenoaks, Kent

Please do not write to these organisations unless you have good cause: they are extremely busy and cannot handle general queries.

Needless to say, there are innumerable other national, regional and local organisations to which the physical geographer can turn for specific advice and data. Most of them are catalogued in the handbooks already listed; and, for those who require information on the firms and organisations dealing with instruments, the following sources are invaluable:

UNESCO *Source Book for Science teaching* (UNESCO, 1956)

Wilson, R. W. *Useful Addresses for Science Teachers* (Edward Arnold, 1968)

Although this great number of publications and organisations may appear overwhelming, our search for the particular book or equipment we require can often be eased by consulting the appropriate digests and abstracts in which the most important new developments are reviewed. To the physical geographer, the most comprehensive reviews appear in *Geo Abstracts* obtainable from the Department of Environmental Sciences, University of East

Anglia, Norwich NOR 88C. The following annual publications are available:

Clayton, K. (ed.)	A.	*Landforms and the Quaternary*
Barkham, J. (ed.)	B.	*Biogeography and climatology*
	E.	*Sedimentology*

The reader should note that the *Biogeography and climatology* publication also includes abstracts of current hydrological research. The *Technical Bulletins* of the British Geomorphological Research Group, a Study Group of the Institute of British Geographers, are also available from the same address. Because these bulletins are devoted to particular instrumental and field techniques recently developed in geomorphology, we have detailed each one under its appropriate chapter in the next section.

The members of the British Geomorphological Research Group have recently contributed a series of well-illustrated articles on 'The Unquiet Landscape' to the *Geographical Magazine*. The series is continuing but back numbers may be available from New Science Publications, 129 Long Acre, London WC2E 9QH.

The Field Studies Council also produce offprints of interest to physical geographers from their journal *Field Studies*. They are retailed by E. W. Classey Ltd, 353 Hanworth Road, Hampton, Middlesex. Titles include:

The Raised Beaches and Strand Lines of South Devon, by A. R. Orme.

Denudation Chronology of parts of South-western England, by D. Brunsden et. al.

The Origin of the Landforms of the Malham area, by K. M. Clayton.

Morphology and Distribution of Features Resulting from Frost Action in Snowdonia, by D. F. Ball and R. Goodier.

Coastal Head Deposits between Start Point and Hope Cove, Devon by D. N. Mottershead.

All are of great value to the reader of this text, especially in following up chapter 8.

By now the reader will have appreciated the physical geographer's preoccupation with instruments capable of measuring particular processes in the field or simulating them in the laboratory. In the final analysis, however, most of us require some mental picture of the process involved whether of radiation exchanges, streamflow, freeze-thaw or whatever. Physical geography in particular has always relied a great deal on the *graphical* portrayal of information, and audio-visual aids have a special place in the teaching of the subject. Those who have grappled with the jig-saw of Continental Drift, for example, and then seen the B.B.C. Television documentary called 'The Restless Earth', will have appreciated the benefits of the animated film as an aid to their studies. Meteorology is especially suited to such techniques, of course, and so we commend appropriate cinefilms from the following organisations:

Services Kinema Corporation, Chalfont Grove, Gerrard's Cross, Buckinghamshire

Longmans Loops in Geography, 48 Grosvenor Street, London, W.1

Other sources will be found in many of the handbooks already listed.

Keeping up-to-date with scientific developments is as important to the physical geographer as any other scientist, particularly now that an increasing awareness of environmental problems attracts workers from related disciplines. Some of the regular weekly and monthly magazines published contain much food for thought; for example,

Nature Macmillan, 4 Little Essex Street, London, W.C.2

New Scientist 128 Long Acre, London, W.C.2

Science Journal and Discovery Dorset House, Stamford Street, London, S.E.1

Scientific American 22 Ryder Street, London, S.W.1

The latter often publishes offprints and collected works like the ones below:

The Biosphere (Freeman, San Francisco, 1970)

Man and the Ecosphere (Freeman, San Francisco, 1971).

Scientific American offprints are available from W. H. Freeman and Co. Ltd, 58 Kings Road, Reading RG1 3AA. Titles of value to the reader are as follows:

No. 803 *Sand* Kuenen

No. 805 *The Changing Level of the Sea* Fairbridge

No. 809 *Glaciers* Field

No. 811 *Radiocarbon dating* Deevey

No. 817 *Erosion by Raindrop* Ellison

No. 821 *Soil* Kellogg

No. 826 *The History of a River* Janssen

No. 845 *Beaches* Bascom

No. 869 *River Meanders* Leopold and Langbein

No. 878 *Water* Revelle

9B 3 Further reading

CHAPTER 1

Beck, S. D. *The Simplicity of Science* (Pelican, 1962)

Bishop, O. N. *Statistics for Biology* (Longmans, 1966)

Chaston, I. *Mathematics for Ecologists* (Butterworth, 1971)

Chatfield, C. *Statistics for Technologists* (Pelican, 1970)

Chorley, R. J. 'The applications of statistical methods to geomorphology' *Essays in Geomorphology* ed. G. H. Dury pp. 275–387 (Heinemann, 1966)

Emmet, E. R. *Learning to Philosophize* (Pelican, 1964)

Gowers, Sir Ernest *The Complete Plain Words* (Pelican, 1962)

Hempel, Carl G. *Philosophy of Natural Science* (Prentice-Hall, 1966)

Moroney, M. J. *Facts from Figures* (Pelican, 1967)

Sawyer, W. W. *The Search for Pattern* (Pelican, 1970)

CHAPTER 2

Barrett, E. C. *Geography from Space* (Pergamon, 1971)

Bruce, J. P. and Clark, R. H. *Introduction to Hydrometeorology* (Pergamon, 1966)

Byers, H. R. *General Meteorology* (Mc Graw-Hill, 1957)

Caborn, J. M. *Shelterbelts and Microclimate* Forestry Commission Bull. 29 (HMSO, 1957)

Chandler, T. J. *Modern Meteorology and Climatology* (Nelson, 1972)

Geiger, R. *The Climate near the Ground* 4th edn (Havard, 1966)

Hare, F. K. *The Restless Atmosphere* 4th edn (Hutchinson, 1968)

Johnson, J. C. *Physical Meteorology* (John Wiley, 1954)

Maunder, W. J. *The Value of the Weather* (Methuen, 1970)

McIntosh, D. H. and Thom, A. S. *Essentials of Meteorology* (Wykeham, 1972)

Meteorological Office *Handbook of Meteorological Instruments Part 1* (HMSO, 1956)

Observer's Handbook 2nd edn (HMSO, 1956)

Stringer, E. T. *Techniques of Climatology* (Freeman, San Francisco, 1972)

Taylor, J. A. and Yates, R. A. *British Weather in Maps* 2nd edn (Macmillan, 1967)

Wiesner, C. J. *Hydrometeorology* (Chapman and Hall, 1970)

Woodall, A. J. *Heat* (English Universities, 1958)

CHAPTER 3

Chandler, T. J. 'London's Urban Climate' *Geographical Jnl* 128, pp. 279–302, 1962

Conrad, V. and Pollak, L. W. *Methods in Climatology* (Havard, 1950)

Crowe, P. R. *Concepts in Climatology* (Longmans, 1971)

George, D. J. 'Temperature Variations in a Welsh Valley' *Weather,* 28, 9, pp. 270–73
Glentilli, J. *Sun, Climate and Life* (Blond, 1972)
Lamb, H. H. *Climate: Fundamentals and Climate now* Vol. 1 (Methuen, 1972)
Leigh, R. (ed.) *The Earth and its place in the Universe* Science No. 8 (Longmans, 1971)
Manley, G. *Climate and the British Scene* (Collins, 1952)
Mattsson, J. O. 'A Simple Dew Recorder' *Weather* 17, 8, pp. 269–72 (1962)
McDonald, J. E. 'Raindrop Spatter-spots' *Weather* 19, 6, pp. 177–9 (1964)
Morris, R. E. and Barry, R. G. 'Soil and Air Temperature in a New Forest Valley' *Weather* 28, 11, pp. 325–31
Neuberger, H. and Cahir, J. *Principles of Climatology: a manual in earth science* (Holt, Rinehart and Wilson, 1969)
Perry, A. J. 'Filtering climatic anomoly fields using principal component analysis' *Trans. Inst. Brit. Geog.* 40, pp. 55–72 (1970)
Sprunt, B. F. 'Geographics: a computer's eye view of terrain' *Area* 4, pp. 55–9 (1970)
Stringer, E. T. *Foundations of Climatology* (Freeman, S. F., 1972)
Taylor, J. A. (ed.) *Weather Forecasting for Agriculture and Industry* (David and Charles, 1972) (Based on the Aberystwyth Annual Symposia in Agricultural and Applied Climatology)
Wooldridge, S. W. *The Geographer as a Scientist* (Nelson, 1956)

CHAPTER 4

Adamson, R. *Pollution: an Ecological approach* (Heinemann, 1972)
Ashby, M. *Introduction to Plant Ecology* (Macmillan, 1961)
Bennett, D. P. and Humphries, D. A. *Introduction to Field Biology* (Edward Arnold, 1965)
Clarke, G. R. and Beckett, P. *The Study of Soil in the Field* 5th edn (Clarendon, 1971)
Dowdeswell, W. H. *Practical Animal Ecology* 2nd edn (Methuen, 1967)
Duffey, E. *The Biotic Effects of Public Pressures on the Environment* (Nature Conservancy, 1967)
Harrison, C. M. 'Recent Approaches to the Description and analysis of Vegetation' *Trans. Inst. Brit. Geog.* 52 (1971)
Kershaw, K. A. *Quantitative and Dynamic Ecology* (Edward Arnold, 1964)
Linford, J. H. *An Introduction to Energetics* (Butterworth, 1966)
McGlashan, N. D. (ed.) *Medical Geography: Techniques and Field Studies* (Methuen, 1972)
Mew, G. and Ball, D. F. 'Grid sampling and air photography in upland soil mapping' *Geographical Jnl.* 138, pp. 8–14 (1972)
Phillipson, J. *Ecological Energetics* (Edward Arnold, 1966) (one of a series from Inst. of Biology)
Rosenak, S. *Soil Mechanics* (Batsford, 1963)
Russell, Sir E. W. *Soil Conditions and Plant Growth* (Longmans, 1961)
Seddon, B. *Introduction to Biogeography* (Duckworth, 1971)
Sheals, J. G. (ed.) *The Soil Ecosystem* (Systematics Assoc., 1969)
Soil Survey Staff Field Handbook *Soil Survey of Great Britain* (HMSO, 1960)
Taylor, J. A. 'Methods of Soil Study' *Geography* 45, pp. 52–67 (1960)
Tivy, J. *Biogeography: a study of plants in the ecosphere* (Oliver and Boyd, 1971)
White, E. M. 'Validity of the transect method for estimating composition of soil-map areas' *Soil Sci. Soc. Amer. Proc.* 30, pp. 129–30 (1966)

CHAPTER 5

Chorley, R. J. (ed.) *Water, Earth and Man* (Methuen, 1969), specifically:
R. J. Chorley 'The drainage basin as the fundamental geomorphic unit' pp. 77–99
G. H. Dury 'Hydraulic Geometry' pp. 319–30
M. J. Kirkby 'Infiltration, throughflow and overland flow' pp. 215–27
J. C. Rodda 'The Assessment of Precipitation' pp. 130–4

J. C. Rodda 'The Flood Hydrograph' pp. 405–18
Dury, G. H. 'Rivers in Geographical Teaching' *Geography* 48, pp. 18–30 (1963)
Drew, D. P. and Smith, D. I. 'Techniques for the Tracing of Subterranean Drainage' *Tech. Bull B.G.R.G.* 2 (1969)
Gregory, K. J. and Walling, D. E. 'Field Measurements in the Drainage Basin' *Geography* 56, pp. 277–92 (1971)
HMSO *British Rainfall* 1964 (With Supplement for 1961–5)
HMSO *Surface Water Yearbook* 1964–5 (With supplement containing details of gauging stations whose streamflow data are given in the main volume)
Horton, R. E. 'Erosional Development of Streams: Quantitative Physiographic Factors' *Rivers and River Terraces* ed. G. H. Dury pp. 117–65 (Macmillan, 1970)
More, R. J. 'Hydrological Models in Geography' *Models in Geography* ed. R. J. Chorley and P. Haggett (Methuen, 1969)
Pegg, R. K. and Ward, R. C. 'What Happens to the Rain?' *Weather* 26, pp. 88–97 (1971)
Penman, H. L. *Vegetation and Hydrology* (Commonwealth Bureau of Soils, Rothamsted, 1963)
Vallentine, H. R. *Water in the Service of Man* (Penguin, 1967)
Ward, R. C. 'Measuring Potential Evapotranspiration' *Geography* 48, pp. 49–55 (1963)
Ward, R. C. 'Some aspects of the recent growth of "geographical" Hydrology' *Geography* 55, pp. 390–406 (1970)

CHAPTER 6

Carson, M. A. and Kirkby, M. J. *Hillslope Form and Process* (Cambridge University Press, 1972)
Chorley, R. J. 'A Re-evaluation of the Geomorphic System of W. M. Davis' *Frontiers in geographical teaching* ed. R. J. Chorley and P. Haggett pp. 21–38 (Methuen, 1965)
Chorley, R. J. 'The Role of Water in Rock Disintegration' *Water, Earth and Man* pp. 135–55 (Methuen, 1969)
Douglas, I. 'Field Methods of Water Hardness Determination' *Tech. Bull. B.G.R.G.* 1 (1968)
Dury, G. H. *Rivers and River Terraces* (Macmillan, 1970), specifically:
G. K. Gilbert 'Land Sculpture in the Henry Mts.' pp. 95–116
W. B. Langbein 'River Meanders and the L. B. Leopold theory of minimum variance' pp. 238–63
High, C. and Hanna, F. K. 'A Method for the Direct Measurement of Erosion on Rock Surfaces' *Tech. Bull. B.G.R.G.* 5
Kidson, C. and Gifford, J. 'The Exmoor Storm of 15 August 1952' *Geography* 38, pp. 1–9 (1953)
King, C. A. M. *Techniques in Geomorphology* (Edward Arnold, 1966)
Kirkby, M. J. 'Measurement and Theory of Soil Creep' *Journal of Geology* 75, pp. 359–78 (1967)
Leopold, L. B. and Langbein, W. B. 'River Meanders' *Scientific American* 214, pp. 60–70
Leopold, L. W., Wolman, M. G. and Miller, J. P. *Fluvial Processes in Geomorphology* (Freeman, San Francisco, 1964)
Ollier, C. D. and Thomasson, A. J. 'Asymmetrical Valleys of the Chiltern Hills' *Geographical Jnl.* 123, pp. 71–80 (1957)
Pitty, A. F. 'A Simple Device for Measuring Hillslopes' *Journal of Geology* 76, pp. 717–20 (1968)
Small, R. J. 'The new geomorphology and the sixth-former' *Geography* 54, pp. 308–18 (1969)
Waters, R. S. 'Morphological Mapping' *Geography* 43, pp. 10–17 (1958)
Young, A. 'Present Rate of Land Erosion' *Nature* 224, pp. 851–2 (1969)
Young, A. *Slopes* (Oliver and Boyd, 1972)

CHAPTER 7

Carr, A. P. 'Shingle spit and river mouth: short term dynamics' *Trans. Inst. Brit. Geog.* 36, pp. 117–29
Carr, A. P. 'Size grading along a pebble beach: Chesil Beach, England' *Journal of Sedimentary Petrology* 39, pp. 297–311

Doornkamp, J. C. and King, C. A. M. *Numerical Analysis in Geomorphology* (Edward Arnold, 1971)
Gresswell, R. K. *Geology for Geographers* (Hulton, 1963)
Geographical Association *Notes on apparatus for demonstrating processes of erosion (together with plans of the tanks)* Designed by Northgate Grammar School for Boys, Ipswich (1970)
Kidson, C., Carr, A. P. and Smith, D. B. 'Further Experiments using radioactive methods to detect the movement of shingle over the sea bed and alongshore' *Geographical Jnl* 124, pp. 210–18
May, V. J. 'The Retreat of Chalk Cliffs' *Geographical Jnl* 137, pp. 203–6 (1971)
Steers, J. A. *The Sea Coast* (Collins, 1963)
Steers, J. A. (ed.) *Introduction to Coastline Development* (Macmillan, 1971), specifically:
C. Kidson and A. P. Carr 'Marking beach materials for tracing experiments' pp. 69–93
C. A. M. King 'The relationship between wave incidence, wind direction and beach changes at Marsden Bay, Co. Durham' pp. 117–32
Per Bruun 'Coastal Research and its economic justification' pp. 204–25
Steers, J. A. (ed.) *Applied Coastal Geomorphology* (Macmillan, 1971), specifically:
H. Valentin 'Land Loss at Holderness' 1852–1952 pp. 116–37
M. A. Arber 'The coastal landslips of south-east Devon' pp. 138–54

CHAPTER 8

Andrews, J. T. 'Techniques of till fabric analysis' *Tech. Bull. B.G.R.G.* 6 (1971)
Brown, E. H. 'Man Shapes the Earth' *Geographical Jnl* 136, pp. 73–85 (1970)
Chorley, R. J. 'The Shape of Drumlins' *Journal of Glaciology* 3, 339–44 (1959)
Crabtree, K. 'Pollen Analysis' *Science Progress* 56, pp. 83–101 (1968)
Dury, G. H. 'General Theory of Meandering Valleys and Underfit Streams' *Rivers and River Terraces* ed. G. H. Dury pp. 264–75 (Macmillan, 1970)
Embleton, C. and King, C. A. M. *Glacial and Periglacial Geomorphology* (Edward Arnold, 1969)
Inman, D. L. 'Measures to Describe the size and distribution of sediments' *Journal of Sedimentary Petrology* 27, pp. 3–27 (1952)
James, P. A. 'The measurement of soil frost heave in the field' *Tech. Bull. B.G.R.G.* 8 (1971)
Kellaway, G. A. 'Glaciation and the stones of Stonehenge' *Nature* 233, pp. 30–35 (1969)
King 'Trend-surface analysis of Central Pennine erosion surfaces' *Trans. Inst. Brit. Geog.* 47, pp. 47–60 (1969)
Lewis, W. V. and Miller, M. M. 'Kaolin Model Glaciers' *Journal of Glaciology* 2, 533–8 (1955)
Linton, D. L. 'The Problem of Tors' *Geographical Jnl* 121, pp. 470–87 (1955)
Stevens, N. and Synge, F. M. 'Pleistocene shorelines' *Essays in Geomorphology* ed. G. H. Dury (Heinemann, 1966)
West, R. G. *Pleistocene Geology and Biology* (Longmans, 1968)
Wiman, S. 'A preliminary study of experimental frost weathering' *Geografiska Annaler* 45, pp. 113–21 (1963)

Index

Figures in italic refer to illustrations